# Role of Copper in Lipid Metabolism

Editor

**K. Y. Lei, Ph.D.**
Professor
Department of Nutrition and Food Science
University of Arizona
Tucson, Arizona

Associate Editor

**Timothy P. Carr, Ph.D.**
Research Associate
Department of Comparative Medicine
Bowman Gray School of Medicine
Wake Forest University
Winston-Salem, North Carolina

**CRC Press**
**Boca Raton Ann Arbor Boston**

**Library of Congress Cataloging-in-Publication Data**

Role of copper in lipid metabolism / editor, K. Y. Lei, associate editor, Timothy P. Carr.

p. cm.

Includes bibliographical references.

ISBN 0-8493-5564-8

1. Hypocupremia—Complications and sequelae. 2. Blood lipids--Metabolism. I. Lei, K. Y., 1944- . II. Carr, Timothy P.

RC627.C68R65 1990

612.3'97—dc20 89-38403

CIP

This book represents information obtained from authentic and highly regarded sources. Reprinted material is quoted with permission, and sources are indicated. A wide variety of references are listed. Every reasonable effort has been made to give reliable data and information, but the author and the publisher cannot assume responsibility for the validity of all materials or for the consequences of their use.



Direct all inquiries to CRC Press, Inc., 2000 Corporate Blvd., N.W., Boca Raton, Florida 33431.



International Standard Book Number 0-8493-5564-8

Library of Congress Card Number 89-38403
Printed in the United States

# PREFACE

Severe copper deficiency has been reported in infants and children suffering from generalized malnutrition and prolonged diarrhea, as well as in infants and adults receiving prolonged total parental nutrition. Although severe copper deficiency is rare, the prevalence of marginal deficiency in human populations has not been established. In Japan, it has been suggested that marginal copper deficiency which escapes clinical recognition may be more common in infants nourished with powdered milk alone. The Food and Nutrition Board of the U.S. National Academy of Sciences suggests that 2 mg of copper per day may be adequate for adults. However, a number of studies have demonstrated that diets commonly consumed and hospital diets provide only slightly more than 1 mg/d and may be low in copper. The consumption of diets low in copper for prolonged periods could lead to marginal copper deficiency. Thus, the prevalence of marginal copper deficiency may not be as rare in the U.S. as previously believed. Hypercholesterolemia has been observed in marginal and severely copper-deficient rats as well as other species, including humans. Although hypercholesterolemia is a major risk factor of cardiovascular disease, the contribution of depressed copper status to the vascular health of human populations has not been established. Nevertheless, recent findings, mostly derived from experimental animals, have provided a large body of information on the influence of severe and marginal copper deficiency on lipid metabolism and the cardiovascular system. The first portion of this book reviews the influence of copper deficiency on plasma cholesterol response in various species, in addition to cholesterol and lipoprotein metabolism in the rat. Other chapters summarize the role of copper in lipid digestion as well as the metabolism of long chain fatty acids, prostaglandins, and cardiac lipids. Finally, the interaction of dietary copper and carbohydrates on blood lipids, the effect of copper deficiency on arterial structure, and the copper deficiency theory for ischemic heart disease are also presented.

Kai Yui Lei

# THE EDITOR

**Kai Yui Lei, Ph.D.,** is Professor of Nutrition and Food Science in the College of Agriculture at the University of Arizona at Tucson.

Dr. Lei received his B.Sc. degree from the University of London in 1968. He obtained his M.S. degree in 1970 from the Department of Nutrition, University of Guelph, Ontario, Canada. In 1973, he received his Ph.D. degree from the Department of Food Science and Human Nutrition at Michigan State University, East Lansing. After two years of postdoctoral research in the laboratory of Dr. Ananda S. Prasad at Wayne State University School of Medicine in Detroit, he was appointed Assistant Professor in 1975 and Associate Professor in 1978 in the Nutrition Program at Mississippi State University. In 1980, he moved to the University of Arizona as an Associate Professor of Nutrition and became a Professor in 1988.

Dr. Lei is a member of the American Institute of Nutrition, American Heart Association Council on Arteriosclerosis, American Dietetic Association, and the honorary society Sigma Xi. He has been the recipient of research grants from the National Institutes of Health, the U.S. Department of Agriculture, American Heart Association, and private industry. He has published more than 50 papers and review chapters. Dr. Lei's current research interests are focused on lipoprotein synthesis, intracellular processing and secretion, as well as membrane biogenesis and membrane incorporation and turnover of fatty acids, sterols, and phospholipids.

## CONTRIBUTORS

**Kenneth G. D. Allen, Ph.D.**
Professor
Department of Food Science and Human Nutrition
Colorado State University
Fort Collins, Colorado

**Patsy M. Brannon, Ph.D.**
Associate Professor
Department of Nutrition and Food Science
University of Arizona
Tucson, Arizona

**Timothy P. Carr, Ph.D.**
Research Associate
Department of Comparative Medicine
Bowman Gray School of Medicine
Wake Forest University
Winston-Salem, North Carolina

**Stephen C. Cunnane, Ph.D.**
Assistant Professor
Department of Nutritional Sciences
University of Toronto
Toronto, Ontario, Canada

**Craig A. Hassel, Ph.D.**
Assistant Professor and Extension Specialist
Department of Food Science and Nutrition
University of Minnesota
St. Paul, Minnesota

**Mary Ann Johnson, Ph.D.**
Associate Professor
Department of Foods and Nutrition
University of Georgia
Athens, Georgia

**Leslie M. Klevay, M.D., S.D. in Hyg.**
U.S. Department of Agriculture
Agriculture Research Service
Grand Forks Human Nutrition Research Center
Grand Forks, North Dakota

**Kai Yui Lei, Ph.D.**
Professor
Department of Nutrition and Food Science
University of Arizona
Tucson, Arizona

**Melvin M. Mathias, Ph.D.**
Cooperative State Research Service
U. S. Department of Agriculture
Washington, D. C.

**Denis M. Medeiros, Ph.D., R.D.**
Associate Professor
Department of Human Nutrition and Food Management
The Ohio State University
Columbus, Ohio

# TABLE OF CONTENTS

Chapter 1

# PLASMA CHOLESTEROL RESPONSE IN COPPER DEFICIENCY

**K. Y. Lei**

## TABLE OF CONTENTS

# I. INTRODUCTION

This chapter is intended to review animal and human studies performed to examine the influence of dietary copper (Cu) deficiency on plasma cholesterol levels. Klevay[1] first presented experimental data which suggested that an increased dietary zinc to copper ratio lead to observed hypercholesterolemia. A subsequent hypothesis was formulated by Klevay[2] which stated that a relative or an absolute deficiency of copper, characterized by a high ratio of dietary zinc to copper, results in hypercholesterolemia and constitutes as a major factor in the etiology of coronary heart disease. Since 1973, a large number of studies have been performed in this area with the majority of the work carried out with rats. Among the numerous rat studies existed marked differences in basal diet composition, route of copper and zinc supplementation, duration of experiment, and strain of rats employed. These differences no doubt contributed to discrepancies in plasma cholesterol response. Consistent in these studies, however, was that copper deficiency was primarily induced by semipurified diets deficient or marginal in copper. In addition to semipurified diets, some investigators have also manipulated the copper status by supplementing copper and zinc in the drinking water (Table 1). Hypercholesterolemia has been induced by copper deficiency in species other than rats, including man,[3,4] cynomolgus monkeys,[5] rabbits,[6] and mice.[7]

Animal studies which report hypercholesterolemia as a result of copper deficiency have generally demonstrated a trend of increasing plasma cholesterol levels with increasing severity of copper deficiency. This trend appears to hold true in spite of differences in diet composition and strain of animals used. The extent of copper deficiency is directly related to marked restriction in copper intake and prolonged duration on diet.

# II. HYPERCHOLESTEROLEMIA IN COPPER-DEFICIENT RATS

At least 12 independent laboratories (Table 1) have successfully induced hypercholesterolemia in rats as a result of copper deficiency. Because of the ease by which copper deficiency and hypercholesterolemia can be produced in rats, a number of laboratories including our own have routinely used this animal model to study the mechanisms responsible for the observed hypercholesterolemia.

## A. Male Rats

In 1973, Klevay[1] (Table 1, Lab 1) reported hypercholesterolemia in rats produced by providing a diet containing less than 1 μg Cu/g and drinking water with zinc to copper ratios (microgram per milliliter of zinc and copper) of 10:0.25 and 20:0.5 to weanling male rats as compared with 10:2 to controls. The duration required for the earliest demonstratable difference in plasma cholesterol was determined. Increased zinc to copper ratio of drinking water from 10:2 to higher ratios induced hypercholesterolemia. Alternatively, elevated plasma cholesterol level was produced by reducing the copper level from 2.10 to 0.25 or 0.5 μg/ml. The data indicated that not only was the zinc to copper ratio important in determining plasma cholesterol level, an increase in the absolute amount of copper from 0.25 to 0.5 μg/ml prolonged the time required for hypercholesterolemia to occur. In subsequent studies, hypercholesterolemia was repeatedly demonstrated in young male rats by Klevay and co-workers[8-13] with the combination of a low copper diet and drinking water supply.

Among the early reports on the influence of copper deficiency and dietary zinc to copper ratio on cholesterol metabolism are the two large-scale factorial studies conducted by Murthy and Petering[14] and Petering et al.[15] (Table 1, Lab 2) involving numerous levels of dietary zinc and copper using young male rats. An inverse relationship was demonstrated between dietary and serum copper, and the levels of serum cholesterol, triglyceride, and phospholipids. They suggested that dietary copper governs the metabolic rate of cholesterol more than the zinc to copper ratio. In recent studies these researchers[16,17] have modified their semipurified diets low in copper, with minor changes in protein and/or carbohydrate sources, and have continued to demonstrate the induction of hypercholesterolemia by copper deficiency in young rats.

Soon after the zinc to copper ratio hypothesis[2] was proposed, a series of studies was inititated in our laboratory (Table 1, Lab 3) to examine the influence of copper status on cholesterol metabolism. In the early studies,[18-21] hypercholesterolemia was induced in young male Sprague-Dawley rats by a semipurified diet low in copper (<2 μg/g) and high in zinc (120 μg/g) for 8 to 9 weeks. The basal diet consisted of 20% casein, 64% glucose, 10% corn oil, as well as 1 and 5% of our own vitamin and mineral mixes, respectively. However, after the American Institute of Nutrition (AIN) diet formulation[22] was published, all of our subsequent studies were performed according to the AIN formulation with 20% casein and 64% glucose monohydrate. Hypercholesterolemia was always observed using this basal diet.[23-29] From 1981 to present, casein from U.S. Biochemical (Cleveland, Ohio) was used for all our studies because of its low copper content. The copper content of the basal diet has since been reduced to a range of 0.6 to 0.9 μg/g. Symptoms of copper deficiency and hypercholesterolemia have been observed in young male rats as early as after 6 weeks of dietary treatment. Thus our data suggest that the lower the copper content of the diet, the less time it will require to induce hypercholesterolemia.

Prior to establishing his own research program, Allen (Table 1, Lab 4) initiated his research in the area of copper and cholesterol metabolism in association with Klevay.[30-32] In his earlier studies with Klevay, as well as in studies performed in his own laboratory,[33,34] copper deficiency was induced

## TABLE 1
## Studies of Copper Deficiency-Induced Hypercholesterolemia in Rats

| Lab | Author, year | Strain, sex | Age or initial wt, time on diet | Dietary copper in feed (water)[a]: Cu content | Dietary copper in feed (water)[a]: Zn/Cu | Protein, carbohydrate, fat, fiber, mineral | Physiological measurements (% change from controls): Tissue Cu | Hematocrit | Heart wt | Plasma cholesterol | Comments |
|---|---|---|---|---|---|---|---|---|---|---|---|
| 1 | Klevay, 1973[1] | Carworth, male | Weanling, 5—21 weeks | <1 μg/g (0.25→2.0 μg/ml) | — (40/1→5) | Egg white, 20%<br>Sucrose, 62.4%<br>Corn oil, 9%<br>Cellulose,3%<br>Jones Foster, 4% | — | ↓6—19% | — | ↑13—33% | Increased plasma cholesterol consistent with Zn/Cu of 40 |
| | Klevay and Viestenz, 1981[8] | Sprague-Dawley, male | Weanling, 5 weeks | 0.79, 3.0 μg/g (—) | 15.8, 4.2 (—) | Egg white, 20%<br>Sucrose, 62.4%<br>Corn oil, 9%<br>Cellulose, 3%<br>Jones Foster, 4% | — | — | — | ↑39% | Abnormal electrocardiograms; S-T segment depression and bundle branch block |
| | Lau and Klevay, 1981[9] | Sprague-Dawley, male | Weanling, 4—7 weeks | No data (0, 2.0 μg/ml) | — (10/0, 10/ 2) | Egg white, 20%<br>Sucrose, 62.4%<br>Corn oil, 9%<br>Cellulose, 3%<br>Jones Foster, 4% | ↓37% (kidney) | ↓24—36% | — | ↑21—32% | Plasma LCAT ↓22—32%; LCAT activity not corrected for possible plasma volume changes |
| | Lau and Klevay, 1982[10] | Sprague-Dawley, male | Weanling, 3—6 weeks | No data (0, 2.0 μg/ml) | — (10/0, 10/ 2) | Egg white, 20%<br>Sucrose, 62.4%<br>Corn oil, 9%<br>Cellulose, 3%<br>Jones Foster, 4% | ↓26—42% (heart) | ↓14—19% | — | ↑26—38% | Postheparin plasma lipoprotein lipase ↓40—47%; not corrected for plasma volume changes |
| | Koop et al., 1983[11] | Sprague-Dawley, male | Weanling, 7 weeks | <0.9 μg/g (0, 2.0 μg/ml) | — (10/0, 10/ 2) | Egg white, 20%<br>Sucrose, 62.4%<br>Corn oil, 9% | ↓37% (liver) | ↓39% | — | ↑39% | Depressed His-Purkinje system conductivity and S-T segment; ↓ATP and ADP in tissues |

| | | | | | | | | | | | | |
|---|---|---|---|---|---|---|---|---|---|---|---|---|
| | | | | | | Cellulose, 3%<br>Jones Foster, 4% | | | | | | |
| | Borg et al., 1985[12] | Sprague-Dawley, male | Weanling, 4—6 weeks | No data<br>(0, 2.0 μg/ml) | —<br>(10/0, 10/2) | Egg white, 20%<br>Sucrose, 62.4%<br>Corn oil, 9%<br>Cellulose, 3%<br>Jones Foster, 4% | — | ↓ 17—34% | — | ↑ 23—29% | Gross and microscopic lesions in connective tissue network of cardiac and skeletal muscle |
| | Klevay, 1987[13] | Sprague-Dawley, male | Adult 300 g, 7—12 weeks | No data<br>(0, 2.0 μg/ml) | —<br>(10/0, 10/2) | Egg white, 20%<br>Sucrose, 62.4%<br>Corn oil, 9%<br>Cellulose, 3%<br>Jones Foster, 4% | — | — | — | ↑ 18—27% | Systolic blood pressure ↑ 11—26 mm Hg |
| 2 | Murthy and Petering, 1976[14] | Carworth, male | Weanling, 9 weeks | No data<br>(0.25→2.0 μg/ml) | —<br>(160→1.5) | Egg white, 20%<br>Sucrose, 62%<br>Corn oil, 9%<br>Cellulose, 3%<br>Jones Foster 4% | — | Positive correlation with Cu intake | — | Highest ↑ with lowest Cu intake | Total cholesterol inversely related to dietary Cu and serum Cu |
| | Petering et al., 1977[15] | Sprague-Dawley, male | Weanling, 17 weeks | 0.43 μg/g<br>(0→16 μg/ml) | —<br>(40/ 0→1.5) | Egg white, 20%<br>Corn starch, 62%<br>Corn oil, 9%<br>Cellulose, 3%<br>Wesson, 4% | ↓ Cu (liver) | — | — | Highest ↑ with lowest Cu intake | Total cholesterol inversely related to serum ceruloplasmin or Cu levels |
| | Stemmer et al., 1985[16] | ?, Male | Weanling, 13 weeks | 0.5→8.0 μg/g<br>(0→7.5 μg/ml) | 280→2.5<br>(—) | 3 Proteins, each 20%<br>2 CHO, each 62%<br>Corn oil, 9%<br>Cellulose, 3%<br>Bernhart-Tomarelli, 4% | ↓ Cu (serum) | Slight ↓ | ↑ | Highest ↑ with low Cu casein diet | Total cholesterol ↑ with low Cu and casein or milk as protein source |
| | Petering et al., 1986[17] | Sprague-Dawley, male | Weanling, 13 weeks | 0.5→8.0 μg/g<br>(0→7.5 μg/ml) | 280→2.5<br>(—) | Egg white, 20%<br>3 CHO, 30— 62%<br>Corn oil, 9%<br>Cellulose, 3% | ↓ Cu (serum; Zn/Cu = 40 or 280) | ↓ (Zn/Cu = 280) | ↑ (Zn/Cu = 280) | ↑ (Zn/Cu = 280) | Total cholesterol also ↑ in rats fed 0.5 μg Cu/g diet (Zn/Cu = 40) and sucrose |

## TABLE 1 (continued)
## Studies of Copper Deficiency-Induced Hypercholesterolemia in Rats

| Lab | Author, year | Strain, sex | Age or initial wt, time on diet | Dietary copper in feed (water)[a] | | Protein, carbohydrate, fat, fiber, mineral | Physiological measurements (% change from controls) | | | | Comments |
|---|---|---|---|---|---|---|---|---|---|---|---|
| | | | | Cu content | Zn/Cu | | Tissue Cu | Hematocrit | Heart wt | Plasma cholesterol | |
| | | | | | | Bernhart-Tomarelli, 4% | | | | | |
| 3 | Lei, 1977[18] | Sprague-Dawley, male | Weanling, 9 weeks | 2, 18 μg/g | 6.7, 60 | Casein, 20%<br>Glucose, 63%<br>Corn oil, 10%<br>No added fiber<br>Own mineral, 5% | ↓ 59% (liver) | — | ↑ 25% | ↑ 40% | Cholesterol feeding masked Cu-deficient effect on plasma total cholesterol |
| | Lei, 1978[19] | Sprague-Dawley, male | Weanling, 8 weeks | 2, 18 μg/g | 6.7, 60 | Casein, 20%<br>Glucose, 63%<br>Corn oil, 10%<br>No added fiber<br>Own mineral, 5% | — | — | ↑ 16% | ↑ 35% | ↑ in plasma free and ester cholesterol; rate of cholesterol oxidation and excretion not influenced by Cu |
| | Looney and Lei, 1978[20] | Sprague-Dawley, male | Weanling, 9 weeks | 2, 18 μg/g | 6.7, 60 | Casein, 20%<br>Glucose, 63%<br>Corn oil, 10%<br>Cellulose, 0, 8, 16%<br>Own mineral, 5% | — | — | — | ↑ 35% (0% fiber) | Highest ↑ in total cholesterol in 0% fiber diet; addition of fiber ↑ Cu intake and ↓ total cholesterol |
| | Shao and Lei, 1980[21] | Sprague-Dawley, male | Weanling, 8 weeks | 2, 18 μg/g | 6.7, 60 | Casein, 20%<br>Glucose, 63%<br>Corn oil, 10%<br>No added fiber<br>Own mineral, 5% | — | — | ↑ 42% | ↑ 21% | ↑ in total cholesterol mainly due to ↑ in plasma free cholesterol |

| | | | | | | | | | | | |
|---|---|---|---|---|---|---|---|---|---|---|---|
| | Lin and Lei, 1981[23] | Sprague-Dawley, male | Weanling, 16 weeks | 1.6, 8 μg/g | 5, 25 | Casein, 20%<br>Glucose, 63%<br>Corn oil, 10%<br>Cellulose, 5%<br>AIN 5% | — | — | — | ↑ 16% (avg from 5—16 wks) | Slight ↑ in total cholesterol from 5 to 16 weeks |
| | Lei et al., 1983[24] | Sprague-Dawley, male | Weanling, 7 weeks | 0.85, 8 μg/g | 5, 47 | Casein, 20%<br>Glucose, 63%<br>Corn oil, 10%<br>Cellulose, 5%<br>AIN, 5% | ↓ 86% (liver) | ↓ 42% | ↑ 50% | ↑ 43% | ↑ in total cholesterol mainly in HDL, with slight ↑ in LDL |
| | Croswell and Lei, 1985[25] | Sprague-Dawley, male | Weanling, 7 weeks | 0.85, 8 μg/g | 5.47 | Casein, 20%<br>Glucose, 63%<br>Corn oil, 10%<br>Cellulose, 5%<br>AIN, 5% | ↓ 59% (liver) | — | ↑ 48% | ↑ 48% (in HDL) | ↑ in total HDL cholesterol associated mainly with apo E-rich HDL fraction |
| | Hassel et al., 1987[26] | Sprague-Dawley, male | Weanling, 8 weeks | 0.7, 8 μg/g | 5, 57 | Casein, 20%<br>Glucose, 63%<br>Corn oil, 10%<br>Cellulose, 5%<br>AIN, 5% | ↓ 74% | — | ↑ 36% | ↑ 35% | |
| | Hassel et al., 1988[27] | Sprague-Dawley, male | Weanling, 8 weeks | 0.7, 8 μg/g | 5, 57 | Casein, 20%<br>Glucose, 63%<br>Corn oil, 10%<br>Cellulose, 5%<br>AIN, 5% | ↓ 78% | — | ↑ 62% | ↑ 38% | |
| 4 | Allen and Klevay, 1978[32] | Sprague-Dawley, male | Weanling, 5—9 weeks | 0.57, 5 μg/g | 1.5, 2.5 | Egg white, 20%<br>Sucrose, 51%<br>Cottonseed oil, 9.5%<br>Coconut oil, 9.5%<br>Cellulose, 3%<br>Jones Foster, 4% | ↓ 78% (liver; 9 wks) | ↓ 15% (5 and 9 wks) | — | ↑ 48% (5 wk)<br>↑ 65% (9 wk) | Total cholesterol inversely related to hepatic Cu content |

## TABLE 1 (continued)
## Studies of Copper Deficiency-Induced Hypercholesterolemia in Rats

| Lab | Author, year | Strain, sex | Age or initial wt, time on diet | Dietary copper in feed (water)[a] Cu content | Zn/Cu | Protein, carbohydrate, fat, fiber, mineral | Physiological measurements (% change from controls) Tissue Cu | Hematocrit | Heart wt | Plasma cholesterol | Comments |
|---|---|---|---|---|---|---|---|---|---|---|---|
| | Allen and Klevay, 1978[30] | Sprague-Dawley, male | Weanling, 5—26 weeks | 0.57, 5 μg/g | 1.5, 2.5 | Egg white, 20%<br>Sucrose, 51%<br>Cottonseed oil, 9.5%<br>Coconut oil, 9.5%<br>Cellulose, 3%<br>Jones Foster, 4% | ↓74% (liver; 26 wks) | — | — | ↑62% (5 wk);<br>↑129% (26 wk) | Total bile acids and biliary cholesterol ↑ in Cu deficiency during 1st day, but not during 2nd and 3rd day of collection |
| | Allen and Klevay, 1980[31] | Sprague-Dawley, male | Weanling, 7 weeks | 0.57, 5 μg/g | 1.5, 2.5 | Egg white, 20%<br>Sucrose, 62.4%<br>Corn oil, 9.5%<br>Cottonseed oil, 9.5%<br>Cellulose, 3%<br>Jones Foster, 4% | ↓67% (liver) | — | — | ↑104% | ↑ in total cholesterol mainly due to ↑ in HDL cholesterol |
| | Harvey and Allen, 1981[33] | Sprague-Dawley, male | Weanling, 9 weeks | 0.6, 5.9 μg/g | 1.5, 2.1 | Egg white, 20%<br>Sucrose, 62.4%<br>Corn oil, 9.5%<br>Cottonseed oil, 9.5%<br>Cellulose, 3%<br>Jones Foster, 4% | ↓73% (liver) | — | — | ↑57% | Plasma LCAT activity ↓ 74%; not corrected for plasma volume changes |
| | Harvey and Allen, 1985[34] | Sprague-Dawley, male | Weanling, 9 weeks | 0.6, 5.9 μg/g | 1.5, 2.1 | Egg white, 20%<br>Sucrose, 62.4%<br>Corn oil, 9.5% | ↓63—73% (liver) | ↓17% | — | ↑ 55—91% | Total cholesterol ↑ in both experiments |

| | | | | | | | | | | | |
|---|---|---|---|---|---|---|---|---|---|---|---|
| | | | | | | Cottonseed oil, 9.5%<br>Cellulose, 3%<br>Jones Foster, 4% | | | | | |
| 5 | Lefevre et al., 1985[35] | Sprague-Dawley, male | Young 120—140 g, 6 weeks | 0, 5 μg/g | 1/0, 1000/5 | Egg white, 25%<br>Sucrose, 60%<br>Corn oil, 1.5%<br>No added fiber<br>Own mineral, 6% | ↓ Cu (liver) | — | ↑ | ↑ (free and HDL) | Total cholesterol appeared to be ↑ at Zn/Cu of 100/0 and 1000/0 |
| | Lefevre et al., 1986[36] | Sprague-Dawley, male | Young 120—140 g, 6 weeks | 0, 5 μg/g | 1/0, 1000/5 | Egg white, 25%<br>Sucrose, 60%<br>Corn oil, 1.5%<br>No added fiber<br>Own mineral,6% | ↓ 49% (liver) | ↓ 5% | No change | ↑ 18% | ↑ in total cholesterol mainly due to ↑ in HDL cholesterol |
| 6 | Koo et al., 1988[37] | Fischer-344, male | Weanling, 8 weeks | 0.5, 7.5 μg/g | — | Egg white, 20%<br>Glucose, 36%<br>Corn starch, 31%<br>Corn oil, 4%<br>Cellulose, 3%<br>AIN, 3.5% | ↓ 58% (liver) | — | ↑ 14% | ↑ 30% | Diets contained 1% cholesterol |
| | Lee and Koo, 1988[38] | Fischer-344, male | Weanling, 9—12 weeks | 0.5, 7.5 μg/g | — | Egg white, 20%<br>Glucose, 36%<br>Corn starch, 31%<br>Corn oil, 4%<br>Cellulose, 3%<br>AIN, 3.5% | ↓ 56% (liver) | — | ↑ 55% | ↑ 46% | Diets contained 1% cholesterol |
| 7 | Nielsen et al., 1982[39] | Sprague-Dawley, female | Weanling, 8—11 weeks | 0.5 → 12.4 μg/g | — | Milk powder, 45%<br>Ground corn, 39%<br>Corn oil, 10%<br>No added fiber<br>Own mineral, 5% | ↓ 87% (liver) | — | — | ↑ 18—38% | ↑ in total cholesterol with Cu-deficient diets; highest ↑ with 45 μg Fe and 0 μg Ni/g diet |

**TABLE 1 (continued)**
**Studies of Copper Deficiency-Induced Hypercholesterolemia in Rats**

| Lab | Author, year | Strain, sex | Age or initial wt, time on diet | Dietary copper in feed (water)[a] Cu content | Zn/Cu | Protein, carbohydrate, fat, fiber, mineral | Physiological measurements (% change from controls) Tissue Cu | Hematocrit | Heart wt | Plasma cholesterol | Comments |
|---|---|---|---|---|---|---|---|---|---|---|---|
| 8 | Johnson 1990[40] | Sprague-Dawley, female and male | Weanling, 4 weeks | <1, 5 μg/g | — | Casein, 20%<br>Sucrose or corn starch, 65.3%<br>Corn oil, 5%<br>Cellulose, 5%<br>AIN, 3.5% | — | — | — | ↑ 15—53% | ↑ in total cholesterol after only 4 weeks in Cu-deficient rats fed sucrose |
| 9 | Wu et al., 1984[41] | Long-Evans, male | Weanling, 22 weeks | 0.7, 3.7, 8.7 μg/g | 57, 11, 4.6 | Casein, 20%<br>Sucrose, 50%<br>Corn starch, 15%<br>Corn oil, 5%<br>Cellulose, 5%<br>AIN, 3.5% | — | ↓ 12% | — | ↑ 36% | ↑ in total cholesterol only with diet containing 0.7 μg/g |
| 10 | Reiser et al., 1983[42] | Sprague-Dawley, male | Weanling, 7 weeks | 1 μg/g<br>(0, 5 μg/ml) | —<br>(—) | Egg white, 20%<br>3 CHO, each 62%<br>Corn oil, 9%<br>Cellulose, 3%<br>Jones Foster, 4% | ↓ 44—74% (liver) | ↓ 25—31% (no change when fed starch as source) | ↑ 54—66% | ↑ 36—185% | ↑ in total cholesterol of 85% and 61% in Cu-deficient rats fed fructose and starch, respectively, as CHO source |
| 11 | Sinthusek and Magee, 1984[43] | Sprague-Dawley, ? | Weanling, 10 weeks | 1.5, 6 μg/g | 30/1.5, 7.5/6.0 | Egg white, 28%<br>Starch, 51%<br>Corn or coconut oil, 13%<br>Cellulose, 2%<br>Hawk-Oser, 4% | Slight ↓ (liver) | — | — | Marginal ↑ | ↑ in total cholesterol with marginal Cu intake |
| 12 | O'Dell 1981[45] | ?, ? | ?, ? | <1, 10 μg/g | — | Non-fat milk Based diet | — | — | — | ↑ 24% | |

[a] Supplemental copper provided in water is represented in parentheses.

in rats with a low copper diet alone without mineral supplementation in the drinking water. Hypercholesterolemia was repeatedly observed in young male Sprague-Dawley rats fed a copper-deficient diet containing 0.6 μg Cu and 12.5 μg Zn/g, usually for periods of 5 to 9 weeks. The basal diet was similar in composition to that used in Klevay's laboratory and consisted of 20% egg white, 51% sucrose, 9.5% coconut oil, 9.5% cottonseed oil, 3% cellulose, vitamin and mineral (Jones Foster, 4%) mixes.

Recent studies by two groups of experienced investigators[35-38] in the area of lipoprotein research also reported hypercholesterolemia in young male rats fed semipurified copper-deficient diets for 6 to 8 weeks. Instead of initiating their studies with weanling rats like most investigators, Lefevre et al.[35,36] (Table 1, Lab 5) acclimated older male Sprague-Dawley rats, 120 to 140 g body weight, for at least 7 d with a basal diet supplemented with 5 μg Cu and 10 μg Zn/g diet. The basal diet consisted of 25% egg white, 59.5% sucrose, 8% corn oil, 1.5% vitamin mix, and 6% salt mix. Thereafter, a 6-week dietary treatment with basal diets containing no copper supplement (copper content not listed) and with 1 to 1000 μg zinc supplemented per gram diet resulted in mild copper deficiency and a small elevation of plasma-free cholesterol and high density lipoprotein (HDL) cholesterol.[35] In a follow-up study[36] designed to study HDL subfractions, a small elevation (17.6%) in plasma total cholesterol was observed in these older rats fed the basal diet with no copper added but with 100 μg zinc supplemented per gram diet.

Apart from their elegant studies in the area of dietary zinc and lipoprotein metabolism, Koo and co-workers[37,38] (Table 1, Lab 6) have recently extended their lipoprotein research into the dietary copper area. Hypercholesterolemia was observed in young male Fischer-344 rats fed a semipurified diet containing 0.5 μg Cu/g for 8 weeks.[37] The basal diet consisted of 20% egg white, 36% glucose, 31% corn starch, and 4% corn oil. It was similar to the AIN formulation[22] with the exception that 1% cholesterol was added. In a subsequent study[38] designed to examine the effect of copper deficiency on HDL subfractions, elevated plasma cholesterol level was again observed under similar experimental conditions.

### B. Female Rats

In addition to the numerous studies performed with male rats, the large scale factorial experiments designed by Nielsen et al.[39] (Table 1, Lab 7) to examine the influence of copper status, as well as its interaction with dietary nickel and iron, on plasma and liver lipid parameters provided vital information on female rats. Weanling female Sprague-Dawley rats were fed copper-deficient diets, as low as 0.4 to 0.5 μg/g, with various levels of nickel and iron for periods of 8 to 11 weeks. Marked elevations in plasma cholesterol levels were observed in rats fed the copper-deficient diet containing either 15 or 45 μg/g of nickel supplementation. The researchers concluded that when the copper deficiency was neither very severe or mild, observed increases in plasma cholesterol, total lipids, and lipid phosphorus were made more severe by dietary nickel supplementation. The important findings of the interaction between nickel and copper on plasma lipid parameters notwithstanding, this extensive study demonstrated that, similar to findings with male rats, hypercholesterolemia will result if copper status is depressed in female rats.

Confirmatory data were obtained in a recent study performed by Johnson[40] (Table 1, Lab 8) to examine the interaction between dietary copper and carbohydrates on plasma lipids. Elevated plasma cholesterol levels were observed after only 4 weeks in female, as well as male, Sprague-Dawley rats fed a copper-deficient diet containing sucrose.

### C. Strain of Rats

While the Sprague-Dawley rat is the most often used strain, copper deficiency-induced hypercholesterolemia has also been observed in the Carworth,[1,14] the Fischer-344[37] and the Long-Evans[41] rat. Wu et al.[41] (Table 1, Lab 9) fed male Long-Evans weanling rats semipurified diets (AIN-76) which were low, marginal, and adequate in copper (0.7, 3.7, and 8.7 μg Cu/g, respectively) for 22 weeks. Rats fed the 0.7 μg Cu/g diet were confirmed to be deficient by the reduction in growth and hematocrit. In addition, a 36% elevation in plasma cholesterol was observed in these copper-deficient rats. It is interesting to note that a diet containing 3.7 μg Cu/g was sufficient to maintain normal growth and similar hematocrit and plasma cholesterol levels as in rats provided with 8.7 μg Cu/g diet. The hypercholesterolemic data obtained in Long-Evans rats, as well as in other strains, suggest that elevated plasma cholesterol levels can be induced in different strains of rats provided that the diet is sufficiently deficient in copper.

### D. Copper and Carbohydrate Interaction

Various research groups have extensively studied the interaction of dietary carbohydrate with copper on lipid, carbohydrate, and copper metabolism. An extensive review in the area of dietary copper and carbohydrate interaction on lipid metabolism is presented in Chapter 6. The study by Reiser et al.[42] (Table 1, Lab 10) is a representative study which demonstrated that copper deficiency elevated plasma cholesterol and triglycerides in rats fed different types of carbohydrate. In this study, male weanling Sprague-Dawley rats were provided with basal diet (0.97 μg Cu/g) containing either starch, sucrose, or fructose (62%) as carbohydrate for 7 weeks. The copper-adequate rats received 5 μg Cu/ml and copper-deficient rats received no supplement. Although copper deficiency elevated plasma cholesterol regardless of the source of carbohydrate, the increase was markedly higher and the hepatic copper content much lower in copper-deficient rats fed fructose than those fed starch. Furthermore, the reduction in hematocrit and increase in heart weight observed in the sucrose- and fructose- fed copper-deficient rats were not detected when starch was used. Thus starch-fed rats appeared not to be as adversely affected as compared to sucrose- and fructose-fed rats receiving the same copper intake. Nevertheless, all these carbohydrates resulted in hypercholesterolemia when the dietary copper content was reduced to 0.9 μg/g. Furthermore, these researchers postulated that the lower level of hepatic copper found in copper-deficient rats fed sucrose and fructose as compared to starch may have resulted from a reduction in intestinal absorption of copper.

### E. Copper and Lipid Interaction

The interaction of dietary fat, either corn oil or coconut oil, with dietary copper and zinc was examined by Sinthusek and Magee[43] (Table 1, Lab 11). Weanling rats were provided with a marginal copper diet containing 1.5 μg Cu/g and corn oil for 4 weeks. Thereafter, 6 weeks of dietary treatment was imposed. Increases in plasma total cholesterol were observed in rats fed the marginal copper diets as compared to adequate copper diets containing either coconut or corn oil with 7.5 μg Zn/g. However, at the higher zinc intake of 30μg/g, plasma cholesterol was elevated only in rats fed the marginal copper diet containing corn oil. The inability of the copper-deficient diet containing coconut oil and 30 μg Zn/g to induce hypercholesterolemia remains unexplained and requires substantiation by future studies. In a subsequent report,[44] this group of investigators postulated that the dietary fat composition and level may influence the effect of copper and zinc on lipid parameters.

Finally, O'Dell[45] (Table 1, Lab 12) presented confirmation data from his laboratory in the New Zealand Workshop on Trace Elements, demonstrating hypercholesterolemia in rats fed a copper-deficient diet based on nonfat milk containing <1 μg Cu/g. Thus after the systematic review of data from 12 independent laboratories, the hypercholesterolemic effect of copper deficiency appears to be firmly established in rats.

## III. COPPER DEPLETION AND REPLETION STUDIES

### A. Humans

There are only two human studies specifically designed to study cholesterol and lipid metabolism in subjects provided with diets inadequate in copper. The first study conducted by Klevay et al.[3] (Table 2) examined the plasma cholesterol response of a young male subject during a 15-week depletion and a 6-week repletion period. A diet consisting of a 3-d rotation of menu, using conventional foods and providing only 0.8 mg Cu/d, was used during the depletion period. Although the study involved only one subject, a significant and progressive elevation of plasma total cholesterol level resulting in a 16% increase at the end of the depletion period was observed. The progressive reversal of plasma cholesterol back to the original or predepletion level after a 6-week repletion (4 mg Cu/d) provided credibility to the study.

A more extensive study was performed by Reiser et al.[4] (Table 2) to

**TABLE 2**
**Studies of Copper Deficiency-Induced Hypercholesterolemia in Humans, Monkeys, and Mice**

| Author, year | Strain, sex, age, | Time on diet | Dietary copper in feed (water)[a] Cu content | Zn/Cu | Protein, carbohydrate, fat, fiber, mineral | Physiological measurements (% change from controls) Tissue Cu | Hemato-crit | Heart wt | Plasma cholesterol | Comments |
|---|---|---|---|---|---|---|---|---|---|---|
| Klevay et al., 1984[3] | Human<br>Male,<br>29 yrs | 6 wk control<br>15 wk depletion<br>6 wk repletion | 1.3 mg/d<br>0.8 mg/d<br>4.8 mg/d | 10.5<br>17.1<br>2.9 | *Conventional foods:*<br>Protein, 60 g/d<br>CHO, 343 g/d<br>Fat, 97 g/d<br>Ca Suppl, 0.36 g/d | — | No change | — | ↑ 16% | Total cholesterol progressively ↑ during depletion, returned to normal after repletion |
| Reiser et al., 1987[4] | Human<br>Male,<br>21—57 yrs | Pretest<br>11 wk depletion<br>3 wk repletion | 1.38 mg/d<br>1.02 mg/d<br>4.02 mg/d | —<br>—<br>— | *Conventional foods:*<br>Protein 98 g/d<br>CHO 349 g/d<br>Fat 119 g/d | — | — | — | ↑ 14%<br>(in LDL) | ↑ in total cholesterol only a trend; LDL cholesterol ↑ 14%, HDL cholesterol ↓ 20% |
| Milne et al., 1981[5] | Cynomolgus monkey,<br>Male,<br>Adult | 28 wk depletion<br>4 wk repletion | (0.1 μg/ml)<br>(0.6 μg/ml) | —<br>— | Casein, 41 g/l<br>Dextrin-Maltose, 145 g/l<br>Corn oil, 22 g/l<br>Fiber, 11 g/l<br>Own salt mix, 18 g/l | — | No change | — | ↑ 30% | Total cholesterol gradually ↑ during depletion, returned to normal after Cu repletion on adequate Vit. C intake, but remained elevated on excessive Vit. C intake |
| Jones, 1984[7] | Swiss White mice<br>Female,<br>Weanling | 7—9 weeks | 1.3, 19 μg/g | — | Egg albumin, 25%<br>Sucrose, 57%<br>Vegetable oil, 10%<br>No added fiber<br>Hurley-Bell, 6% | ↓ 35%<br>(liver) | — | — | ↑ 27% | ↓ in plasma Cu (76%) and red cell SOD (24%) confirmed Cu deficiency |

[a] Supplemental copper provided in water is represented in parentheses.

examine the influence of copper status on blood cholesterol concentration and its lipoprotein distribution in 24 male subjects. Diets inadequate to abundant in copper were used in this study and were composed of natural and refined food, organized as a 7-d rotating menu and designed to satisfy all known nutrient requirements except for copper. After an 11-week depletion period, with an average copper intake of 1.02 mg/d, reductions in copper status were demonstrated in the subjects by decreases in erythrocyte superoxide dismutase and serum copper levels. Although the trend for an increase in plasma total cholesterol did not reach significance, a significant elevation in low density lipoprotein (LDL) cholesterol (14% and 24% above pretest and repletion levels, respectively) and a reduction in HDL cholesterol (20% and 31% below pretest and repletion levels, respectively) concentrations were demonstrated. Furthermore, the adverse alterations in LDL and HDL cholesterol levels reverted back to normal after a 3-week repletion period with 4-mg copper intake (3-mg copper supplement and 1-mg dietary copper) per day. Data from these two human studies suggest that, at least in male subjects, plasma cholesterol level may respond to a depressed copper status in a similar manner as in rats. Differences in the distribution of cholesterol among the various lipoprotein fractions may exist mainly because LDL and HDL are the major lipid-transporting lipoproteins in man and rats, respectively. Nevertheless, the human data can be taken as supporting evidence that copper deficiency may lead to hypercholesterolemia in species other than rats.

**B. Monkeys**

Elevated serum cholesterol levels have been observed in nonhuman primates after a prolonged consumption of a diet low in copper. The carefully executed study by Milne et al.[5] (Table 2) on eight adult male cynomolgus (*Macaca fasicularis*) monkeys examined the interaction of dietary vitamin C and copper on cholesterol and copper metabolism. A semipurified liquid diet providing 0.1 μg Cu/ml was used. During the extensive 28-week copper depletion period, a gradual elevation of serum cholesterol was observed in monkeys consuming only 1 mg of ascorbate per day per kilogram body weight. At the end of the depletion, a 30% increase in serum cholesterol was observed over the predepletion level and this elevation was reverted nearly back to normal after a 4-week repletion with an additional 0.5 μg Cu/ml supplemented. A similar magnitude of elevation in serum cholesterol was also observed during depletion in animals supplemented with a pharmacological dose of 25 mg ascorbate per day per kilogram body weight. However, the additional 0.5 μg Cu/ml supplemented during repletion was insufficient to depress and normalize the cholesterol level with the concurrent excessive ascorbate intake. These researchers suggested that the amount of copper (0.6 μg/ml) consumed was relatively unavailable for the regulation of cholesterol metabolism when ascorbic acid intake was high. Although the plasma cholesterol was not partitioned into different lipoprotein fractions, the serum total cholesterol data substantiated and confirmed the findings of the previous human studies.

## IV. OTHER COPPER-DEFICIENT SPECIES WITH HYPERCHOLESTEROLEMIA

### A. Mice

Hypercholesterolemia has also been induced in female Swiss White mice[7] fed from weanling on a semipurified copper-deficient diet containing 1.3 μg Cu/g (Table 2). After a 7 to 9-week period of dietary treatment, the mice fed the copper-deficient diet demonstrated mild reductions in liver copper (35%) and in red cell superoxide dismutase (24%) levels as well as a 76% reduction in plasma copper concentration, when compared to adequate rat fed 19 μg Cu/g diet. In view of the mild changes in indexes of copper status, the researcher considered the mice fed the copper-deficient diet to be subclinically deficient in copper. Nevertheless, a 27% elevation in plasma cholesterol level was observed in these female mice with subclinical copper deficiency.

### B. Rabbits

Rabbits are highly responsive to diets containing high fat and cholesterol, resulting in hypercholesterolemia. However, because of their requirement for high dietary fiber, the use of a natural fiber source to promote good growth responses usually results in copper intake too high to allow for efficient copper depletion. The strategy of using a hypocupremic agent, 2,3,2-tetramine (TETA), in addition to a low copper diet to induce copper deficiency in rabbits was undertaken in our laboratory[6] (Table 2). Male weanling New Zealand White rabbits were administered 100 mg of TETA orally for 10 d and maintained on a low copper diet. The basal diet consisted of 20% soy protein, 54% glucose, 5% corn oil, 15% wheat straw, and vitamin and mineral (USP XIV, 5%) mixes. Immediately after TETA treatment, copper deficiency was observed as evident by cardiac hypertrophy, markedly reduced plasma and tissue copper contents, and hypercholesterolemia associated mainly with an elevation (29%) in LDL cholesterol. Thereafter, rabbits maintained on the low copper diet (2.3 μg Cu/g) retained the copper-deficient symptoms and hypercholesterolemia, 10 weeks post-TETA treatment. Despite the hypercholesterolemia and copper-deficient symptoms observed in this study, further studies will be required to firmly establish that hypercholesterolemia can be induced in rabbits by copper-deficient diets alone. Before this can be accomplished, a semipurified diet containing less than 0.5 μg Cu/g and a fiber source that will sustain relatively good growth response, needs to be developed.

# V. STUDIES WITH NO HYPERCHOLESTEROLEMIC RESPONSE

## A. Rats

A well designed study executed by Sherman[46] to examine the interaction of copper and iron deficiency on serum lipids and tissue minerals in female Sprague-Dawley rats failed to induce changes in serum cholesterol level after 18 weeks of treatment (Table 3). The semipurified copper-deficient diet, containing <1 μg Cu and 15 μg Zn, can be considered to be as deficient in copper as most diets used to induce hypercholesterolemia, but the zinc content was relatively low. Despite the 18 week long treatment period and the low copper intake, the rats were only very mildly depleted in copper as evidenced by the lack of anemia, cardiomegaly, and coarse fur, as well as the small reduction in liver copper stores. The female rats used (87 g initial body weight) were at least 1 week post-weanling in age, and may have acquired a higher copper store than weanling rats used in most studies. In addition, the low zinc intake, which is only slightly higher than a marginal zinc intake of 10 μg/g diet, may have allowed a more efficient intestinal absorption of copper. Thus, the combined effect of a higher initial copper status and a low dietary zinc intake may have contributed to the negative results. Interestingly, a trend for the cholesteryl ester to be higher than the controls appeared to be evident even though the difference is not significant. It is reasonable to speculate that had dietary zinc been supplemented at 30 μg/g, as recommended by AIN,[22] an increase in serum cholesterol may have occurred. In view of the fact that hypercholesterolemia has been observed in copper-deficient female rats previously, it would appear that sex of animals exerted negligible influence on the outcome of this study.

The large-scale factorial study of Fischer et al.,[47] (Table 3) which involved various levels of dietary copper and zinc, was also unable to demonstrate an elevation in serum cholesterol with the lowest copper intake of 1.95 μg/g (basal level 0.45 μg, and 1.5 μg supplemented). At this marginal copper level and the highest zinc intake of 45 μg/g, only a small reduction of about 25% in liver copper content was attained. A general trend of gradual increments of liver copper content as time progressed was observed for rats fed the marginal copper diets. This trend suggested that the copper status of these rats was being improved rather than being depleted with time. Thus the lack of change in serum cholesterol can be best explained by the inability of the marginal diets to deplete copper stores to levels comparable to those observed in studies which demonstrated hypercholesterolemia.

The importance of the experimental duration required to induce copper deficiency and hypercholesterolemia was demonstrated in the study of Caster and Doster.[48] No alteration in plasma cholesterol was observed after a 3-week consumption of a copper-deficient diet containing only 0.5 μg Cu/g (Table 3). (Since no tissue copper data were provided, the copper status of these

**TABLE 3**
**Studies Unable to Demonstrate Hypercholesterolemia in Copper-Deficient Animals**

| Author, year | Strain, sex | Age or initial wt, time on diet | Dietary copper in feed (water)[a] | | Protein, carbohydrate, fat, fiber, mineral | Physiological measurements (% change from controls) | | | | Comments |
|---|---|---|---|---|---|---|---|---|---|---|
| | | | Cu content | Zn/Cu | | Tissue Cu | Hemato-crit | Heart wt | Plasma cholesterol | |
| Sherman, 1981[46] | Sprague-Dawley rat, male | Young 87 g, 18 weeks | <1.0, 10 μg/g | — | Casein, 20%<br>Sucrose, 30%<br>Corn starch, 31%<br>Corn oil, 10%<br>Cellulose, 2%<br>NRC, 3.7% | ↓32% (liver) | No change | No change | No change | Total cholesterol tended to be higher, but not significantly |
| Fischer et al. 1980[47] | Wistar rat, male | Weanling, 15 weeks | 1.95, 3.45, 6.45 μg/g | 3.1 → 23.4 | Egg white, 20%<br>Glucose, 38%<br>Starch, 25%<br>Corn oil, 10%<br>Fiber, 3%<br>AIN, 3.5% | ↓ Cu (liver) | — | — | No change | Total cholesterol data highly variable from week 5 —15; largest ↓ in liver Cu in rats fed 45.6 μg Zn/g diet |
| Caster and Doster, 1979[48] | White rat, male | Weanling, 4 weeks | 0.5, 1.8, 5.6 μg/g | 2 → 220 | Casein, 20%<br>Sucrose, 20%<br>Starch, 48%<br>Corn oil, 4%<br>Cellulose, 4%<br>Own mineral, ?% | — | — | — | No change | Total cholesterol data highly variable |
| Prohaska et al., 1985[49] | C57BL mice, male | Weanling, 4 weeks | 0.5 μ/g (0, 20 μg/ml) | — | AIN-76A diet | ↓80% (liver) | — | — | No change | |

[a] Supplemental copper provided in water is represented in parentheses.

young rats could not be evaluated. In view of the average time required to induce hypercholesterolemia — approximately 7 to 9 weeks — elevated plasma cholesterol level may have occurred had this study continued for 4 additional weeks.

## B. Mice

An interesting supposition suggesting that the age of the animal when copper deficiency is initiated as an important factor in the development of hypercholesterolemia was postulated by Prohaska et al.,[49] after finding no hypercholesterolemia in a copper-deficient mice study. Elevated plasma cholesterol was not observed in 12-d-old male offspring from dams fed a copper-deficient diet during gestation and lactation, as well as in 7-week-old male offspring with the period of copper deficiency covering the entire period of lactation and the 4 weeks after weaning. The basal diet was a modified AIN-76A formulation containing 0.5 μg Cu/g diet. Liver copper data were not available for the 12-d-old offspring, but liver copper was reduced 80% in the copper-deficient 7-week-old offspring as compared to controls. These investigators were surprised that hypercholesterolemia was not observed in these older mice, in view of the marked reduction in liver copper, which is indicative of severe copper deficiency, as well as the data from other studies demonstrating elevated plasma cholesterol levels in copper-deficient mice.[7]

Although these studies involving both rats and mice did not provide supportive evidence for the induction of hypocholesterolemia by copper deficiency, Prohaska et al.[49] emphasized the importance of experimental duration, dietary zinc and copper content, and tissue mineral data for the evaluation of a degree of deficiency as factors in the design of experiments to study the influence of dietary copper on cholesterol metabolism. In addition, it would appear that the age or stage of development when copper deficiency is initiated as being critical in the induction of hypercholesterolemia is an attractive supposition and deserves further consideration. During the first 4 to 5 weeks post-weaning, the animals are rapidly laying down new tissues which require a tremendous quantity of cholesterol. The demand of cholesterol for rapid growth may be sufficiently large to have masked the hypercholesterolemic effect of copper deficiency. Thus, in studies which demonstrated little or no difference in final weight gains between dietary treatments, hypercholesterolemia may become evident only after the rapid growth had slowed down. However, hypercholesterolemia may also develop during this rapid growth period when the rate of growth and cholesterol deposition into new tissue are markedly retarded by extreme severity of copper deficiency.[9] Furthermore, in studies which initiated copper deficiency with older animals, hypercholesterolemia may develop when the animals are sufficiently copper-deficient.[35] In addition, when these matured animals are depleted to a lesser extent, an adverse alteration in cholesterol distribution among the lipoprotein fraction may be observed rather than a significant elevation in total cholesterol level in the plasma.[36]

## VI. CONCLUSIONS

A large number of studies and a great deal of information have been generated since the copper deficiency-hypercholesterolemia relationship was first proposed. The following trends are now apparent from the available data:

1. Results from twelve independent laboratories have firmly established that copper deficiency affects cholesterol metabolism, resulting in hypercholesterolemia in both male and female rats.
2. Hypercholesterolemia has also been induced in other species, including humans and nonhuman primates, by dietary copper deficiency.
3. Results from rat studies indicate that hypercholesterolemia was observed when copper intake was markedly restricted or when marginal copper intake was accompanied by a relatively high zinc supplement.
4. In most studies, elevated plasma cholesterol was observed after an average of 7 to 9 weeks of dietary treatment, in 10 to 12-week-old copper-deficient rats, which had slowed down in growth by this age.
5. Rat diets containing $<1$ $\mu$g Cu/g produced hypercholesterolemia of greater magnitude in a short duration and required no excessive zinc supplementation. However, diets with $>2$ $\mu$g Cu/g may require high zinc supplements to promote hypercholesterolemia.

As more studies are performed in the future, the hypercholesterolemic response as a result of copper deficiency may be established as a universal phenomena regardless of strains or even species. In view of the limited number of human studies performed so far, additional human depletion and repletion studies especially designed to study cholesterol and lipoprotein metabolism are needed to firmly establish the copper deficiency-hypercholesterolemia relationship in humans. In addition, animal studies should be performed to elucidate the mechanism(s) responsible for the hypercholesterolemia observed in copper deficiency.

## REFERENCES

1. **Klevay, L. M.,** Hypercholesterolemia in rats by an increase in the ratio of zinc to copper ingested, *Am. J. Clin. Nutr.*, 26, 1060, 1973.
2. **Klevay, L. M.,** Coronary heart disease: the zinc/copper hypothesis, *Am. J. Clin. Nutr.*, 28, 764, 1975.
3. **Klevay, L. M., Inman, L., Johnson, L. K., Lawler, M., Mahalko, J. R., Milne, D. B., Lukaski, H. C., Bolonchuk, W., and Sandstead, H. H.,** Increased cholesterol in plasma in a young man during experimental copper depletion, *Metabolism,* 33, 1112, 1984.

4. **Reiser, S., Powell, A., Yang, C. Y., and Canary, J. J.,** Effect of copper intake on blood cholesterol and its lipoprotein distribution in men, *Nutr. Rep. Int.,* 36, 641, 1987.
5. **Milne, D. B., Omaye, S. T., and Amos, W. H., Jr.,** Effect of ascorbic acid on copper and cholesterol in adult cynomolgus monkeys fed a diet marginal in copper, *Am. J. Clin. Nutr.,* 34, 2389, 1981.
6. **Hing, S. A. O. and Lei, K. Y.,** Copper deficiency and hypercholesterolemia induced by *N, N'*-bis-(2-amino ethyl)-1,3-propane diamine (2,3,2-tetramine) in rabbits, *Fed. Proc. Fed. Am. Soc. Exp. Biol.,* 45, 357, 1986.
7. **Jones, D. G.,** Effects of dietary copper depletion on acute and delayed inflammatory responses in mice, *Res. Vet. Sci.,* 37, 205, 1984.
8. **Klevay, L. M. and Viestenz, K. E.,** Abnormal electrocardiogram in rats deficient in copper, *Am. J. Physiol.,* 240, H185, 1981.
9. **Lau, B. W. C. and Klevay, L. M.,** Plasma lecithin: cholesterol acyltransferase in copper-deficient rats, *J. Nutr.,* 111, 1698, 1981.
10. **Lau, B. W. C. and Klevay, L. M.,** Postheparin plasma lipoprotein lipase in copper-deficient rats, *J. Nutr.,* 112, 928, 1982.
11. **Kopp, S. J., Klevay, L. M., and Feliksik, J. M.,** Physiological and metabolic characterization of a cardiomyopathy induced by chronic copper deficiency, *Am. J. Physiol.,* 245, H855, 1983.
12. **Borg, T. K., Klevay, L. M., Gay, R. E., Siegel, R., and Bergin, M. E.,** Alteration of the connective tissue network of striated muscle in copper deficient rats, *J. Mol. Cell Cardiol.,* 17, 1173, 1985.
13. **Klevay, L. M.,** Hypertension in rats due to copper deficiency, *Nutr. Rep. Int.,* 35, 999, 1987.
14. **Murthy, L. and Petering, H. G.,** Effect of dietary zinc and copper interrelationships on blood parameters of the rat, *J. Agric. Food Chem.,* 24, 808, 1976.
15. **Petering, H. G., Murthy, L., and O'Flaherty, E.,** Influence of dietary copper and zinc on rat lipid metabolism, *J. Agric. Food Chem.,* 25, 1105, 1977.
16. **Stemmer, K. L., Petering, H. G., Murthy, L., Finelli, V. N., and Menden, E. E.,** Copper deficiency effects on cardiovascular system and lipid metabolism in the rat; the role of dietary proteins and excessive zinc, *Ann. Nutr. Metab.,* 29, 332, 1985.
17. **Petering, H. G., Murthy, K. L., Stemmer, K. L., Finelli, V. N., and Menden, E. E.,** Effect of copper deficiency on the cardiovascular system of the rat. The role of the dietary sucrose and excessive zinc, *Biol. Trace Elem. Res.,* 9, 251, 1986.
18. **Lei, K. Y.,** Cholesterol metabolism in copper-deficient rats, *Nutr. Rep. Int.,* 15, 597, 1977.
19. **Lei, K. Y.,** Oxidation, excretion and tissue distribution of [26-$^{14}$C]cholesterol in copper-deficient rats, *J. Nutr.,* 108, 232, 1978.
20. **Looney, M. A. and Lei, K. Y.,** Dietary fiber, zinc and copper: effects on serum and liver cholesterol levels in the rat, *Nutr. Rep. Int.,* 17, 329, 1978.
21. **Shao, M. J. S. and Lei, K. Y.,** Conversion of [2-$^{14}$C]mevalonate into cholesterol, lanosterol and squalene in copper-deficient rats, *J. Nutr.,* 110, 859, 1980.
22. **American Institute of Nutrition,** Report of the AIN ad hoc committee on standards for nutritional studies, *J. Nutr.,* 107, 1340, 1977.
23. **Lin, I. M. and Lei, K. Y.,** Cholesterol kinetic analyses in copper-deficient rats, *J. Nutr.,* 111, 450, 1981.
24. **Lei, K. Y., Hassel, C. A., and Allen, D. K.,** Alterations in plasma lipid, lipoprotein and apolipoprotein concentrations in copper-deficient rats, *J. Nutr.,* 113, 2178, 1983.
25. **Croswell, S. C. and Lei, K. Y.,** Effect of copper deficiency on the apolipoprotein E-rich high density lipoproteins in rats, *J. Nutr.,* 115, 473, 1985.
26. **Hassel, C. A., Carr, T. P., Marchello, J. A., and Lei, K. Y.,** Apolipoprotein E-rich HDL binding to liver plasma membranes in copper-deficient rats, *Proc. Soc. Exp. Biol. Med.,* 187, 296, 1988.

27. **Hassel, C. A., Lei, K. Y., Carr, T. P., and Marchello, J. A.,** Lipoprotein receptors in copper-deficient rats: apolipoprotein E-free high density lipoprotein binding to liver membranes, *Metabolism,* 36, 1054, 1987.
28. **Carr, T. P. and Lei, K. Y.,** High density lipoprotein cholesteryl ester and protein catabolism in hypercholesterolemic rats induced by copper deficiency, *Metabolism,* in press, 1990.
29. **Zhang, J., Lei, K. Y., and Carr, T. P.,** Uptake and degradation of apo E-free high density lipoproteins in cultured hepatocytes of copper-deficient rats, *FASEB J.,* 2, A1206, 1988.
30. **Allen, K. G. D. and Klevay, L. M.,** Copper deficiency and cholesterol metabolism in the rat, *Atherosclerosis,* 31, 259, 1978.
31. **Allen, K. G. D. and Klevay, L. M.,** Hyperlipoproteinemia in rats due to copper deficiency, *Nutr. Rep. Int.,* 22, 295, 1980.
32. **Allen, K. G. D. and Klevay, L. M.,** Cholesterolemia and cardiovascular abnormalities in rats caused by copper deficiency, *Atherosclerosis,* 29, 81, 1978.
33. **Harvey, P. W. and Allen, K. G. D.,** Decreased plasma lecithin: cholesterol acyltransferase activity in copper-deficient rats, *J. Nutr.,* 111, 1855, 1981.
34. **Harvey, P. W. and Allen, K. G. D.,** Lipoproteins and liver lipids in copper-deficient rats, *Nutr. Res.,* 5, 511, 1985.
35. **Lefevre, M., Keen, C. L., Lönnerdal, B., Hurley, L. S., and Schneeman, B. O.,** Different effects of zinc and copper deficiency on composition of plasma high density lipoproteins in rats, *J. Nutr.,* 115, 359, 1985.
36. **Lefevre, M., Keen, C. L., Lönnerdal, B., Hurley, L. S., and Schneeman, B. O.,** Copper deficiency-induced hypercholesterolemia: effects on HDL subfractions and hepatic lipoprotein receptor activity in the rat, *J. Nutr.,* 116, 1735, 1986.
37. **Koo, S. I., Lee, C. C., and Norvell, J. E.,** Effect of copper deficiency on the lymphatic absorption of cholesterol, plasma chylomicron clearance and postheparin lipase activities, *Proc. Soc. Exp. Biol. Med.,* 188, 410, 1988.
38. **Lee, C. C. and Koo, S. I.,** Effect of copper deficiency on the composition of three high-density lipoprotein subclasses as separated by heparin-affinity chromatography, *Biochim. Biophys. Acta,* 963, 278, 1988.
39. **Nielsen, F. H., Zimmerman, T. L., and Shuler, T. R.,** Interactions among nickel, copper and iron in rats. Liver and plasma content of lipids and trace elements, *Biol. Trace Elem. Res.,* 4, 125, 1982.
40. **Johnson, M. A.,** Effects of dietary carbohydrate and other factors in blood lipids during copper deficiency, in *Role of Copper in Lipid Metabolism,* Lei, K. Y. and Carr, T. P., Eds., CRC Press, Boca Raton, 1990, chap. 6.
41. **Wu, B. N., Medeiros, D. M., Lin, K-N., and Thorne, B. M.,** Long term effects of dietary copper and sodium upon blood pressure in the Long-Evans rat, *Nutr. Res.,* 4, 305, 1984.
42. **Reiser, S., Ferretti, R. J., Fields, M., and Smith, J. C.,** Role of dietary fructose in the enhancement of mortality and biochemical changes associated with copper deficiency in rats, *Am. J. Clin. Nutr.,* 38, 214, 1983.
43. **Sinthusek, G. and Magee, A. C.,** Relationships of dietary zinc and copper ratios to plasma cholesterol and liver trace mineral deposition in young rats fed saturated and unsaturated fats, *Nutr. Res.,* 4, 841, 1984.
44. **Frimpong, N. A. and Magee, A. C.,** Effects of dietary copper and zinc on serum lipid parameters of young male rats, *Nutr. Rep. Int.,* 35, 551, 1987.
45. **O'Dell, B. L.,** Copper-Zinc interaction, effect of excess zinc on copper status, in *Proc. N.Z. Workshop on Trace Elements,* University of Otago, Dunedin, 1981, 157.
46. **Sherman, A. R.,** Copper and iron deficiencies: effects on serum lipids and tissue minerals, *Nutr. Res.,* 1, 363, 1981.

47. **Fischer, P. W. F., Giroux, A., Belonje, B., and Shah, B. G.,** The effect of dietary copper and zinc on cholesterol metabolism, *Am. J. Clin. Nutr.*, 33, 1019, 1980.
48. **Caster, W. O. and Doster, J. M.,** Effect of the dietary zinc/copper ratio on plasma cholesterol level, *Nutr. Rep. Int.*, 19, 773, 1979.
49. **Prohaska, J. R., Korte, J. J., and Bailey, W. R.,** Serum cholesterol levels are not elevated in young copper-deficient rats, mice or brindled mice, *J. Nutr.*, 115, 1702, 1985.

Chapter 2

# CHOLESTEROL METABOLISM IN COPPER-DEFICIENT RATS

**K. Y. Lei**

## TABLE OF CONTENTS

## I. INTRODUCTION

Following the induction of hypercholesterolemia in the copper-deficient rat model, a number of investigators have extended their work to examine the possible mechanisms responsible for the observed increase in plasma cholesterol. Some of the earlier studies addressed the changes in tissue distribution of free, ester and total cholesterol. Other studies were designed to examine the influence of copper deficiency on cholesterol synthesis, degradation, and absorption. These studies are reviewed in this chapter to provide an understanding of the changes in cholesterol metabolism associated with copper deficiency. Studies designed to establish the effect of copper deficiency on the *in vivo* plasma clearance and tissue uptake of high density lipoproteins, as well as the distribution of cholesterol among lipoprotein fractions and subfractions, are reviewed fully in Chapters 3 and 4, but will be discussed briefly in the present chapter.

## II. OVERVIEW OF CHOLESTEROL METABOLISM

Intestinal absorption of dietary cholesterol is about 50% efficient in humans and tends to be depressed as intake is increased.[1] The absorbed cholesterol is packaged into intestinal lipoproteins, mostly as chylomicrons. Remnants of chylomicrons are produced after the particles are delivered to muscle and adipose tissues and their triglycerides removed following hydrolysis by endothelial lipoprotein lipase (EC 3.1.1.34). Chylomicron remnants are then cleared rapidly by the liver via a specific receptor-mediated process which recognizes apolipoprotein E.[2] This process is not down regulated by high cholesterol intakes,[3] and prolonged consumption of diets high in cholesterol may result in cholesterol accumulation in the liver. However, the low density lipoprotein receptor (LDL or apo B,E receptor) is down regulated by cholesterol accumulation in the liver, which may lead to the elevation of plasma LDL levels. Accumulation of cholesterol in the liver and LDL in the plasma may occur when the ability of the liver to eliminate cholesterol directly in the bile or to convert cholesterol to bile acid is exceeded by high cholesterol intake.[4] These changes may also be observed when bile acid synthesis is compromised, as in hypothyroidism.[5]

Apart from the dietary source, significant amounts of cholesterol are synthesized *de novo* by tissues. By measuring the rate of [$^3$H]water incorporation into digitonin-precipitable sterols, Spady and Dietschy[6] compared the *in vivo* rates of sterol synthesis in animals that carry most of their cholesterol in high density lipoproteins (HDL; rat and hamster) or LDL (guinea pig) or equally in HDL and LDL (monkey and rabbit). Of the total body sterol synthesis, the liver contributes 50% in rats, 40% in monkeys, 27% in hamsters, 18% in rabbits, and 16% in guinea pigs. Thus the major site of synthesis for the rat is the liver, but the major sites of synthesis for the other species appear to be the gastrointestinal tract, skin, and carcass (mainly muscle). The latter pattern may also apply to baboons and humans.[7] As dietary cholesterol is increased, hepatic cholesterol synthesis tends to diminish. Studies performed on a number of species demonstrate the presence of a negative feedback mechanism that regulates the activity of 3-hydroxy-3-methylglutaryl coenzyme A (HMG-CoA) reductase (EC 1.1.1.34), which catalyzes the rate-limiting step of cholesterol synthesis.[8]

Various metabolic fates are available for hepatic cholesterol. In addition to functioning as an important component of cellular membranes, cholesterol is assembled into lipoproteins that are secreted and transported to peripheral tissues. In view of the large quantity of cholesterol secreted in lipoproteins by the liver, a quantity far greater than total body cholesterol synthesis, a substantial amount of lipoprotein cholesterol must be returned to the liver and recycled to sustain hepatic lipoprotein production. Excess cellular cholesterol can be converted by acyl coenzyme A:cholesterol acyltransferase (ACAT; EC 2.3.1.26) to cholesterol ester and later stored.[9]

Hepatic cholesterol may be eliminated by being secreted into the bile; that not reabsorbed by the enterohepatic circulation is eventually excreted. In addition, cholesterol is converted to bile acid, which aids the intestinal absorption of lipids. The loss of bile acid through this pathway is small since intestinal reabsorption and hepatic uptake is very efficient.[10] A negative feedback mechanism also regulates bile acid synthesis by controlling the activity of cholesterol 7α-hydroxylase (EC 1.14.13.17), which catalyzes the rate-limiting step in bile acid formation.[11,12] For example, bile acid synthesis is depressed by cholic or chenodeoxycholic acid feeding, resulting in an enlarged bile acid pool,[13] but production is enhanced by cholestyramine therapy or ileal exclusion which increase elimination.[14] Finally, a significant amount of cholesterol is lost through the sloughing of cells from skin, hair, and gastrointestinal tract.

## III. TISSUE CHOLESTEROL CONCENTRATIONS

Although numerous laboratories have reported increased plasma total cholesterol in copper-deficient rats (Chapter 1), only a few studies have partitioned the total cholesterol into free and ester components, and even fewer studies have analyzed cholesterol levels in other tissues. In addition, the use of different analytical methods, the presentation of data by bar graphs, and the involvement of large numbers of treatments in multifactorial designs in the various studies created difficulties in the evaluation of the available data. Consequently, selected studies will be presented to provide an overview of the changes in cholesterol concentrations observed in the plasma and various tissues.

### A. Plasma Free and Esterified Cholesterol Concentrations

In addition to the established increase in plasma total cholesterol level

(Chapter 1), increases in both free and esterified cholesterol levels have been observed by Lei[15] in the plasma of copper-deficient rats (Table 1). Similar findings have been reported by Harvey and Allen.[16] The magnitude of the changes was similar even though different analytical procedures were used. The free cholesterol to cholesteryl ester ratio appeared to remain constant,[15,16] indicating that there was no disproportional increase in either cholesterol component. The overall distribution among lipoproteins may also remain relatively constant. The observed increase in plasma cholesterol concentration indicates that the plasma cholesterol pool is enlarged, provided the plasma volume does not decrease. Since an *increase* in plasma volume[17] has been observed in copper-deficient rats, a twofold increase in plasma cholesterol pool size was established when plasma volume enlargement was taken into consideration.[18,19] In a recent study[20] designed to examine the developmental sequence of hypercholesterolemia and anemia in copper-deficient rats, increases in plasma volume as well as plasma cholesterol pool size appeared to precede the actual increase in plasma cholesterol concentration. The enlarged plasma cholesterol pool may have resulted from a reduction of plasma clearance while the amount of cholesterol entering the plasma pool remained constant. Alternatively, the observed hypercholesterolemia may be due to an inability of the clearance processes to handle an increased amount of cholesterol entering the plasma pool. Available data appear to support the latter contention.

### B. Cholesterol Concentrations in Hepatic and Peripheral Tissues

In copper-deficient rats, a small reduction in hepatic total cholesterol concentration has been repeatedly observed in our laboratory,[15,18,19,21-23] as well as confirmed by other investigators[24-26] (Table 1). Reductions in both free and esterified cholesterol levels in the liver of copper-deficient rats have also been observed by us.[15] Similar findings have been reported by Lefevre et al.[26] Among these studies, the magnitude of the changes were similar in spite of the different analytical methods used. The ratio of free cholesterol to cholesteryl ester also remained relatively constant, as in the plasma, indicating that there was no disproportional reduction in either cholesterol components.

Studies using cholestyramine to enhance excretion of endogenous steroids have demonstrated that a reduction in liver cholesterol is accompanied by an increase in hepatic HMG-CoA reductase activity and an enhanced expression of lipoprotein receptors.[27] Indeed, an increase in hepatic HMG-CoA reductase activity was found to be associated with reduction in liver cholesterol level observed in copper-deficient rats.[28] Thus the liver exhibits increased HMG-CoA reductase activity and cholesterol synthesis, possibly in response to the reduced hepatic cholesterol level attributed to a net efflux of liver cholesterol to the plasma pool.

Cholesterol concentrations have also been determined in peripheral tissues

**TABLE 1**
**Effect of Copper Deficiency on Plasma and Tissue Cholesterol Concentrations**

| Investigators | Tissue | Chemical form | Dietary copper: Adequate | Marginal | Deficient | Observed change | Methods |
|---|---|---|---|---|---|---|---|
| Lei (1978)[15] | Serum (mg/dl) | Free | 18.4 ± 1.4 | — | 25.6 ± 1.9 | ↑ | Solvent extraction, saponification, digitonin precipitation, and colorimetric assay |
| | | Ester | 64.8 ± 5.1 | — | 86.9 ± 5.0 | ↑ | |
| | | F/E ratio | 0.28 | | 0.30 | — | |
| | Liver (mg/g) | Free | 1.80 ± 0.07 | — | 1.55 ± 0.08 | ↓ | Same as above |
| | | Ester | 0.75 ± 0.09 | — | 0.59 ± 0.04 | ↓ | |
| | | F/E ratio | 2.4 | | 2.6 | — | |
| Harvey and Allen (1981)[16] | Plasma (mg/dl) | Free | 15.1 ± 1.7 | — | 22.2 ± 3.2 | ↑ | Cholesterol oxidase in the presence and absence of esterase |
| | | Total | 71.2 ± 3.9 | — | 110.8 ± 10.9 | ↑ | |
| | | F/E ratio | 0.27 | | 0.25 | — | |
| | Liver (mg/g dry) | Free | 4.1 ± 0.6 | — | 4.1 ± 0.6 | — | Solvent extraction, enzymatic assay |
| | | Ester | 2.4 ± 0.3 | — | 1.7 ± 0.3 | ↓ | |
| | | F/E ratio | 1.7 | | 2.4 | — | |
| Lefevre et al. (1985)[26] | Liver (mg/g) | Free | 2.58 ± 0.11 | — | 2.22 ± 0.17 | ↓ | Enzymatic assay (Note: Data selected from Large factorial experiment indicating reduction) |
| | | Ester | 0.83 ± 0.13 | — | 0.74 ± 0.13 | — | |
| | | F/E ratio | 3.11 | | 3.00 | — | |
| Nielsen et al. (1982)[24] | Liver (mg/g) | Total | 3.14 | 3.03 | 2.74 | ↓ | Solvent extraction, enzymatic assay (Note: Data selected from large factorial experiment indicating reduction) |
| Hassel et al. (1988)[22] | Liver (mg/g) | Total | 2.24 ± 0.05 | — | 1.65 ± 0.06 | ↓ | Solvent extraction, enzymatic assay |
| | Heart (mg/g) | Total | 0.85 ± 0.03 | — | 0.87 ± 0.03 | — | |
| | Kidney (mg/g) | Total | 2.96 ± 0.22 | — | 3.08 ± 0.18 | — | |
| Hunsaker et al. (1984)[29] | Aorta (mg/g) | Total | 2.1 ± 0.4 | — | 1.9 ± 0.5 | — | Solvent extraction, enzymatic assay |

such as the kidney,[22,23] heart,[22,23] and aorta[29] of copper-deficient rats. Unlike the cholesterol concentrations in the liver and plasma, no consistent trend of alteration was detected in these tissues as a result of copper deficiency.

## IV. CHOLESTEROL DEGRADATION STUDIES

A number of *in vivo* studies, each providing valuable information, has been performed to establish the influence of copper deficiency on the process of cholesterol degradation: (A) bile acid production was estimated by measuring the oxidation of [26-$^{14}$C]cholesterol to $^{14}CO_2$; (B) biliary steroid production and excretion was determined by assaying the amount of bile acids and cholesterol collected from bile duct cannulae over 72 h; (C) the pool size and half-life of a rapidly exchangeable pool, which consisted of cholesterol from plasma, liver, and erythrocytes as well as some cholesterol in viscera tissues, were determined by a 2-pool kinetic analysis of serum cholesterol decay curve; and (D) kinetics of plasma clearance and tissue uptake of HDL ester were established by the use of [$^{3}$H]cholesteryl linoleyl ether as a non-degradable marker.

### A. Oxidation, Excretion, and Tissue Distribution of [26-$^{14}$C]cholesterol

In order to examine the influence of dietary copper on cholesterol catabolism, Lei[15] employed [26-$^{14}$C]cholesterol to label the tissue cholesterol pools. Since the three-carbon fragment derived from the oxidation of the cholesterol side chain is mainly used for $CO_2$ production, the determination of exhaled $^{14}CO_2$ provided a measure of cholesterol oxidation and bile acid formation. The study involved four treatments in a factorial arrangement with two levels of dietary copper ($<$2 and 18 mg/kg of diet) and two levels of dietary cholesterol (0 and 20 g/kg of diet). The basal diet consisted of casein, glucose, corn oil, vitamin and mineral mixes, as well as $<$2 mg copper and 120 mg Zn/kg (Chapter 1, Table 1). After 8 weeks of treatment, the 11-week-old male rats were injected intraperitoneally with 2 $\mu$Ci of [26-$^{14}$C]cholesterol per 300 g body weight. Respiratory $^{14}CO_2$ was collected 10 d later as established by Ginter et al.[30] A 20-min $CO_2$ collection was used as a measure of cholesterol catabolism. In addition, feces were collected on days 7 to 10 postinjection to provide a measure of [26-$^{14}$C]cholesterol excretion as fecal lipids. The rates of cholesterol oxidation and excretion were calculated from the specific activity of serum total cholesterol and the amounts of radioactivity exhaled in respiratory $CO_2$ and excreted in fecal lipids, respectively.

Tissue cholesterol data suggested that a shift of cholesterol from the liver to serum pool may be responsible for the hypercholesterolemia observed in copper-deficient rats.[15] The liver total cholesterol concentration in rats fed the copper-deficient diets appeared to be reduced but did not reach significance as in other studies (Table 2). However, a significant reduction (15%) of liver free cholesterol concentration was observed. A significant reduction (24%)

**TABLE 2**
**Influence of Dietary Copper and Cholesterol on Liver Cholesterol Concentration and Specific Activity[a]**

| Liver cholesterol | Control (n = 12) | Copper deficient (n = 10) | Control +2% cholesterol (n = 11) | Copper deficient +2% cholesterol (n = 10) | ANOVA[b] |
|---|---|---|---|---|---|
| | | Concentration (mg/100 g) | | | |
| Free | 180 ± 7 | 155 ± 8 | 179 ± 22 | 151 ± 11 | Cu |
| Ester | 75 ± 9 | 59 ± 4 | 907 ± 100 | 892 ± 91 | Ch |
| Total | 255 ± 11 | 214 ± 10 | 1086 ± 89 | 1042 ± 86 | Ch |
| | | Specific activity (dpm × $10^3$/mg) | | | |
| Free | 8.34 ± 0.27 | 8.42 ± 0.44 | 3.29 ± 0.19 | 3.05 ± 0.25 | Ch |
| Ester | 5.38 ± 0.22 | 4.09 ± 0.12 | 2.77 ± 0.23 | 3.04 ± 0.28 | Ch, X |
| Total | 7.50 ± 0.19 | 7.21 ± 0.32 | 2.87 ± 0.22 | 3.03 ± 0.27 | Ch |

[a] Values represent means ± SEM.
[b] Two-way analysis of variance using dietary copper (Cu) and cholesterol (Ch) as factors. Presence of symbol indicates significant effect at the $p < 0.05$ level; X = interaction.

From Lei, K. Y., *J. Nutr.*, 108, 232, 1978. With permission.

in specific activity of liver cholesteryl ester was also observed in rats fed the copper-deficient diet with no added cholesterol. As expected, cholesterol supplementation markedly increased (13.4-fold) hepatic cholesteryl ester concentration to the same extent in both copper-deficient and copper-adequate rats.

In addition, cholesterol supplementation significantly increased the concentration and reduced the specific activities of free, esterified, and total cholesterol in the serum[15] (Table 3). In the rats fed the copper-deficient diet with no added cholesterol, serum free and esterified cholesterol levels were increased to the same extent and slightly lower, respectively, as in cholesterol-supplemented rats. Furthermore, small reductions in specific activities of free, esterified, and total cholesterol in the serum were observed in rats fed the copper-deficient diets. These reductions may have resulted from an increase in the serum cholesterol pool size or an increase in turnover of these metabolites or both.

As expected, cholesterol supplementation significantly increased the rates of cholesterol oxidation (3.25-fold) and excretion (1.9-fold).[15] In contrast, copper deficiency exerted no effect on cholesterol oxidation and excretion (Table 4). The data demonstrated that the copper-deficient rats receiving no cholesterol supplement possessed the capacity to increase cholesterol oxida-

**TABLE 3**
**Influence of Dietary Copper and Cholesterol on Serum Cholesterol Concentration and Specific Activity[a]**

| Serum cholesterol | Control (n = 12) | Copper deficient (n = 10) | Control +2% cholesterol (n = 11) | Copper deficient +2% cholesterol (n = 10) | ANOVA[b] |
|---|---|---|---|---|---|
| | | | Concentration (mg/100 g) | | |
| Free | 18.4 ± 1.4 | 25.6 ± 1.9 | 26.8 ± 0.8 | 25.4 ± 1.4 | Ch, X |
| Ester | 64.8 ± 5.1 | 86.9 ± 5.0 | 92.6 ± 5.8 | 100.8 ± 5.1 | Cu, Ch |
| Total | 83.2 ± 5.7 | 112.4 ± 5.9 | 119.4 ± 6.1 | 126.2 ± 5.6 | Cu, Ch |
| | | | Specific activity (dpm × $10^3$/mg) | | |
| Free | 8.25 ± 0.33 | 7.13 ± 0.25 | 2.83 ± 0.23 | 2.61 ± 0.15 | Cu, Ch |
| Ester | 8.55 ± 0.49 | 7.77 ± 0.50 | 2.92 ± 0.34 | 2.75 ± 0.18 | Cu, Ch |
| Total | 8.37 ± 0.35 | 7.57 ± 0.39 | 2.89 ± 0.30 | 2.71 ± 0.16 | Cu, Ch |

[a] Values represent means ± SEM.
[b] Two-way analysis of variance using dietary copper (Cu) and cholesterol (Ch) as factors. Presence of symbol indicates significant effect at the $p < 0.01$ level; X = interaction.

From Lei, K. Y., *J. Nutr.*, 108, 232, 1978. With permission.

**TABLE 4**
**Influence of Dietary Copper and Cholesterol on Oxidation and Excretion of [26-$^{14}$C]Cholesterol[a]**

| | Control (n = 12) | Copper deficient (n = 10) | Control +2% cholesterol (n = 11) | Copper deficient +2% cholesterol (n = 10) | ANOVA[b] |
|---|---|---|---|---|---|
| Cholesterol converted to $CO_2$ (mg/d) | 5.92 ± 0.46 | 7.15 ± 0.97 | 21.18 ± 2.40 | 21.31 ± 1.79 | Ch |
| Cholesterol excreted in fecal lipids (mg/d) | 1.87 ± 0.43 | 2.88 ± 0.49 | 5.06 ± 0.67 | 3.98 ± 0.86 | Ch |

[a] Values represent means ± SEM.
[b] Two-way analysis of variance using dietary copper (Cu) and cholesterol (Ch) as factors. Presence of symbol indicates significant effect at the $p < 0.01$ level.

From Lei, K. Y., *J. Nutr.*, 108, 232, 1978. With permission.

tion and excretion when compared to those receiving cholesterol supplement. However, this mechanism was not utilized by the rats fed the copper-deficient diet with no added cholesterol as a means to normalize the hypercholesterolemia. The possibility of reduced cholesterol degradation and elimination being responsible for the observed hypercholesterolemia in copper deficiency appears to be unlikely.

### B. Biliary Steroid Excretion

Biliary excretion of bile acids and cholesterol represents the major pathway for cholesterol elimination. An impairment in this process may be a contributory factor for hypercholesterolemia observed in copper deficiency. The mixed-function oxygenase (monooxygenase) which hydroxylates the steroid nucleus in bile acid synthesis demonstrates certain similarities to other monooxygenases requiring copper.[31] Thus, Allen and Klevay[32] quantitated bile acids and cholesterol collected from biliary cannulae to determine whether the hypercholesterolemia in copper-deficient rats is a consequence of reduced bile acid synthesis and excretion. Male weanling Sprague-Dawley rats were allotted to two treatments, copper-deficient (0.57 μg Cu/g diet) and pair-fed control (5.57 μg Cu/g diet). The basal diet consisted of 51% sucrose, 20% egg white, 9.5% coconut oil, and 9.5% cottonseed oil (Chapter 1, Table 1). After 36 d, pairs of rats were fasted for 12 h, anesthetized with ether, and the bile ducts cannulated. The cannulae exited at the interscapular region, sutured to the skin, and suspended inside a restraint spring attached to the top of the cage to allow the rat free access to feed and water. Biliary excretions were collected at 0 to 8, 8 to 24, 24 to 48, and 48 to 72 h after cannulation. A copper-deficient rat and its pair-fed control were processed on subsequent days from day 36 until all ten pairs of rats were cannulated by day 63. Total biliary cholesterol was determined enzymatically with cholesterol esterase and oxidase. Ether extracts of the saponifiable fraction derived from bile samples were used for total bile acid determination. A modified procedure of the method of Turnberg and Anthony-Mote,[33] using 3α-hydroxysteroid dehydrogenase, was employed.

In general, total biliary bile acid and cholesterol excretion, as well as bile volume for both treatments, tended to be lower during 0 to 8 h and 8 to 24 h, because of the shorter duration, as compared to later collection periods[32] (Table 5). It should be noted that these earlier periods were preceded by surgery and a 12-h fast. A steady state was attained by the 24 to 48 h and 48 to 72 h collection periods for both treatment groups. In the controls, these parameters were significantly lower during 0 to 8 h and 8 to 24 h and were comparable by the second and third day as compared to the deficient rats. The initially depressed level of total bile acid and cholesterol excretion appeared to reflect the lower volume of bile produced in the controls immediately following cannulae placement. These differences may have resulted from a more intensive surgical trauma suffered by the controls. However, the controls

**TABLE 5**
**Biliary Steroid Excretion in Copper-Deficient and Pair-Fed Control Rats With Biliary Cannulae[a]**

| | Time period (h) | Pair-fed control | Copper deficient | *p*-value[b] |
|---|---|---|---|---|
| Bile volume (ml) | 0—8 | 4.35 ± 0.16 | 6.16 ± 0.35 | <0.005 |
| | 8—24 | 6.72 ± 0.87 | 8.76 ± 0.51 | NS |
| | 24—48 | 9.91 ± 0.31 | 11.07 ± 0.62 | NS |
| | 48—72 | 9.83 ± 0.11 | 10.33 ± 0.57 | NS |
| Total biliary bile acids (μmol 3α-hydroxysteroid) | 0—8 | 59.91 ± 2.27 | 93.66 ± 6.76 | <0.001 |
| | 8—24 | 38.34 ± 3.03 | 70.87 ± 6.72 | <0.005 |
| | 24—48 | 74.73 ± 4.67 | 91.57 ± 7.16 | NS |
| | 48—72 | 110.05 ± 5.82 | 116.48 ± 7.98 | NS |
| Total biliary cholesterol (mg) | 0—8 | 0.60 ± 0.07 | 0.80 ± 0.06 | NS |
| | 8—24 | 0.56 ± 0.06 | 1.08 ± 0.15 | <0.025 |
| | 24—48 | 1.67 ± 0.14 | 1.86 ± 0.08 | NS |
| | 48—72 | 1.72 ± 0.13 | 1.88 ± 0.13 | NS |

[a] Values represent means ± SEM.
[b] NS = not significant ($p > 0.05$).

From Allen, K. G. D. and Klevay, L. M., *Atherosclerosis*, 31, 259, 1978. With permission.

rapidly recovered and a steady state was attained by days 2 and 3. Alternatively, the transient increase in bile acid collected from copper-deficient rats may reflect the presence of an enlarged enteroheptic bile acid pool which was rapidly depleted during the first day of collection. Subsequent collections on days 2 and 3 may be more meaningful since they represent steady-state production to replace the diverted bile acids. The overall profile of biliary excretion indicates that the *steady-state* excretion of bile acids and cholesterol was not impaired in copper deficiency. Furthermore, the inability of the copper-deficient rats to markedly increase biliary steroid excretion in order to counteract the copper deficiency-induced hypercholesterolemia would suggest that cholesterol from the enlarged plasma cholesterol pool is unavailable for bile acid production and elimination.

### C. Cholesterol Kinetic Analysis

The influence of copper deficiency on the kinetics of cholesterol metabolism was examined by Lin and Lei[34] using a two-pool kinetic analysis of the disappearance curve of serum cholesterol-specific activity following [4-$^{14}$C]cholesterol administration. The study involved two treatment groups, copper-deficient and copper-adequate (1.6 and 8 mg/kg of diet, respectively). The basal diet consisted of casein, glucose, corn oil, and cellulose, as well as vitamin and mineral mixes (Chapter 1, Table 1). It was similar to that recommended by the American Institute of Nutrition[35] except that no copper supplement was included in the mineral mix. Weanling male Sprague-Dawley rats were provided with their respective diets *ad libitum* for 16 weeks. After 5 weeks of treatment, rats were injected intraperitoneally with 6.35 μCi [4-$^{14}$C]cholesterol per kilogram body weight. A disappearance curve of the serum cholesterol specific activity (SA) was obtained for each treatment group by killing one animal at 1, 2, 7, 14, 28, 42, 56, 70, and 77 d after tracer injection. The study was designed to generate five decay curves of serum cholesterol SA for each treatment. These curves were subjected to a kinetic two-pool analysis[36,37] to derive a number of kinetic parameters (Table 6).

The turnover of plasma cholesterol in humans conforms to a two exchangeable-pool model.[37] A report by Avigan et al.[38] also indicated that the turnover of serum cholesterol in the rat could be described by a two-compartment system. The conformation of serum cholesterol data to a two-pool model in these studies suggested that the various tissue pools of cholesterol could be divided into two groups in terms of the rates at which they equilibrated with serum cholesterol.[37] The first group of pools equilibrated rapidly (i.e., hours to days) with serum cholesterol, whereas the second group of pools equilibrated slowly (i.e., days to weeks) with serum cholesterol. When the rates of equilibration of the different pools with plasma were sufficiently similar, this group of pools was assumed to behave as a single pool.

In the studies by Lin and Lei,[34] curve analysis indicated that the turnover of serum cholesterol in rats could be described by a two-pool model. Copper deficiency significantly increased the pool size ($M_A$) of the rapidly turning over pool A, reduced the rate constant ($K_{AA}$) for the total removal of cholesterol from pool A, lowered the rate constant ($K_A$) for the irreversible removal of cholesterol from pool A, and reduced the rate constant for the transport of cholesterol from pool A to pool B[34] (Table 6). However, copper deficiency did not alter the production rate ($PR_A$) of cholesterol in pool A, but did result in a significant increase in the half-life ($t_{1/2A}$) for the "fast curve".

Field et al.,[39] Chobanian and Hollander,[40] Goodman,[41] and Gould et al.[42] have indicated that the plasma-free cholesterol, cholesteryl ester, red blood cell cholesterol, and liver cholesterol all equilibrated with each other rapidly. Therefore, the plasma, red blood cell, and liver cholesterol were all part of the fast turning over pool (pool A). Chobanian and Hollander[40] reported that some of the cholesterol in several other viscera such as spleen, kidney, lung, and intestine also equilibrated with plasma cholesterol sufficiently fast to comprise part of pool A. The remainder of the cholesterol in these viscera and the cholesterol in peripheral tissues (such as skeletal muscle) equilibrated with plasma cholesterol slowly to comprise the slow turning over pool, pool B.[39,40]

In the interpretation of the enlarged pool size and prolonged half-life of the rapidly equilibrating pool as a result of copper deficiency, it is important

## TABLE 6
## Turnover of Serum Cholesterol in Rats[a]

| | Copper-adequate | Copper deficient | *p*-value[b] |
|---|---|---|---|
| $C_A$, dpm/mg (Y-intercept of "fast" curve) | 12129 ± 500 | 7771 ± 1114 | <0.01 |
| $C_B$, dpm/mg (Y-intercept of "slow" curve) | 8246 ± 423 | 8247 ± 873 | NS |
| $t_{1/2A}$ (half-life of "fast" curve) | 2.61 ± 0.30 | 4.13 ± 0.61 | <0.05 |
| $t_{1/2B}$ (half-life of "slow" curve) | 18.94 ± 0.72 | 18.54 ± 0.70 | NS |
| α, $day^{-1}$ (slope of "fast" curve) | 0.278 ± 0.026 | 0.172 ± 0.015 | <0.05 |
| β, $day^{-1}$ (slope of "slow" curve) | 0.0368 ± 0.0014 | 0.0376 ± 0.0015 | NS |
| $K_{AA}$, $day^{-1}$ (rate constant for cholesterol removal from pool A) | −0.1800 ± 0.0149 | −0.1008 ± 0.0071 | <0.005 |
| $K_{BB}$, $day^{-1}$ (rate constant for cholesterol removal from pool B) | −0.1350 ± 0.0132 | −0.1083 ± 0.0160 | NS |
| $M_A$, mg (size of pool A) | 203.21 ± 9.97 | 255.56 ± 10.30 | <0.01 |
| $PR_A$, mg/d (rate of entry of new cholesterol to pool A) | 15.31 ± 0.27 | 15.40 ± 0.31 | NS |
| $K_{BA}$, $day^{-1}$ (rate constant for cholesterol transport from B → A) | 0.1350 ± 0.0132 | 0.1083 ± 0.0160 | NS |
| $K_A$, $day^{-1}$ (rate constant for cholesterol removal from pool A) | 0.0758 ± 0.0022 | 0.0607 ± 0.0034 | <0.01 |
| $K_{AB}$, $day^{-1}$ (rate constant for cholesterol transport from A → B) | 0.1042 ± 0.0131 | 0.0401 ± 0.0038 | <0.005 |

[a] Values represent means ± SEM.
[b] NS = not significant ($p < 0.05$).

From Lin, I. M. and Lei, K. Y., *J. Nutr.*, 111, 450, 1981. With permission.

to bear in mind that this pool of cholesterol was comprised not only of cholesterol from serum but also that of liver and erythrocytes, as well as some of the cholesterol from several other visceral tissues. A number of possibilities exist which may explain the alterations associated with pool A. The most promising possiblity is that the prolonged half-life of cholesterol within this rapidly exchangeable pool may have resulted from the cholesterol being reused for an enhanced hepatic lipoprotein synthesis, an increase in hepatic cholesterol synthesis, and a prolonged net efflux of cholesterol from the liver to the plasma pool, leading to hypercholesterolemia. This contention is particularly attractive in view of the data described in the following section.

**D. Plasma Clearance and Tissue Uptake of [$^3$H]cholesteryl Ether**

A detailed description of the experimental design, the methodology and discussion in the use of [$^3$H]cholesteryl linoleyl ether as a nondegradable marker for HDL cholesteryl ester, as well as the use of $^{125}$I-HDL to label HDL protein in a dual label approach to examine plasma clearance and tissue uptake of HDL cholesteryl ester and protein is provided in Chapter 4. Hence, only a brief description of these studies[18,19] is presented here to provide an overview of HDL cholesterol metabolism.

In these experiments[18,19] the HDL fraction was selected to be labeled and studied since greater than 80% of the plasma cholesterol of the rat is carried by HDL and both HDL cholesterol and protein are selectively elevated as a result of copper deficiency. Copper-deficient (0.8 μg Cu/g) and control (6.5 μg Cu/g) diets were fed 6 weeks to induce copper deficiency in male weanling Sprague-Dawley rats. In the copper-deficient rats marked enlargement of the plasma pools of HDL cholesterol and protein were accompanied by reductions in hematocrits and liver cholesterol level, which are established signs of copper deficiency. HDL were isolated and purified by ultracentrifugation and column chromatography. The HDL cholesteryl ester and protein components were labeled with [$^3$H]cholesteryl linoleyl ether ([$^3$H]CLE) and with $^{125}$I, respectively. Freshly labeled HDL were administered i.v. to rats from the same treatment from which it was originally derived. Blood and tissue samples were obtained at various time intervals up to 12 h postinjection for plasma clearance and tissue uptake studies. Plasma kinetic analysis indicated that cholesterol ester was cleared 1.6-fold faster than HDL protein in controls and 2.5-fold faster in copper-deficient rats. Absolute mass clearance and tissue uptake of both HDL components were increased in copper deficiency. Hepatic uptake of HDL cholesteryl ester was increased 2.1-fold and the liver was responsible for virtually all of the increased removal of HDL cholesteryl ester from the plasma of copper-deficient animals. In both treatment groups, the skin and muscle were major sites of HDL protein uptake. These bulk tissues were responsible for most of the increased removal of plasma HDL protein in copper-deficient rats. Since previous studies demonstrated that copper deficiency exerted little or no change in bile acid and cholesterol excretion,[15,32] the increased uptake of HDL cholesteryl ester by the liver may be used to sustain an enhanced synthesis and secretion of nascent lipoproteins.

## V. CHOLESTEROL SYNTHESIS STUDIES

A logical explanation for the hypercholesterolemia in copper-deficient rats is a prolonged, sustained increase in cholesterol production. The influence of copper deficiency on the process of cholesterol synthesis has been examined by *in vitro* and *in vivo* studies. The *in vitro* incubation of liver slices and [$^{14}$C]acetate provided evidence that fatty acid and cholesterol synthesis may be increased in copper deficiency.[43] An increase in hepatic HMG-CoA reductase activity confirmed the presence of an increased cholesterol synthetic capacity in copper-deficient rats.[28] In addition, results of an *in vivo* study using [$^{3}$H]water to measure cholesterol synthesis also provided evidence for an enhanced cholesterol synthesis in copper-deficient rats.[44] Furthermore, an accelerated efflux of newly synthesized cholesteryl ester from the liver to the plasma confirmed the secretion process is enhanced in copper deficiency.[21] Thus, several lines of evidence derived from different experimental approaches support the contention that hepatic cholesterol synthesis is increased, leading to an increased net efflux from liver to plasma, subsequently resulting in hypercholesterolemia.

### A. *In Vitro* Hepatic Cholesterogenesis and Lipogenesis From [$^{14}$C]acetate

An *in vitro* incubation system using freshly prepared liver slices was employed by Lei[43] to examine the effect of copper deficiency on the rates of cholesterol and fatty acid synthesis from [1-$^{14}$C]acetate. The experiment consisted of four treatments in a factorial arrangement, with two levels of dietary copper (<2 and 18 mg/kg diet) and two levels of dietary cholesterol (0 and 20 g of cholesterol supplemented per kilogram diet). The basal diet, deficient in copper, contained less than 2 mg Cu/kg and 120 mg Zn/kg (Chapter 1, Table 1). Copper was added to the adequate diet to provide 18 mg Cu/kg. Weanling male Sprague-Dawley rats were fed their respective diets *ad libitum* for 9 weeks. Thereafter, rats were killed and livers were rapidly excised. Slices of the left lateral lobe of liver, weighing about 200 mg, were prepared with a Stadie-Riggs hand microtome and incubated for 3 h at 37° in 3 ml Krebs-Ringer bicarbonate buffer containing per milliliter: 0.1 U porcine insulin, 5 μmol glucose, 10 μmol acetate and 0.2 μCi [1-$^{14}$C]acetate. The incubation procedures and methods for isolation and counting the radioactive fatty acids and digitonin-precipitable sterols (DPS) were described previously.[45] Incorporation of [1-$^{14}$C]acetate into DPS was assumed to represent rates of cholesterol synthesis. Since digitonin precipitates mainly 3-β-hydroxysterols, the contribution of other newly synthesized 3-β-hydroxysterols (mainly lanosterol) is small in comparison to cholesterol.

Established signs of copper deficiency such as cardiac hypertrophy, growth depression, reduction in plasma copper level and depleted hepatic copper stores were observed in rats fed the copper-deficient diets.[43] In addition, reductions in liver lipid and cholesterol levels were observed in copper-deficient rats, especially those supplemented with dietary cholesterol. A significant increase in the rate of synthesis for fatty acids, and not for cholesterol (Table 7), was observed when the data were expressed as synthetic rate per 100 mg liver tissue (nanomoles of $^{14}$C-acetate incorporated into DPS and fatty acids per 100 mg wet liver tissue per 3 h). In view of the marked differences in body and liver weights between the treatments, the synthesis rates were expressed as nanomoles of $^{14}$C-acetate incorporated per liver organ per 100 g body weight in 3 h. When the organ and body weights were taken into consideration, a small but significant trend of increased cholesterol synthesis was established in copper-deficient rats. In addition, a twofold increase in hepatic fatty acid synthesis was still evident in the deficient groups when the synthetic rate was recalculated.

A dramatic change in energy metabolism of the myocardium of severely copper-deficient rats was reported by Kelly et al.[46] Increases in lactic dehydrogenase and mitochondrial oxidative enzymes, including NAD diaphorase as well as succinic, malic, and glutamic dehydrogenases, were observed in myocardium of copper-deficient rats. The feedback control of NAD dehydrogenase of the citric acid cycle is achieved primarily through changes of the oxidation-reduction state of the pyridine nucleotide system.[47] Thus, some or all of the increased enzyme activities observed by Kelly et al.[46] may represent a loss of coordinated control in the presence of high NADH levels. The liver may also respond to copper deficiency by increases in enzyme activity of the glycolytic and citric pathways and thus increase the flow of substrates through these pathways. This may explain the observed increased rate of conversion of [$^{14}$C]acetate into fatty acids.[43]

Increases in the observed synthetic rates for fatty acids and possibly cholesterol indicate that the liver of the copper-deficient rat has an increased synthetic capacity capable of providing lipids to sustain and enhance production of lipoproteins. This *in vitro* finding supports the hypothesis that increases in the production of fatty acids and cholesterol, incorporation of lipids into lipoproteins, and lipoprotein secretion are responsible for the hypercholesterolemia and hyperlipoproteinemia observed in copper-deficient rats.

### B. *In Vivo* Cholesterogenesis From [$^{14}$C]mevalonate

An *in vivo* study was performed by Shao and Lei[21] to examine the influence of copper deficiency on the conversion of mevalonate into cholesterol by hepatic and renal tissues in rats. Mevalonate, a more immediate precursor of cholesterol than acetate, was selected since it is mainly committed to cholesterol synthesis. However, in view of the extremely small concentration of mevalonate in tissues and the difficult analytical method, an estimation of this precursor pool is often not possible. In metabolic studies involving the use of radioactive mevalonate as a tracer, the mevalonate pool was assumed

**TABLE 7**
**Influence of Dietary Copper and Cholesterol on Serum and Liver Cholesterol Concentration, Liver Lipids, and Liver Cholesterol and Fatty Acid Synthesis[a]**

| | Control (n = 9) | Copper deficient (n = 9) | Control +2% cholesterol (n = 9) | Copper deficient +2% cholesterol (n = 6) | ANOVA[b] |
|---|---|---|---|---|---|
| Serum cholesterol (mg/dl) | 80.6 ± 5.3 | 113.0 ± 10.8 | 100.8 ± 11.0 | 95.1 ± 10.9 | X |
| Liver cholesterol (mg/g) | 5.13 ± 0.33 | 4.28 ± 0.25 | 55.6 ± 6.2 | 39.8 ± 4.0 | Cu, Ch, X |
| Liver lipids (mg/g) | 45.3 ± 2.1 | 39.4 ± 2.3 | 158.8 ± 11.2 | 108.9 ± 8.3 | Cu, Ch, X |
| Cholesterol synthesis[c] | 1.19 ± 0.29 | 1.17 ± 0.28 | 0.55 ± 0.07 | 0.76 ± 0.15 | Ch |
| Fatty acid synthesis[c] | 8.37 ± 2.37 | 14.21 ± 3.89 | 4.22 ± 0.87 | 9.74 ± 3.50 | Cu |

[a] Values represent means ± SEM.
[b] Two-way analysis of variance using dietary copper (Cu) and cholesterol (Ch) as factors. Presence of symbol indicates significant effect at the $p < 0.05$ level; X = interaction.
[c] Cholesterol and fatty acid synthesis expressed as nmoles of $^{14}C$-acetate incorporated into digitonin-precipitable sterols and fatty acids, respectively, per 100 mg liver per 3 h.

From Lei, K. Y., *Nutr. Rep. Int.*, 15, 597, 1977. With permission.

to remain constant between treatments. The experiment consisted of two levels of dietary copper, namely deficient and adequate. The basal diet, deficient in copper, contained less than 2 mg Cu/kg and 120 mg Zn/kg (Chapter 1, Table 1). Copper was added to form the adequate diet which contained 18 mg Cu/kg. Weanling male Sprague-Dawley rats were fed their respective diets *ad libitum* for 8 weeks. Five fasted rats from each treatment were killed at 40, 80, and 120 min after injection of [2-$^{14}$C]mevalonate (0.5 μCi/100 g body weight; 46 μCi/μmol) via the femoral vein. The dose and time periods were similar to those reported by Hellström et al.;[48] the half-life and turnover of blood radioactivity, after the i.v. injection of 7 μg of dl-[5-$^{3}$H]mevalonate was found to be 28 and 40 min, respectively. The half-life of blood mevalonate labeled with $^{14}$C was reported to be 7.4 min by Hagenfeldt and Hellström.[49] The dose of 3.2 μg of [$^{14}$C]mevalonate for a 200-g rat was considered to be a pulse. Lipids from serum and tissue samples were extracted with chloroform-methanol (2/1; v/v) and partitioned by thin layer chromatography (TLC).[50] Separation of squalene, lanosterol, and cholesterol of kidney tissue was also performed by TLC.[48]

A significant elevation of serum-free cholesterol level was observed in rats fed the copper-deficient diet.[21] Serum cholesteryl ester levels appeared to be higher in the rats fed the copper-deficient diet than in those fed the copper-adequate diet, but the difference was not significant. In a previous study Lei[15] observed a significant elevation of serum free and esterified cholesterol levels in rats fed a similar copper-deficient diet. The free cholesterol SA of the serum (Figure 1A) appeared to increase rapidly from 0 to 40 min, slow down from 80 to 120 min, and approach a peak around 120 min. The data suggest that the free cholesterol newly synthesized from [$^{14}$C]mevalonate in liver and extrahepatic tissues gradually enters the serum pool during the time span of the study. For the rats fed the copper-deficient diet, the serum-free cholesterol SA was not different from that of the controls at each time interval. The fact that the serum-free cholesterol concentration was significantly elevated in rats fed the copper-deficient diet suggests that the actual amount of $^{14}$C-labeled cholesterol was increased. However, the increase in $^{14}$C-labeled serum-free cholesterol was proportional to the overall increase in serum-free cholesterol concentrations, or the serum-free cholesterol SA would be different among the two treatments. The results suggest that the turnover of the serum free cholesterol may not be adversely affected by copper deficiency.

The serum cholesteryl ester SA of rats fed the copper-deficient diet was significantly elevated at all time periods after [$^{14}$C]mevalonate injection (Figure 1B). It would take the controls 80 to 90 min postinjection to reach the same value of serum ester SA attained by the deficient group at 40 min. Thus, in the rats fed the copper-deficient diet, the esterified cholesterol newly synthesized from [$^{14}$C]mevalonate may be entering the serum pool at an increased rate or leaving this pool at a reduced rate.

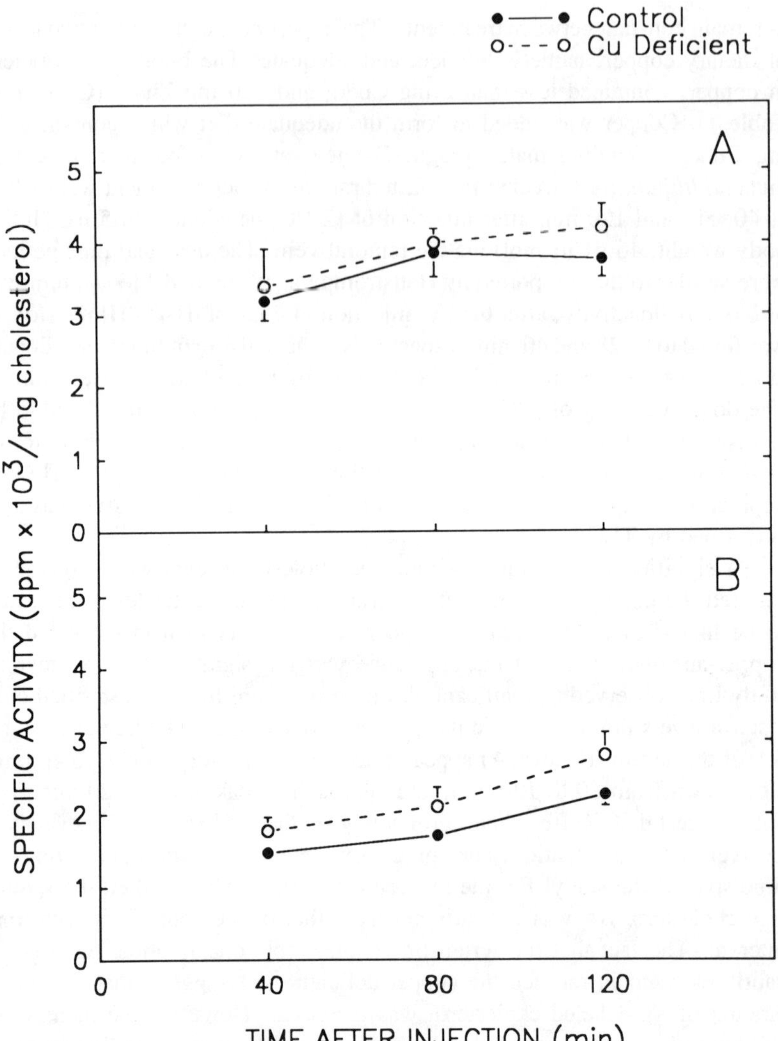


FIGURE 1. Specific activity of free cholesterol (*panel A*) and cholesteryl ester (*panel B*) in the serum after injection of [$^{14}$C]mevalonate. Data were subjected to two-way analysis of variance using dietary copper and time after injection as factors. Each point represents the mean ± SEM of five observations. *Panel A:* A linear time effect was observed ($p < 0.05$). *Panel B:* A dietary copper effect ($p < 0.05$) and a linear time effect ($p < 0.005$) were observed. (From Shao, M. J. S. and Lei, K. Y., *J. Nutr.*, 110, 859, 1980. With permission.)

No significant differences in the free cholesterol level of the liver were observed between the two treatments.[21] However, a significant reduction of liver cholesteryl ester concentration was observed in rats fed the copper-deficient diet. Lei also observed a significant reduction in the concentration of liver total cholesterol[15,43] and free cholesterol[15] in rats fed a copper-deficient diet. A shift of the cholesterol from the liver pool to the serum pool was proposed by Lei[43] as a possible mechanism contributing to the hypercholesterolemia in copper-deficient rats. No difference in the SA of free cholesterol in the liver was observed between the two treatments (Figure 2A). The data indicate that the peak incorporation may have occurred before 40 min. The serum free cholesterol SA appeared to have peaked between 80 and 120 min and thus lent support to the interpretation that the liver-free cholesterol SA may have peaked before 40 min.

The liver cholesteryl ester SA of the rats fed the copper-deficient diet appeared to have reached a peak at or before 40 min (Figure 2B), but that of the control rats seemed to be increasing progressively during the time intervals of the study. The results suggest that the turnover rate of the liver cholesteryl ester pool may be increased in the rats fed the copper-deficient diet. The serum and liver data indicate that esterified cholesterol, newly synthesized from [$^{14}$C]mevalonate by liver and extrahepatic tissues, may be entering the serum pool at an increased rate and the turnover of the serum cholesteryl ester pool may be reduced in rats fed the copper-deficient diet. However, the metabolism of free cholesterol may also be influenced by copper deficiency since the concentration of free cholesterol was elevated in the serum and reduced in the liver[15] in rats fed the copper-deficient diet.

The concentrations of total cholesterol, the radioactivity (RA) of squalene and lanosterol, and the SA of total cholesterol of the kidney were not significantly different between the rats fed the two diets.[21] The RA of kidney squalene appeared to have peaked before 40 min and the RA of kidney lanosterol seemed to have reached a peak at or before 40 min after [$^{14}$C]mevalonate injection. In rats, the kidney converts a significant portion of mevalonate to form squalene and lanosterol. However, the capacity of the kidney to convert squalene and lanosterol to form cholesterol is less than that of the liver.[47,51] In addition, lanosterol and squalene in the kidney may be carried via the blood circulation to the liver to be converted to cholesterol.[49] The SA of kidney total cholesterol was increased progressively after [$^{14}$C]mevalonate injection. The results suggest that squalene and lanosterol, the immediate precursors of cholesterol, newly synthesized from [$^{14}$C]mevalonate within the kidney, were entering the total cholesterol pool in the kidney during the time intervals of the experiment. In addition, the results indicate that mevalonate metabolism in renal tissue was not different between the rats fed the copper-deficient diet and those fed the adequate diet.

In contrast, Allen and Klevay[52] reported a reduction of both plasma and liver cholesteryl ester SA in copper-deficient rats at 4 h after a subcutaneous injection of 3 $\mu$Ci of [5-$^{3}$H]mevalonate per 100 g body weight. They suggested that the reduction in plasma cholesteryl ester SA was the result of an increase in plasma cholesteryl ester pool associated with copper deficiency. In their study Allen and Klevay[52] limited the feed intake of their control rats by pair feeding them against the copper-deficient rats. A 20% increase in body weight

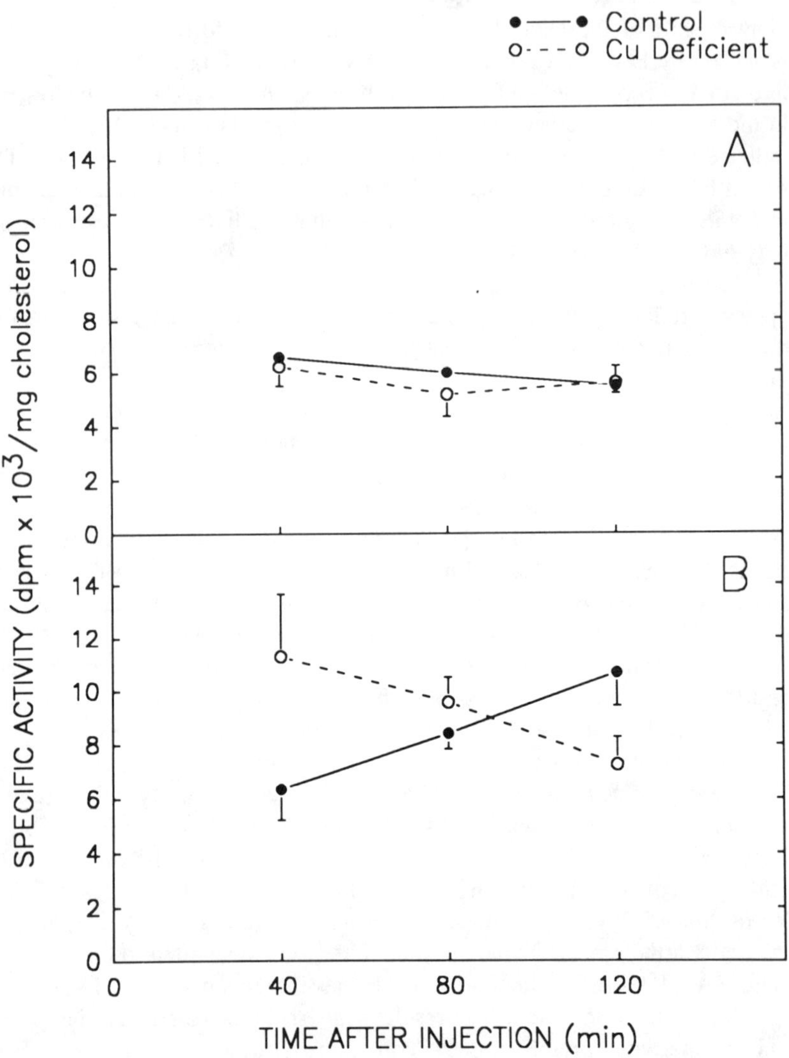


FIGURE 2. Specific activity of free cholesterol (*panel A*) and cholesteryl ester (*panel B*) in the liver after injection of [$^{14}$C]mevalonate. Data were subjected to two-way analysis of variance using dietary copper and time after injection as factors. Each point represents the mean ± SEM of five observations. *Panel B:* An interaction between the copper effect and the linear time effect was observed ($p$ <0.005). (From Shao, M. J. S. and Lei, K. Y., *J. Nutr.*, 110, 859, 1980. With permission.)

gain in the control rats was observed despite the pair-feeding treatment. Thus, the control rats were utilizing the feed more efficiently for growth than the copper-deficient rats.

Significant metabolic adaptions have been reported by Leveille[45] and

Fabry et al.[53] in meal-fed rats. The access of these meal-fed rats to feed and caloric intake was restricted and subsequently they became more efficient in their ability to absorb and store nutrients. In addition, hepatic cholesterogenesis was observed to be much higher in meal-fed rats than in *ad libitum*-fed rats.[45,53]

Since Allen and Klevay[52] restricted the feed intake of rats fed the copper-adequate diet on a pair-fed basis, these rats may have adapted to a meal-feeding pattern similar to that induced in the study of Leveille.[45] Thus a stimulation of cholesterogenesis may be produced in rats fed the copper-adequate diet on a pair-fed basis. This may account for the observation of Allen and Klevay,[52] that the liver and serum cholesteryl ester SA were lower in rats fed the copper-deficient diet as compared to those pair fed the copper-adequate diet. In a subsequent paper, Allen and Klevay[32] converted the SA data published previously[52] into percent dose incorporation data (i.e., the percent of the injected [$^3$H]mevalonate which was converted into plasma or liver cholesterol and lipids). This extrapolation does not take the size of the particular pool into consideration and may be invalid in this case since the serum cholesterol pool is enlarged as a result of copper-deficiency. We feel that in metabolic studies involving the use of isotopes, SA data or the expression of data as a function of enlarged pool size should be used in the interpretation of the results.

### C. *In Vivo* Cholesterogenesis From [$^3$H]water

Recently, Yount et al.[44] estimated cholesterol synthesis by determining the rate of [$^3$H]water incorporation into DPS *in vivo*.[6] Weanling male Sprague-Dawley rats were fed a semipurified diet[35] with or without added copper for a period of 5 weeks. Various organs were removed and analyzed 1 h after injection of labeled substrate. While this technique measures [$^3$H]water incorporated into DPS present within the tissues, it does not account for newly synthesized sterols secreted from the tissues. Therefore, determination was made of DPS from the plasma to represent newly synthesized cholesterol cleared to the plasma compartment during the experiment. The majority of whole-body cholesterol synthesis in either treatment group was attributable to the skin and, to a lesser extent, the muscle. Significant differences in cholesterol synthesis in the kidney, small intestine, and liver were not detected between treatment groups, although synthesis in these organs of copper-deficient rats tended to be greater. On the other hand, the rate of accumulation of newly synthesized cholesterol in the plasma of copper-deficient animals was at least twofold greater compared to controls, implying an increased flux of newly synthesized cholesterol to the plasma compartment during copper deficiency.

Compelling evidence suggests an increased flux of hepatic cholesterol to the plasma in this animal model. Firstly, increased clearance of mevalonate-derived cholesterol by the liver was observed in the form of cholesteryl ester.[21]

Secondly, the activity of 3-hydroxy-3-methylglutaryl coenzyme A (HMG-CoA) reductase, the rate-limiting enzyme in cholesterol synthesis, is significantly increased in fasted and nonfasted copper-deficient rats.[28] It is possible that [$^3$H]water incorporated into cholesterol is also cleared from the liver at an increased rate, but would not have been detected by measuring the amount of radioactivity remaining in the liver at a given time point. That greater DPS-associated radioactivity recovered in the liver was not detected under similar copper-deficient conditions in which increased HMG-CoA reductase activity was observed, further suggests greater clearance of hepatic cholesterol to the plasma. Finally, total hepatic cholesterol concentration is lower in copper-deficient rats,[15,22,23,43] also indicating increased cholesterol flux from the liver. Thus, the present study[44] of [$^3$H]water incorporation into DPS suggests an increase in cholesterol synthesis in copper-deficient rats which is cleared to the plasma at an increased rate. Supportive evidence implies that the majority of increased synthesis may be attributable to the liver, although the contribution of other organs in this model should not be underestimated.

## VI. ACTIVITY OF RELATED ENZYMES

### A. 3-Hydroxy-3-Methylglutaryl Coenzyme A Reductase

3-Hydroxy-3-methylglutaryl coenzyme A (HMG-CoA) reductase (EC 1.1.1.34) catalyzes the formation of mevalonate from HMG-CoA, which is the committed step and the rate-limiting reaction in cholesterol biosynthesis.[8] The enzyme is a transmembrane glycoprotein[54] confined to smooth, ribosome-poor endoplasmic reticulum,[9] and appears to be highly regulated. Inhibition of cholesterol synthesis by cholesterol feeding was found to be exerted at the level of HMG-CoA reductase.[8] In addition, the enhanced hepatic cholesterogenesis to replace bile acids depleted by cholestryramine treatment may also be exerted at this level.[55]

Yount et al.[28] recently examined the influence of copper deficiency on the hepatic HMG-CoA reductase activity in copper-deficient rats. Male weanling Sprague-Dawley rats were fed semipurified diets, consisting of casein, glucose, and cellulose, with two levels of dietary copper, 0.9 μg/g (deficient) and 6.5 μg/g. After 7 weeks of treatment, cholesterol level was elevated in the plasma, and cholesterol as well as copper levels were reduced in the liver of rats fed the deficient diet. These data are consistent with previous findings observed in copper-deficient rats.[15,16,21-26] The method of Shapiro et al.[56] was used to assay for HMG-CoA reductase activity. The active form of HMG-CoA reductase accounted for 45 to 80% of the activity in the liver of fasted adequate rats, but the active form of the enzyme amounted to only 30% of the total activity in nonfasted adequate rats. Hepatic active and total activities were about three- and eightfold higher, respectively, in the nonfasted relative to the fasted adequate rats. In the fasted state, total activity was elevated two-fold with copper deficiency, while the activity of the active form remained unchanged. However, in the fed state, both total and active forms of the enzyme were increased about two fold in the liver of deficient animals compared to controls. This increase in reductase activity indicates the hepatic capacity for cholesterol synthesis is enhanced in copper deficiency, which corresponds well with the expected induction response to the reduction in hepatic cholesterol concentration observed in copper-deficient rats. The increased cholesterol synthetic capacity supports the hypothesis that the hypercholesterolemia and hyperlipoproteinemia associated with copper deficiency result from increased cholesterol and lipoprotein production.

### B. Lecithin:Cholesterol Acyltransferase

Lecithin:cholesterol acyltransferase (LCAT; EC 2.3.1.43) catalyzes the esterification of free cholesterol to form cholesteryl esters.[57] Fatty acids from the C-2 position of lecithin are transferred to the hydroxyl group of unesterified cholesterol. Apolipoprotein A-I of HDL is an activator of LCAT,[58] and the free cholesterol of HDL is a preferential substrate of the enzyme.[57] Apart from tissue uptake of HDL cholesterol by receptor- or nonreceptor-mediated pathways, the transfer of HDL cholesterol esters to LDL and VLDL aided by LCAT would also allow for their removal from the plasma pool when these lower density lipoproteins are taken up by liver and extrahepatic tissues.[58] Thus LCAT appears to play an important role in the transfer of cholesterol from peripheral tissues to the liver for degradation and elimination. The activity of plasma LCAT in copper-deficient rats has been determined by three independent laboratories.[16,26,59] Two earlier studies demonstrated a reduction in activity, but a later study indicated that LCAT was not affected in a consistent manner by copper status.

Lau and Klevay[59] observed a significant decrease (22 to 32% reduction in three experiments) in plasma LCAT activity in copper-deficient rats. Copper deficiency was induced by providing an egg white and sucrose semipurified diet to male weanling Sprague-Dawley rats (Chapter 1, Table 1). After 26 to 46 d, copper deficiency was verified by the presence of anemia, hypercholesterolemia, and reduced kidney and skeletal muscle copper concentrations. LCAT was determined in freshly obtained plasma by the method of Stokke and Norum[60] and expressed as a percentage of acylation of labeled cholesterol per hour per 100 μl plasma. Unfortunately, plasma-free and esterified cholesterol levels were not determined to provide an estimate of the influence of the reduced LCAT activity on the circulating levels of free and esterified cholesterol. These investigators[59] postulated that the reduction in LCAT activity may have resulted from an impaired enzyme synthesis or that LCAT may be a metalloenzyme. Furthermore, they indicated that the reduction in LCAT activity appeared to correlate closely with the reduction in hematocrit (24 to 36% reduction).

In view of the established increase in plasma volume in copper-deficient rats, as the hematocrit is reduced,[17,19] the expression of plasma enzyme activities or metabolite concentrations should take the expanded plasma volume into consideration. Little or no meaningful reduction in plasma LCAT activity is apparent when the expanded plasma volume is accounted for in the expression of LCAT activity.

A similar study was performed by Harvey and Allen[16] to examine the effect of dietary copper deficiency on rat plasma LCAT activity. Male Sprague-Dawley rats were provided with a casein base semipurified diet for 9 weeks. Three groups of rats were used; *ad libitum*-fed copper-deficient group (0.6 μg Cu/g diet), paired-fed control group (5.9 μg Cu/g diet), and *ad libitum*-fed control group (5.9 μg Cu/g diet). The presence of a marked reduction in liver copper (73% reduction) and hypercholesterolemia (57% increase) established the rats were copper-deficient. Plasma LCAT activity was performed by a modification of the method of Stokke and Norum.[60] Plasma LCAT fractional (percent esterified per hour) and molar activity (micromoles esterified per liter per hour) were decreased 74% and 64% respectively, in copper-deficient as compared to pair-fed controls. No difference in activity was observed between the pair-fed and *ad libitum*-fed controls. These investigators[16] postulated that the decrease in LCAT activity in copper deficiency may be due to a direct role for copper in the LCAT enzymic process either as a possible cofactor, regulator, or in the association of LCAT with HDL. Unfortunately, hematocrit and plasma volume were not determined in this study. As a result, the data could not be evaluated precisely on a total plasma volume basis. Similar to results reported by other researchers,[33] the ratio of plasma cholesteryl ester to cholesterol appeared not to be altered by copper deficiency in this study. These findings are inconsistent with the reported reduction in plasma LCAT activity observed. Since this study represents the only LCAT experiment performed by this group of investigators, additional studies, including plasma volume measurements, are clearly needed to confirm their results.

Plasma LCAT activity was also measured by Lefevre et al.[26] in male Sprague-Dawley rats. Animals with initial body weights of 120 to 140 g were maintained for 6 weeks on an egg white and sucrose basal diet supplemented with 0 or 5 μg Cu/g diet and with 1, 10, 100, or 1000 μg Zn/g diet (Chapter 1, Table 1). Rats fed copper-deficient diets with no copper supplement demonstrated increases in heart weight, plasma free cholesterol, HDL cholesterol, and apolipoprotein concentrations, as well as a reduction of about 50% in liver copper level. Plasma LCAT activity was determined by the method of Stokke and Norum.[60] When LCAT activity was expressed in terms of absolute amount of cholesterol esterified per unit time, a significant zinc and copper interaction effect was observed. However, when LCAT activity was expressed in terms of the percent of free cholesterol esterified per unit time, a reduction in activity was observed only in rats fed diets containing 1 μg Zn/g diet. The possible influence of enlarged plasma volume on LCAT activity was not considered in the expression of enzyme activity in this study. Nevertheless, these authors concluded that LCAT activity was not affected in a consistent manner by copper status. Furthermore, they suggested that the elevation in plasma free cholesterol in copper-deficient rats appeared to be due to a selective elevation in plasma HDL rather than to alterations in plasma LCAT activity. Their conclusion was based on the following findings: (1) the ratio of plasma cholesteryl ester to free cholesterol or the percentage of HDL-free cholesterol was not altered;[16,26,33] (2) a consistent reduction of LCAT activity was not observed in copper deficiency when the data were expressed on a percentage or molar basis; (3) the plasma free cholesterol concentration correlated more closely with HDL cholesterol than with LCAT activity.

At present, the available LCAT data appear to be inconsistent and do not support the contention that a reduction in LCAT activity may contribute significantly to the hypercholesterolemia observed in copper-deficient rats.

### C. Lipoprotein and Hepatic Lipases

Lipoprotein lipase (EC 3.1.1.34) is located on the vascular endothelial surface of skeletal muscle, adipose tissue, heart, lung, and mammary gland.[61] It is responsible for the hydrolysis of lipoprotein-associated triglycerides into mono- and diacylglycerols, as well as free fatty acids.[62] Lipoprotein lipase is activated by apolipoprotein C-II and divalent cations such as calcium, magnesium, and manganese,[63] The ability of copper to activate this enzyme has not been established. Since metals are important to the structure and function of at least 27% of all known enzymes,[64] copper may play a major role in the function of this enzyme. Lipoprotein lipase is tightly bound to the endothelial surface by a glycosaminoglycan associated with cell membrane.[65] In order to assay for its activity, heparin, which possesses a high affinity for the enzyme, is administered i.v. to displace the enzyme into the plasma. The overall lipolytic activity contributed by hepatic lipase, also released by heparin administration, can be eliminated by the use of sodium dodecyl sulfate, a specific inhibitor for hepatic lipase.[66]

Following the hydrolysis of triglycerides by lipoprotein lipase, chylomicrons and very low density lipoproteins are converted to smaller chylomicron remnants and intermediate density lipoproteins, respectively. A reduction in lipoprotein lipase activity may lead to delayed formation and clearance of these particles. Thus, Lau and Klevay[67] postulated that a reduction in lipoprotein lipase may contribute to the hypertriglyceridemia and hypercholesterolemia observed in copper-deficient rats.

Significant reductions in post-heparin plasma lipoprotein lipase activity (40 to 47%) were observed in copper-deficient rats.[67] Copper deficiency was induced by providing an egg white and sucrose base semipurified diet (Chapter 1, Table 1) to male weanling Sprague-Dawley rats for 3 to 6 weeks. After hypercholesterolemia was confirmed, rats were fasted overnight, sodium-heparin (60 IU/kg body weight) was injected i.v., and post-heparin blood samples were collected from the vena cava 5 min later. Plasma samples were treated with equal volumes of 50 m*M* SDS (in 0.2 *M* Tris-HCl, pH 8.2) and used as the source of lipoprotein lipase. Assay of lipoprotein lipase was performed by the method of Baginsky and Brown.[66] Copper deficiency was confirmed by significant reductions of cardiac copper (34.4%) and hematocrit (17%) in the copper-deficient rats as compared to the pair-fed controls. In view of the reduction in hematocrit, the enlarged plasma volume may have accounted for a large part of the decrease in lipoprotein lipase activity. Furthermore, the pair-feeding treatment imposed on the control rats is capable of inducing meal-feeding, which has a marked effect on lipid metabolism [45,53] Thus, the meal feeding imposed may elevate the lipoprotein lipase activity in the controls. Nevertheless, these researchers postulated that the decreased lipoprotein lipase activity may be responsible for the hypertriglyceridemia and hypercholesterolemia of copper deficiency.

A similar study performed by Koo et al.[68] has also demonstrated reductions in post-heparin lipoprotein lipase (reduced 28.5%) as well as hepatic lipase (reduced 26.6%) activities in copper-deficient rats. In this study male weanling Fischer-344 rats were fed *ad libitum* an egg white, glucose, and corn starch base semipurified diet containing 1% cholesterol. The copper-deficient group and controls were fed diets containing 0.5 ppm and 7.5 ppm copper, respectively. After 8 weeks of dietary treatment, rats were fasted for 18 h and injected i.v. with 150 IU heparin per kilogram body weight. Blood samples were collected 6 min later through the orbital sinus. Post-heparin lipoprotein lipase and hepatic lipase activities were assayed according to established methods.[69,70] Although lipoprotein lipase activity was reduced, a transient initial increase in the *in vivo* rate of chylomicron removal, without altering the overall clearance, was observed in copper-deficient rats. These investigators[68] suggested that the apo E-dependent hepatic binding and uptake during the initial phase of chylomicron metabolism is facilitated in copper deficiency. Since hepatic lipase is capable of hydrolyzing HDL phospholipids at the surface of hepatic endothelial cells,[71] a reduction in hepatic lipase may hinder the delivery of HDL cholesterol to hepatocytes. Thus, Koo et al.[68] further postulated that the hepatic uptake of cholesterol mediated by hepatic lipase may be reduced and subsequently contribute to the hypercholesterolemia observed in copper deficiency. However, this postulation may not hold true in view of the increased *in vivo* hepatic uptake of HDL cholesteryl ester observed in copper-deficient rats.[19]

## VII. LYMPHATIC ABSORPTION OF CHOLESTEROL

The influence of copper deficiency on the metabolism of intestinally derived cholesterol and its contribution to plasma cholesterol pool was examined by Koo et al.[68] The use of weanling male Fischer-344 rats, the method for the induction of copper deficiency by a copper-deficient diet containing 1% cholesterol, and the duration and severity of dietary treatment were similar to that presented in their lipoprotein and hepatic lipase study.[68] Intestinal absorption of cholesterol was measured by collecting lymph from cannulated major mesenteric lymph ducts at hourly intervals for 8 h following the intraduodenal infusion of a dose of [4-$^{14}$C]cholesterol (0.13 μCi, 10 mg cholesterol, 50 mg Na-taurocholate, and 1 ml of Intralipid®). In copper-deficient rats cumulative $^{14}$C-cholesterol absorption and output of lymphatic cholesterol were reduced at 4 h and thereafter until 8 h. During the 8 h period, a total of only 7.3% and 9.2% of the cholesterol load was absorbed by copper-deficient and copper-adequate animals, respectively. This suggests that most of the large cholesterol load of 10 mg provided may eventually be excreted. No treatment difference was observed for the distribution of $^{14}$C-radioactivity between free and esterified cholesterol in the lymph collected at each time interval. Thus, mucosal esterification of cholesterol appeared not to be compromised by copper deficiency. These researchers suggested that a general reduction in absorptive capacity may be responsible for the reduced cholesterol absorption since partial atrophy of the duodenal and jejunal villi has been observed in copper deficiency.[72] In addition, they postulated that the reduction in cholesterol absorption may have resulted from an impairment in the energy requiring processes involved in the synthesis and secretion of chylomicrons, in view of the marked reduction in the copper-dependent cytochrome oxidase activity and morphological changes in mitochondria of intestinal epithelium of copper-deficient rats.

The effect of copper deficiency on the rate of plasma clearance of chylomicrons was also examined by Koo et al.[68] [$^{3}$H]Retinol was used as chylomicron marker since it is incorporated as the ester form into the particle core and is not readily exchangeable with lipoproteins of higher densities or taken up by extrahepatic tissues. A dose of 27.8 mg of [$^{3}$H]chylomicrons derived from either deficient or adequate rats was injected into both deficient and adequate recipient rats with closely matched body weights in a crossover design. The plasma [$^{3}$H]chylomicron clearance, expressed as percent dose, was faster in copper-deficient rats during the first 3 min. However, the percentage clearance of the dose was not different at 6 and 10 min between treatments. In addition, the half-life of [$^{3}$H]chylomicrons and total $^{3}$H-radioactivity taken up by the liver were not influenced by copper deficiency, irrespective of the source of chylomicrons. Previous studies have established that the rate of plasma clearance of chylomicrons is governed by the rates of chylomicron remnant formation by peripheral lipolysis and their subsequent uptake by the liver. These authors suggested that the initial transient increase in clearance may be due to an initial increase in binding and uptake of remnants by liver apo E receptors. This contention is supported by the elevated *in vitro* binding of apo E-rich HDL by purified hepatic plasma membranes derived from copper-deficient rats.[22] Furthermore, Koo et al.[68] pointed out that, despite the reduction in post-heparin plasma lipoprotein lipase activity,[67,68] the overall plasma chylomicron clearance was not impaired by copper deficiency. Thus, the small changes observed in intestinal cholesterol absorption and

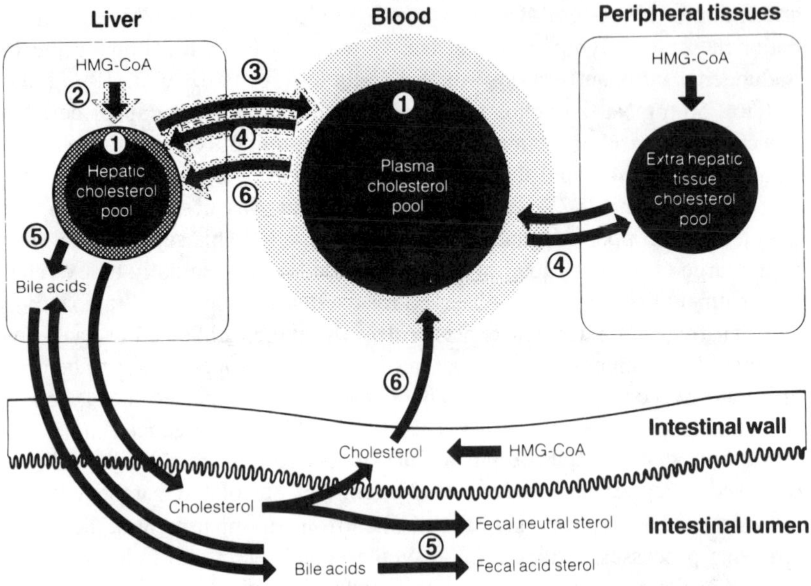


FIGURE 3. Overview of cholesterol metabolism in normal and copper-deficient rats. Established metabolic pathways are represented by solid arrows. Alterations resulting from dietary copper deficiency are represented by dashed lines. (1) Increased plasma and decreased liver cholesterol pools. (2) Increased hepatic HMG-CoA reductase activity. (3) Increased hepatic clearance of newly synthesized cholesteryl ester to the plasma. (4) Increased HDL cholesteryl ester uptake specifically in the liver. (5) No change in bile acid production or fecal cholesterol excretion. (6) Reduction in lymphatic cholesterol absorption and transient increase in chylomicron clearance.

plasma chylomicron clearance may be unrelated to the hypercholesterolemia observed in copper deficiency.

## VIII. CONCLUSION

An overview of cholesterol metabolism in copper-deficient rats is depicted in Figure 3. The following changes are listed by number in Figure 3:

1. Elevations in plasma and reductions in hepatic free and esterified cholesterol concentrations are indicative of an enlarged plasma and a slightly depleted liver cholesterol pool. Since the cholesterol contents of the heart, kidney, and aorta were not affected, the size of these cholesterol pools was not altered.
2. An increased capacity of the liver to synthesize cholesterol was demonstrated by a twofold increase in hepatic HMG-CoA reductase activity *in vitro*.
3. Hepatic cholesterol ester, newly synthesized from [$^{14}$C]mevalonate, was

leaving the liver and entering the serum pool at a markedly increased rate.

4. The absolute mass of HDL cholesterol ester clearance from the plasma was increased twofold. An enhanced hepatic uptake was responsible for most of the increase in plasma clearance.
5. Bile acid production, determined by measuring [26-$^{14}$C]cholesterol oxidation, was not altered. Total biliary steroid and cholesterol excretion, measure by direct quantitation of bile acids and cholesterol collected from bile cannulae, was increased during the first day post-cannulation but was not altered during days 2 and 3 when steady state was reached. Thus, bile acid production and biliary cholesterol excretion were not impaired.
6. A small reduction in intestinal absorption and a transient increase in plasma chylomicron clearance were established.
7. Kinetic analysis of serum [$^{14}$C]cholesterol decay curve established the enlargement of pool size and prolongation of half-life for the rapidly exchangeable cholesterol pool, which is composed of cholesterol from serum, erythrocytes, and liver, as well as some cholesterol from other viscera tissues. This may have resulted from a recycling of cholesterol between the serum and tissue pools and the inability of the cholesterol degradation process to normalize the hypercholesterolemia.

When the above data are considered together, a meaningful overview of altered metabolic processes leading to the observed hypercholesterolemia in copper-deficient rats is apparent. Available data support the concept that cholesterol derived from an increased hepatic cholesterol synthesis is packaged into lipoprotein to facilitate an enhanced clearance of newly synthesized cholesterol ester from the liver to plasma. The increased uptake of HDL cholesteryl ester by the liver indicates that an elevated amount of plasma cholesterol is recycled back to the liver to sustain the net efflux of cholesterol from the liver. Cholesterol derived from the increased hepatic cholesterol synthesis and uptake of plasma HDL cholesteryl ester is unavailable to enhance bile acid production and cholesterol elimination. A prolonged net efflux of cholesterol from the liver results in a depletion of liver cholesterol content, as well as an increased plasma cholesterol concentration and enlarged plasma cholesterol pool.

## REFERENCES


1. **Felig, P., Havel, R. J., and Smith, L. H., Jr.,** Metabolism and nutrition, in *The Biological Principles of Disease,* Smith, L. H., Jr. and Thier, S. O., Eds., W. B. Saunders, Philadelphia, 1981, 479.
2. **Mahley, R. W., Hui, D. Y., Innerarity, T. L., and Weisgraber, K. H.,** Two independent lipoprotein receptors on hepatic membranes of dog, swine and man: apo-B,E and apo-E receptors, *J. Clin. Invest.,* 68, 1197, 1981.
3. **Angelin, B., Ravida, C. A., Innerarity, T. L., and Mahley, R. W.,** Regulation of hepatic lipoprotein receptors in the dog, *J. Clin. Invest.,* 71, 816, 1983.
4. **Havel, R. J.,** The formation of LDL: mechanisms and regulation, *J. Lipid Res.,* 25, 1151, 1984.
5. **Havel, R. J.,** Functional activities of hepatic lipoprotein receptors, *Annu. Rev. Physiol.,* 48, 119, 1986.
6. **Spady, D. K. and Dietschy, J. M.,** Sterol synthesis in vivo in 18 tissues of the squirrel monkey, guinea pig, rabbit, hamster, and rat, *J. Lipid Res.,* 24, 303, 1983.
7. **Turley, S. D. and Dietschy, J. M.,** Cholesterol metabolism and excretion, in *The Liver, Biology and Pathobiology,* Vol. 1, Arias, I., Popper, H., Schacter, D., and Shafritz, D. A., Eds., Raven Press, New York, 1982, 467.
8. **Rodwell, V. W., Nordstrom, J. L., and Mitschelen, J. J.,** Regulation of HMG-CoA reductase, *Adv. Lipid Res.,* 14, 1, 1976.
9. **Balasubramaniam, S., Mitropoulos, K. A., Venkatesan, S., Myant, N. B., Peters, T. J., Postiglione, A., and Mancini, M.,** Analytical fraction of human liver microsomal fractions: localization of cholesterol and of the enzymes relevant to its metabolism, *Clin. Sci.,* 60, 435, 1981.
10. **Angelin, B., Björkhem, I., Einarsson, K., and Ewerth, S.,** Hepatic uptake of bile acids in man: fasting and postprandial concentrations of individual bile acids in portal venous and systemic blood serum, *J. Clin. Invest.,* 70, 724, 1982.
11. **Danielsson, H. and Sjovall, J.,** Bile acid metabolism, *Annu. Rev. Biochem.,* 44, 233, 1975.
12. **Myant, N. B. and Mitropoulos, K. A.,** Cholesterol 7α-hydroxylase, *J. Lipid Res.,* 18, 135, 1977.
13. **Kallner, M.,** The effect of chenodeoxycholic acid feeding on bile acid kinetics and fecal neutral steroid excretion in patients with hyperlipoproteinemia, *J. Lab. Clin. Med.,* 86, 595, 1975.
14. **Grundy, S. M., Ahrens, E. H., Jr., and Salen, G.,** Interruption of the enterohepatic circulation of bile acids in man: comparative effects of cholestyramine and ileal exclusion on cholesterol metabolism, *J. Lab. Clin. Med.,* 78, 94, 1971.
15. **Lei, K. Y.,** Oxidation, excretion and tissue distribution of [26-$^{14}$C]cholesterol in copper-deficient rats, *J. Nutr.,* 108, 232, 1978.
16. **Harvey, P. W. and Allen, K. G. D.,** Decreased plasma lecithin:cholesterol acyltransferase activity in copper-deficient rats, *J. Nutr.,* 111, 1855, 1981.
17. **Lei, K. Y., Hassel, C. A., and Allen, D. K.,** Alterations in plasma lipid, lipoprotein and apolipoprotein concentrations in copper-deficient rats, *J. Nutr.,* 113, 2178, 1983.
18. **Carr, T. P. and Lei, K. Y.,** In vivo apoprotein catabolism of high density lipoproteins in copper-deficient, hypercholesterolemic rats, *Proc. Soc. Exp. Biol. Med.,* 191, 370, 1989.
19. **Carr, T. P. and Lei, K. Y.,** High density lipoprotein cholesteryl ester and protein catabolism in hypercholesterolemic rats induced by copper deficiency, *Metabolism,* in press, 1990.
20. **Al-Othman, A. A. and Lei, K. Y.,** Time course development of enlarged plasma volume and cholesterol pool size in copper-deficient rats, *FASEB J.,* 3, A358, 1989.
21. **Shao, M. J. S. and Lei, K. Y.,** Conversion of [2-$^{14}$C]mevalonate into cholesterol, lanosterol and squalene in copper-deficient rats, *J. Nutr.,* 110, 859, 1980.

22. **Hassel, C. A., Carr, T. P., Marchello, J. A., and Lei, K. Y.,** Apolipoprotein E-rich HDL binding to liver plasma membranes in copper-deficient rats, *Proc. Soc. Exp. Biol. Med.,* 187, 296, 1988.
23. **Hassel, C. A., Lei, K. Y., Carr, T. P., and Marchello, J. A.,** Lipoprotein receptors in copper-deficient rats: apolipoprotein E-free high density lipoprotein binding to liver membranes, *Metabolism,* 36, 1054, 1987.
24. **Nielsen, F. H., Zimmerman, T. L., and Shuler, T. R.,** Interactions among nickel, copper and iron in rats. Liver and plasma content of lipids and trace elements, *Biol. Trace Elem. Res.,* 4, 125, 1982.
25. **Harvey, P. W. and Allen, K. G. D.,** Lipoproteins and liver lipids in copper-deficient rats, *Nutr. Res.,* 5, 511, 1985.
26. **Lefevre, M., Keen, C. L., Lönnerdal, B., Hurley, L. S., and Schneeman, B. O.,** Different effects of zinc and copper deficiency on composition of plasma high density lipoproteins in rats, *J. Nutr.,* 115, 359, 1985.
27. **Brown, M. S. and Goldstein, J. L.,** HMG-CoA reductase is induced in response to removal of cholesterol from cells, *Science,* 191, 150, 1976.
28. **Yount, N. Y., McNamara, D. J., Al-Othman, A. A., and Lei, K. Y.,** The effect of copper deficiency on rat hepatic 3-hydroxy-3-methylglutaryl coenzyme A reductase activity, *J. Nutr. Biochem.,* in press, 1990.
29. **Hunsaker, H. A., Morita, M., and Allen, K. G. D.,** Marginal copper deficiency in rats, aortal morphology of elastin and cholesterol values in first-generation adult males, *Atherosclerosis,* 51, 1, 1984.
30. **Ginter, E., Nemec, R., Gerven, J., and Mikus, L.,** Quantification of lowered cholesterol oxidation in guinea pigs with latent vitamin C deficiency, *Lipids,* 8, 135, 1973.
31. **Whale, K. W. J. and Davies, N. T.,** Effect of dietary deficiency in the rat on fatty acid composition of adipose tissue and desaturase activity of liver microsomes, *Br. J. Nutr.,* 34, 105, 1975.
32. **Allen, K. G. D. and Klevay, L. M.,** Copper deficiency and cholesterol metabolism in the rat, *Atherosclerosis,* 31, 259, 1978.
33. **Turnberg, L. A. and Anthony-Mote, A.,** The quantitative determination of bile salts in bile using thin layer chromatography and 3α-hydroxysteroid dehydrogenase, *Clin. Chim. Acta,* 24, 253, 1969.
34. **Lin, I. M. and Lei, K. Y.,** Cholesterol kinetic analyses in copper-deficient rats, *J. Nutr.,* 111, 450, 1981.
35. **American Institute of Nutrition,** Report of the AIN ad hoc committee on standards for nutritional studies, *J. Nutr.,* 107, 1340, 1977.
36. **Gurpide, E., Mann, J., and Sandberg, E.,** Determination of kinetic parameters in a two-pool system by administration of one or more tracers, *Biochemistry,* 3, 1250, 1964.
37. **Goodman, D. S. and Noble, R. P.,** Turnover of plasma cholesterol in man, *J. Clin. Invest.,* 47, 231, 1968.
38. **Avigan, J., Steinberg, D., and Berman, M.,** Distribution of labeled cholesterol in animal tissues, *J. Lipid Res.,* 3, 216, 1962.
39. **Field, H., Jr., Swell, L., Schools, P. E., Jr., and Treadwell, C. R.,** Dynamic aspects of cholesterol metabolism in different areas of the aorta and other tissues in man and their relationship to atherosclerosis, *Circulation,* 22, 547, 1960.
40. **Chobanian, A. V. and Hollander, W.,** Body cholesterol metabolism in man. I. The equilibration of serum and tissue cholesterol, *J. Clin. Invest.,* 41, 1732, 1962.
41. **Goodman, D. S.,** The in vivo turnover of individual cholesterol esters in human plasma lipoprotein, *J. Clin. Invest.,* 43, 2026, 1964.
42. **Gould, R. G., LeRoy, G. V., Okita, G. T., Kabara, J. J., Keegan, P., and Bergenstal, D. M.,** The use of $C^{14}$-labeled acetate to study cholesterol metabolism in man. *J. Lab. Clin. Med.,* 46, 372, 1955.
43. **Lei, K. Y.,** Cholesterol metabolism in copper-deficient rats, *Nutr. Rep. Int.,* 15, 597, 1977.

44. **Yount, N. Y., Carr, T. P., McNamara, D. J., and Lei, K. Y.,** Copper deficiency-induced hypercholesterolemia: tissue cholesterol synthesis in the rat using [$^3$H]water, *Biochim. Biophys. Acta,* submitted, 1989.
45. **Leveille, G. A.,** Glycogen metabolism in meal-fed rats and chicks and the time sequence of lipogenic and enzymatic adaptive changes, *J. Nutr.,* 90, 449, 1966.
46. **Kelly, A. W., Kesterson, J. W., and Carlton, W. W.,** Myocardial lesions in the offspring of female rats fed a copper deficient diet, *Exp. Mol. Pathol.,* 20, 40, 1974.
47. **Lanoue, K., Nicklas, W. J., and Williamson, J. R.,** Control of citric acid cycle activity in rat heart mitochondria, *J. Biol. Chem.,* 245, 102, 1970.
48. **Hellström, K. H., Siperstein, M. D., Bricker, L. A., and Luby, L. J.,** Studies of the in vivo metabolism of mevalonic acid in the normal rat, *J. Clin. Invest.,* 52, 1303, 1973.
49. **Hagenfeldt, L. and Hellström, K.,** Blood concentration and turnover of circulating mevalonate in the rat, *Life Sci.,* 11, 669, 1972.
50. **Skipski, V. P. and Barclay, M.,** Thin layer chromatography of lipids, *Methods Enzymol.,* 14, 530, 1969.
51. **Raskin, P. and Siperstein, M. D.,** Mevalonate metabolism by renal tissue in vitro, *J. Lipid Res.,* 15, 20, 1974.
52. **Allen, K. G. D. and Klevay, L. M.,** Cholesterol metabolism in copper deficient rats, *Life Sci.* 22, 1691, 1978.
53. **Fabry, P., Poledne, R., Kazdova, L., and Braun, T.,** The effect of feeding frequency and type of dietary carbohydrate on hepatic lipogenesis in the albino rat, *Nutr. Dieta,* 10, 81, 1968.
54. **Chin, D. J., Gil, G., Russell, D. W., Liscum, L., Luskey, K. L., Basu, S. K., Okayama, H., Berg, P., Goldstein, J. L. and Brown, M. S.,** Nucleotide sequence of 3-hydroxy-3-methyl-glutaryl coenzyme A reductase, a glycoprotein of endoplasmic reticulum, *Nature (London),* 308, 613, 1984.
55. **Brown, M. S., Kovanen, P. T., and Goldstein, J. L.,** Regulation of plasma cholesterol by lipoprotein receptors, *Science,* 212, 628, 1981.
56. **Shapiro, D. J., Nordstrom, J. L., Mitschelen, J. J., Rodwell, V. W., and Schimke, R. T.,** Micro assay for 3-hydroxy-3-methylglutaryl-CoA reductase in rat liver and in L-cell fibroblasts, *Biochim. Biophys. Acta,* 370, 369, 1974.
57. **Gjone, E., Norum, K. R., and Glomset, J. A.,** Familial lecithin:cholesterol acyltransferase deficiency, in *The Metabolic Basis of Inherited Disease,* 4th ed., Stanbury, J. B., Wyngaarden, J. B., and Fredrickson, D. S., Eds., McGraw-Hill, New York, 1978, 589.
58. **Glomset, J. A.,** Lecithin:cholesterol acyltransferase, *Prog. Biochem. Pharmacol.,* 15, 41, 1979.
59. **Lau, B. W. C. and Klevay, L. M.,** Plasma lecithin:cholesterol acyltransferase in copper-deficient rats, *J. Nutr.,* 111, 1698, 1981.
60. **Stokke, K. T. and Norum, K. R.,** Determination of lecithin-cholesterol acyltransferase in human blood plasma, *Scand. J. Clin. Lab. Invest.,* 27, 21, 1971.
61. **Tan, M. H.,** The lipoprotein lipase system: new understanding, *Can. Med. Assoc. J.,* 118, 675, 1978.
62. **Fielding, C. J. and Havel, R. J.,** Lipoprotein Lipase, *Arch. Pathol. Lab. Med.,* 101, 225, 1977.
63. **Strinivasan, S. R., Radhakrishnamurthy, B., and Berenson, G. S.,** Studies of the interaction of heparin with serum lipoproteins in the presence of $Ca^{++}$, $Mg^{++}$ and $Mn^{++}$, *Arch. Biochem. Biophys.,* 170, 334, 1975.
64. **Mildvan, A. S.,** Metals in enzyme catalysis, in *The Enzymes,* Boyer, P. D., Ed., Academic Press, New York, 1970, 445.
65. **Staprans, I., Andrews, C. D., Lurz, F. W., and Felts, J. M.,** Separation of a lipoprotein lipase cofactor from the $\alpha_1$-acid glycoprotein fraction from the urine of nephrotic patients, *Biochim. Biophys. Acta,* 617, 514, 1980.

66. **Baginsky, M. L. and Brown, W. V.,** A new method for the measurement of lipoprotein lipase in postheparin plasma using sodium dodecyl sulfate for the inactivation of hepatic triglyceride lipase, *J. Lipid Res.*, 20, 548, 1979.
67. **Lau, B. W. C. and Klevay, L. M.,** Postheparin plasma lipoprotein lipase in copper-deficient rats, *J. Nutr.*, 112, 928, 1982.
68. **Koo, S. I., Lee, C. C., and Norvell, J. E.,** Effect of copper deficiency on the lymphatic absorption of cholesterol, plasma chylomicron clearance and postheparin lipase activities, *Proc. Soc. Exp. Biol. Med.*,, 188, 410, 1988.
69. **Nilsson-Ehle, P. and Ekman, R.,** Rapid, simple and specific assays for lipoprotein lipase and hepatic lipase, *Artery,* 3, 194, 1977.
70. **Nilsson-Ehle, P.,** Measurements of lipoprotein lipase activity, in *Lipoprotein Lipase,* Borensztajn, J., Ed., Evener Publishers, Chicago, 1987, 59.
71. **Bamberger, M., Glick, J. M., and Rothblat, G. H.,** Hepatic lipase stimulates the uptake of high density lipoprotein cholesterol by hepatoma cells, *J. Lipid Res.*, 24, 869, 1983.
72. **Fell, B. F., Dinsdale, D., and Mills, C. F.,** Changes in enterocyte mitochondria associated with deficiency of copper in cattle, *Res. Vet. Sci.*, 18, 274, 1975.

Chapter 3

# PLASMA LIPOPROTEINS AND APOLIPOPROTEINS IN COPPER DEFICIENCY

**K. Y. Lei and Timothy P. Carr**

## TABLE OF CONTENTS

## I. INTRODUCTION

This chapter is intended to provide an overview on lipoprotein metabolism and to review the changes in lipoprotein profiles observed in copper-deficient experimental animals and human subjects. Most of the studies have concentrated on the distribution of major lipid classes and apolipoproteins among high density lipoproteins (HDL) in rats. In the early studies with copper-deficient rats, lipoprotein classes were partitioned by the precipitation technique. The use of ultracentrifugation alone or in combination with gel permeation chromatography was involved in later studies to separate the various lipoprotein fractions. HDL subfractions have also been isolated by heparin-Sepharose affinity column chromatography and characterized in the recent copper-deficient rat studies. In addition, alterations in the distribution of cholesterol among lipoproteins have been established in human subjects fed copper-deficient diets as well as in copper-deficient rabbits. Thus the changes in plasma lipoprotein induced by copper deficiency are compared among rats which carry plasma cholesterol mostly in HDL, rabbits which transport cholesterol equally in low density lipoproteins (LDL) and HDL, and humans which carry cholesterol mostly in LDL and with a large component in HDL. A detailed review on the comparative aspects of lipoproteins among the three species has been presented by Chapman.[1]

## II. OVERVIEW OF LIPOPROTEIN METABOLISM

Plasma lipoproteins are soluble macromolecules comprised of various lipids and protein that function to transport cholesterol, triglycerides, and phospholipids in the circulation. Certain lipoproteins serve as vehicles for dietary lipids entering the plasma from the intestine, whereas other lipoproteins function to transport lipids from one cell type to another. The lipids may be utilized as a source of energy, as structural components of cells, or as precursors of cell products. Other than albumin-bound fatty acids and a few other specialized carriers, such as retinol-binding protein, the plasma lipoproteins are the only recognized transport forms of lipids in the circulation.

Lipoprotein particles are spherical, pseudomicellar structures.[2] Neutral lipids, particularly triglycerides and cholesteryl esters, comprise the hydrophobic core. Phospholipids, mainly phosphatidylcholine (lecithin), are amphipathic molecules, having both a nonpolar region and a polar region. The phospholipids form an outer surface monolayer that surrounds the nonpolar lipids. Contained within the monolayer is unesterified cholesterol and protein (apolipoproteins). Since the distribution of lipids between the surface monolayer and the particle core is a function of their phase behavior, appreciable amounts of unesterified cholesterol may be distributed into the core and small amounts of cholesteryl esters and triglycerides may be located in the surface monolayer.[3]

The plasma lipoproteins may be separated into five major classes according to their density, size, and net surface charge: (1) chylomicrons, (2) very low density lipoproteins (VLDL), (3) intermediate density lipoproteins (IDL), (4) low density lipoproteins (LDL), (5) and high density lipoproteins (HDL). Each class is heterogeneous in size and composition, and several subclasses have been identified.[4] Chylomicrons are the largest and least dense of the lipoproteins, containing as much as 90% triglyceride in their hydrophobic core. The chylomicrons are synthesized in the intestine and are the vehicle by which dietary triglycerides and cholesterol are transported from the site of absorption to various tissues throughout the body. Most of the small intestine appears capable of secreting chylomicrons, although the majority is produced by the jejunum.[5] Very low density lipoproteins (VLDL) are synthesized in the liver and function as carriers of endogenously synthesized triglycerides, as well as cholesterol, for distribution to peripheral tissues. VLDL and chylomicrons are generally referred to as triglyceride-rich lipoproteins, although VLDL have proportionally more cholesteryl esters and are, hence, more dense than chylomicrons. Intermediate density lipoproteins (IDL) are actually VLDL ''remnant'' particles, generated through the progressive catabolism of VLDL core triglycerides.[6] IDL particles are therefore smaller and more dense than VLDL. Low density lipoproteins (LDL) are the major carriers of cholesterol and cholesteryl esters in human plasma. They are derived from VLDL through the catabolic cascade that first generates IDL. The plasma concentration of LDL is directly correlated with the pathogenesis of coronary artery disease.[7] High density lipoproteins (HDL) are the smallest and most dense of the lipoproteins, containing comparatively little triglyceride. HDL are much richer in protein (approximately one half protein by weight), while the lipids are primarily phosphatidylcholine and cholesteryl esters. The synthesis of HDL or HDL precursors appears to occur in the liver and intestine, as well as within the plasma compartment during the lipolytic processing of the triglyceride-rich lipoproteins.[8] It has been proposed[9] that HDL are involved in transporting excess cholesterol from peripheral tissues to the liver for excretion during a process known as ''reverse cholesterol transport''. The plasma concentration of HDL are inversely correlated to the development of atherosclerosis.[10,11]

The plasma lipoproteins contain on their surface one or more protein constituents called apolipoproteins. The ability of apolipoproteins to form an association with lipids is based on the observation that polar and nonpolar amino acid residues form amphipathic helical regions.[12] The nonpolar regions penetrate into the lipid surface, whereas the polar residues on the opposite side of the helix remain in contact with the surrounding aqueous phase. It has been proposed[4] that the role of the polar amino acid residues is to keep the apolipoproteins at the surface of the particle. Several functions have been attributed to specific apolipoproteins; in addition to maintaining the structure and integrity of the lipoprotein particle, apolipoproteins regulate lipoprotein metabolism by modulating the activity of enzymes and by mediating the cellular uptake of lipoproteins by specific receptors.

The primary function of plasma lipoproteins is to transport triglycerides and cholesterol. The transport mechanisms by which lipoproteins function involve complex interactions between lipoproteins and various cells and molecules, as well as interactions between different lipoprotein particles. A discussion of lipoprotein metabolism may be separated into three distinct metabolic systems, depending upon the origin or source of lipid being transported: (A) dietary lipids, absorbed in the intestine, are packaged in chylomicrons for delivery to peripheral tissues and the liver in the ''exogenous'' lipid transport pathway; (B) triglycerides of hepatic origin and cholesterol are transported from the liver to the periphery in VLDL, as well as other LDL precursor particles, to comprise the ''endogenous'' pathway; (C) in the third lipid transport system, commonly referred to as ''reverse cholesterol transport'', cholesterol from the peripheral tissues are packaged in HDL and delivered to the liver for excretion.

**A. Exogenous Lipid Transport**

The production of chylomicrons by intestinal epithelial cells is a direct function of the rate of fat absorption. Dietary triglycerides are hydrolyzed, absorbed, and reformed into droplets visible in the smooth endoplasmic reticulum.[13] Cholesterol in the lumen of the intestine is derived from both dietary and endogenous (biliary) sources. Cholesterol is absorbed and esterified by the action of acyl coenzyme A:cholesterol acyltransferase (ACAT; EC 2.3.1.26) also within the smooth endoplasmic reticulum.[14] The growing lipid droplet is subjected to a series of additions of specific apolipoproteins, phospholipids, and carbohydrate before leaving the intestinal cell.[15]

The secretion of chylomicrons from the intestinal mucosal cell appears to be an exocytotic process; once released, intercellular chylomicrons pass through the basement membrane,[16] and finally gain access to the lymphatics through gaps between adjacent endothelial cells.[17] The chylomicrons are delivered directly to the bloodstream by the mesenteric lymph via the thoracic duct. Circulating chylomicrons are rapidly metabolized, initiated by the action of an extrahepatic enzyme, lipoprotein lipase (clearing factor lipase; EC 3.1.1.34). Lipoprotein lipase (LPL) is responsible for the hydrolysis of the core triglycerides, which have a half-life of 5 min in the circulation.[18] LPL activity is found primarily in adipose, muscle, and lung tissue,[19] as well as lactating mammary gland.[20] An obligatory cofactor for the enzyme is apolipoprotein (apo) C-II, located on the surface of the chylomicron particle, although the nature of the interaction between apo C-II and active sites on LPL are unknown.[18] Hydrolysis is initially very rapid and the lipolytic products are efficiently utilized by the tissue cells.[21] As the particle becomes smaller, the A apolipoproteins are rapidly transferred to HDL, together with most of the phospholipids and C apolipoproteins.[22] Chylomicrons may go through a number of attachment-hydrolysis-detachment cycles during lipolysis,[23] until as much as 90% of the core triglycerides have been hydrolyzed. As a result of this process, smaller chylomicron "remnants" are formed, greatly depleted in triglycerides and having little affinity for LPL.[24]

Chylomicron remnants, once formed, are cleared rapidly from the circulation by the liver. The remnant particles, now containing almost exclusively apo B-48 and apo E as their protein constituents, leave the peripheral capillary bed and travel to the liver where they bind to specific receptors on the surface of hepatocytes.[25] The receptors recognize apo E, which appears to be the major determinant of chylomicron remnant uptake by the liver.[26,27] Binding to the receptor is followed by endocytosis of the particle via coated pits, which pinch off to form a primary endosome.[28] Lysosomes derived from the Golgi apparatus fuse with the endosomes, releasing acid hydrolases.[29] This process leads to virtually complete lipolysis and proteolysis of remnant components, which are made available for metabolic processes with the cell. The cholesterol component, transported in chylomicrons mainly as cholesteryl esters, is delivered nearly quantitatively to the liver in chylomicron remnants. The cholesterol is made available for the formation of bile acids[14] or the biosynthesis of lipoproteins.[4]

**B. Endogenous Lipid Transport**

The liver also secretes triglycerides and cholesterol into the circulation as VLDL, a process analogous to chylomicron production by the intestine. Whereas triglycerides of chylomicrons are derived from the diet, the triglycerides of VLDL are synthesized by the liver or are derived from fatty acids transported to the liver from adipose tissue.[22] Cholesteryl esters of VLDL are also synthesized locally by the action of ACAT, or are derived from the uptake of chylomicron remnants.[4] The activity of ACAT in human hepatocytes, however, is low,[30] and newly secreted VLDL are thought to contain relatively few cholesteryl esters. This component of VLDL is mainly acquired after secretion by transfer from HDL,[31] which will be discussed below. Unlike chylomicrons, newly secreted VLDL contain apo E, which are synthesized in the hepatocyte,[32] and a B apolipoprotein (apo B-100).[33] Studies using inhibitors of protein synthesis indicate that apo B-100 is essential for secretion of VLDL particles.[34] Thus, secreted VLDL contain endogenous triglyceride, cholesteryl ester, phospholipid, apo B-100, apo E, and C apolipoproteins.[4]

The metabolism of circulating VLDL is initiated by binding to endothelial LPL, followed by the hydrolysis of core triglycerides. The action of LPL on VLDL particles occurs more slowly than on chylomicrons, apparently because fewer enzyme molecules associate with the smaller VLDL particles.[22] A second enzyme, hepatic lipase, also acts on VLDL triglycerides but does not require apo C-II as a cofactor.[35] Some of the surface proteins, particularly the C apolipoproteins, are transferred to HDL during the hydrolytic process. A portion of the surface phospholipids are also transferred to HDL, while some is hydrolyzed by LPL.[36] Hydrolysis continues until 80 to 90% of the core triglycerides have been removed,[37] resulting in VLDL remnant particles (IDL). The IDL formed during particle interaction with LPL is unstable and the excess surface material (mostly phospholipids and unesterified cholesterol) is transferred to HDL.[38] The resulting IDL contain primarily cholesteryl ester in the hydrophobic core and apo B-100, apo E, and C apolipoproteins.[39] After lipolysis, the IDL particles are released from the capillary wall into the circulation. Thus, by this process, VLDL functions to transport endogenous triglycerides from the liver to the peripheral tissues.

The subsequent metabolism of VLDL remnants (IDL) differs from that of chylomicron remnants. Whereas the latter are rapidly and completely taken up by the liver, a substantial fraction of IDL is normally converted to LDL in humans.[40] Some IDL are catabolized by the liver via "LDL" receptors before being converted to LDL. Evidence suggests that both receptor-mediated uptake and conversion to LDL occur in all mammals but to different relative extents.[40] IDL particles which bind to the liver appear to be mediated primarily by their apo E content. The LDL receptor recognizes both apo E and apo B-100, but the affinity of apo E is much greater.[41,42] IDL not taken up by the liver eventually lose all of their surface apoproteins except apo B-100 by transfer to HDL, as occurs during initial lipolysis. Because of the lower affinity of apo B-100 for the LDL receptor, the resulting LDL particles have a circulating half-life of several days, compared to 2 to 6 h for IDL.[43]

Newly formed LDL particles contain apo B-100 as their sole protein constituent and are rich in cholesteryl esters, thus being the major transport vehicle of cholesterol in humans. These particles are removed from the circulation by two pathways; about two thirds of the LDL is cleared by binding to LDL receptors and about one third is through nonreceptor mechanisms.[7] A variety of tissues are known to express LDL receptors.[44,45] Delineation of the LDL receptor pathway was initiated by Brown and Goldstein[46] in studies of individuals who genetically lacked the LDL receptor (familial hypercholesterolemia). In normal humans, LDL particles are removed from the plasma with a fractional catabolic rate of about 45% of the plasma pool per day.[47] Delivery of cholesterol is achieved when LDL particles bind to the receptor protein, located on the surface of cell membranes in pockets called coated pits.[48] Receptor-bound LDL is then internalized by endocytosis forming vesicles. LDL receptors may be recycled after endocytosis and are reutilized many times before they are eventually catabolized.[49] Lysosomes derived from the Golgi apparatus fuse with the vesicles, releasing hydrolytic enzymes that degrade apo B-100 and hydrolyze the cholesteryl esters. The cholesterol leaves the lysosome and is utilized in membrane synthesis or other cellular functions.[38]

In addition to degradation via the LDL receptor, LDL can be removed from the plasma by nonreceptor mechanisms. Since the latter processes are not dependent upon the synthesis and metabolism of the LDL receptor, degradation by these mechanisms increases when plasma levels of LDL rise.[38] Nearly every organ in the body is capable of LDL uptake by receptor-independent mechanisms,[44,45] and must account for essentially all LDL degradation when LDL receptors are lacking or nonfunctional. At least three categories of degradative mechanisms may be considered independent of the LDL receptor: (1) fluid or bulk endocytosis, (2) nonspecific adsorptive endocytosis, and (3) high affinity uptake by alternative receptor-mediated mechanisms.[50] Additionally, macrophages and other scavenger cells of the reticuloendothelial system contribute to receptor-independent uptake of LDL.[38]

The cholesterol derived from LDL mediates a series of regulatory events which maintain cholesterol homeostasis in the cell.[46] First, increasing cellular cholesterol suppresses the rate-controlling enzyme in cholesterol biosynthesis, 3-hydroxy-3-methylglutaryl coenzyme A (HMG-CoA) reductase (EC 1.1.1.34). Second, intracellular ACAT is activated, which esterifies excess cholesterol for storage as cholesteryl ester droplets. Finally, the synthesis of LDL receptors is suppressed, which prevents an overaccumulation of cellular cholesterol via the receptor mechanism.

Thus, the metabolism of VLDL and LDL are the vehicles by which endogenous triglyceride and cholesterol, respectively, are transported in the body. A very important feature of the VLDL-IDL-LDL cascade is the change of biological activity associated with each particle. Furthermore, it should be noted that while the VLDL-IDL-LDL cascade is generally regarded as the primary source of circulating LDL, VLDL may not be the sole precursor to LDL; direct hepatic secretion of lipoprotein particles resembling LDL has been described.[51] The significance of direct LDL secretion by the liver has not been fully delineated, although modification of the particles appears to occur in the plasma.[52,53] The possibility that several types of LDL precursors are secreted by the liver should not be precluded.

### C. Reverse Cholesterol Transport

An important function of HDL is the transport of cholesterol from peripheral tissues to the liver for excretion from the body in a process known as reverse cholesterol transport.[9] It should be noted that HDL can also deliver cholesterol to steroidogenic tissues.[54] Several recent reviews[8,55-57] describe in detail the current understanding of HDL metabolism and reverse cholesterol transport and, hence, only the briefest review is present herein.

Synthesis and secretion of precursor HDL particles have been demonstrated in the liver and small intestine.[14] These "nascent" particles are discoidal lipoproteins composed of phospholipids, apolipoproteins, and small amounts of free cholesterol. It was suggested by Hamilton et al.[58] that once in the bloodstream, they are transformed to mature spherical HDL particles after interaction with lecithin:cholesterol acyltransferase (LCAT; EC 2.3.1.43). LCAT esterifies free cholesterol, allowing cholesterol to be stored in the core of HDL particles. The HDL originating in the intestine mainly contain newly synthesized apo A-I, with relatively smaller amounts of apo E and C apolipoproteins.[59,60] Hepatic HDL exhibit higher amounts of apo E, compared to apo A-I and C apolipoproteins.[60] Although discoidal HDL structures have been identified in liver perfusates and intestinal lymph, they have not been identified along the intracellular secretory pathways as the triglyceride-rich lipoproteins.[32,58,61] (Chylomicrons and VLDL are abundant in intracellular organelles and can readily be isolated from the Golgi apparatus.) It has been suggested by Eisenberg[8] that the cells produce free apolipoproteins that become associated with phospholipids at a later stage of the secretory process or even after secretion.

A second potential source of HDL precursors is from the catabolism of triglyceride-rich lipoproteins. The apolipoproteins normally found in mature HDL are also secreted with chylomicrons and VLDL, which may be transferred to preexisting HDL particles in the plasma during the course of lipolysis.[22] The transfer mechanisms are unclear, although one mechanism may be the release or exclusion of entire fragments of surface components from shrinking chylomicrons and VLDL particles.[8] Although the relative contribution of this process to the formation of HDL is not fully understood, it is well established that transformation of HDL precursors to mature spherical HDL particles is dependent on LCAT activity.[62] HDL are the locus for LCAT activity and are responsible for the synthesis of essentially all plasma cholesteryl esters, except those that represent dietary cholesterol transported in chylomicrons.[22] The enzyme transfers a fatty acyl residue from lecithin to cholesterol to yield lysolecithin and cholesteryl esters as products. Apo A-I appears to be the most effective activator of LCAT,[63] although apo A-IV and apo C-I may also serve as cofactors.[22]

Reverse cholesterol transport is initiated by the efflux of unesterified (free) cholesterol from cells to HDL particles.[55,64] The mechanisms by which HDL accepts cholesterol from peripheral cells is poorly understood, and there is disagreement on whether specific binding of HDL to the cell surface membrane facilitates cholesterol transfer.[65-72] Nevertheless, when cholesterol is transferred to HDL, cholesterol enrichment in the cell membrane apparently must exceed that of the lipoprotein.[73] Because of the rapid consumption of unesterified cholesterol by the LCAT reaction, net movement of cholesterol from peripheral cells to HDL is therefore favored and reverse cholesterol transport ensues.

The existence of a specific HDL membrane receptor, either in peripheral or hepatic cells, is also in question. A putative "HDL receptor" has apparently been identified and partially characterized in various tissues using ligand blotting techniques. The molecular mass of this HDL-binding protein is uncertain, and values have been reported to be 110 kDa in human fibroblasts,[74] 120 kDa in human placenta,[75] and 78 kDa in rat adrenal, liver, and kidney membranes.[76,77] Conversely, Mendel et al.[78,79] have presented evidence that binding sites for HDL are not "classical" receptors, in that they are kinetically heterogeneous and small in molecular mass.

As Roheim[57] points out, much information concerning the mechanisms of cholesterol transfer from peripheral cells to HDL *in vivo* is limited. Most studies on reverse cholesterol transport and cellular cholesterol transfer have been performed *in vitro* using tissue culture systems, whereas virtually no reports are available on *in vivo* cholesterol transfer. Conversely, much information exists about how cholesterol is transported to the cell; once at the liver cell, HDL cholesterol can be delivered by endocytosis of the intact particle, transfer of free cholesterol, or selective uptake of cholesteryl esters.

# III. RAT STUDIES

## A. Early Studies

Three early studies were performed to examine the influence of copper deficiency on plasma lipoprotein composition in rats using the precipitation method. In 1980, Allen and Klevay[80] used the heparin/$MnCl_2$ method[81] to precipitate VLDL plus LDL, and the HDL cholesterol concentration in the supernatant was determined fluorimetrically.[82] Plasma VLDL cholesterol concentration was estimated as plasma triglyceride concentration divided by 5. Cholesterol concentration of the combined VLDL plus LDL fraction was obtained from the difference between total plasma and HDL cholesterol levels. LDL cholesterol level was then derived from the difference between the calculated cholesterol values of the combined VLDL plus LDL fraction and that of VLDL fraction. Similar to previous findings in rats,[1] 85% of the total plasma cholesterol was found to be carried by HDL (61.7 mg/dl) in the copper-adequate rats. However, this percentage was reduced to 77% in copper-deficient rats. When the data were expressed as plasma concentration, a marked increase (85%) in HDL cholesterol level (114.2 mg/dl) was observed in the copper-deficient rats. Moreover, the amount of cholesterol associated with the combined VLDL plus LDL fraction was also elevated (56%) in copper deficiency. Since the arbitrary formula used for the calculation of VLDL cholesterol may not be justifiable, the individual VLDL and LDL data will not be addressed here.

Conflicting data on the influence of copper deficiency on the distribution of cholesterol among HDL and VLDL plus LDL fractions were reported by Harvey and Allen[83] in a subsequent study using the same precipitation technique. Apart from the marked increase (55 to 91%) in plasma total cholesterol level, the cholesterol distribution among lipoproteins in copper-deficient rats was completely different as compared to their previous study[80] and all other studies[84-93] reported so far. Instead of the usual increase in HDL cholesterol, Harvey and Allen[83] observed a 13 to 17% reduction in copper-deficient rats in two experiments; HDL cholesterol level was reduced from 42.4 to 36.9, and 48.4 to 40.1 mg/dl as a result of copper deficiency. Furthermore, all of the increase in plasma cholesterol of copper-deficient rats was found to reside in the VLDL plus LDL fraction. These major discrepancies may have resulted from an incomplete extraction of HDL cholesterol with hexane or from analytical problems associated with the use of their newly acquired enzymatic assay.[94] In their previous study, HDL cholesterol was determined, without hexane extraction, directly with their established fluorimetric assay.[82]

Using a similar precipitation technique,[95] with phosphotungstic acid and magnesium chloride to remove VLDL and LDL, Lei and Lin[84] observed a 33% increase in total and a 30% increase in unesterified cholesterol of HDL in copper-deficient rats. The magnitude of the increase in total HDL cholesterol (adequate, 36.4 mg/dl; deficient, 48.4 mg/dl) as a result of copper deficiency was smaller than that reported by Allen and Klevay.[80] In both treatment groups, the amount of unesterified cholesterol accounted for 29% of the total cholesterol in serum HDL. The HDL cholesterol data were determined from plasma obtained from rats fed a copper-deficient diet for 5 to 15 weeks in duration. Contrary to the suggestion of Harvey and Allen[83] that prolonged copper deficiency produces reduction in HDL cholesterol level, increases in HDL cholesterol concentration were observed throughout this prolonged study. Furthermore, in all subsequent studies involving the copper-deficient rat model, increases in HDL cholesterol levels accounting for most of the increases in total plasma cholesterol have always been observed.[84-93]

In view of the conflicting results presented by Allen and co-workers[80,83] and the limitations associated with the precipitation method, researchers in this area initiated the use of ultracentrifugation or a combination of ultracentrifugation and gel permeation chromatography for the isolation of plasma lipoproteins. These latter methods not only allow for the quantitative analyses of lipoprotein composition but also supplied ample amounts of purified lipoproteins for both *in vivo* and *in vitro* metabolic studies.

### B. Composition of Lipoprotein Classes

The first study[85] to examine the influence of copper deficiency on the composition of individual lipoprotein classes was accomplished by partitioning the lipoproteins with the method of Rudel et al.[96] using ultracentrifugation and gel permeation column chromatography. This method is gentle as compared to sequential ultracentrifugation, which subjects lipoproteins to high gravitational force for prolonged periods and may lead to the loss of apolipoproteins.[97] In addition, cross contamination between lipoprotein fractions is very small, as judged by apolipoprotein composition and the absence of albumin, when compared to gradient or sequential ultracentrifugation. A typical lipoprotein profile prepared by agarose gel permeation is depicted in Figure 1.

In the copper-deficient rats, the mean values of plasma lipids appeared to be elevated in all the lipoprotein fractions, with the exception of VLDL cholesterol. However, significant increases were observed only for LDL cholesterol and HDL cholesterol and for LDL triglyceride concentrations (Table 1). When expressed as a percentage, the VLDL, LDL, and HDL in the controls

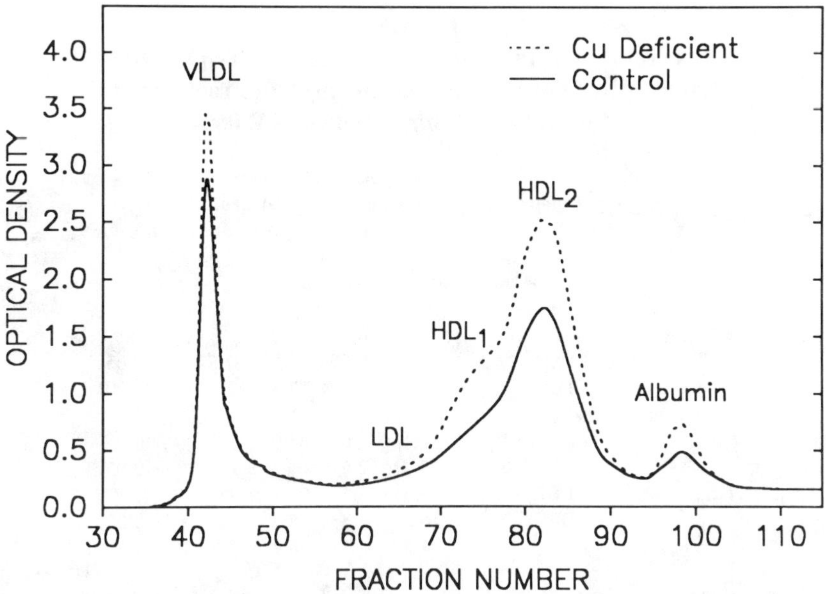


FIGURE 1. Gel permeation chromatographic elution profile of plasma lipoproteins. Plasma was first adjusted to density 1.225 g/ml and centrifuged at 100,000 × *g* for 24 h at 15°C to obtain lipoproteins. Samples were eluted at 20 ml/h with 150 m*M* NaCl, 0.01% EDTA, and 0.02% $NaN_3$ at pH 7.4.

were found to carry 7, 31, and 62% of the total plasma cholesterol, respectively. In copper deficiency, the percentage was reduced to 2% for VLDL, increased to 41% for LDL, and remained almost unchanged at 57% for HDL. However, the absolute amounts of HDL cholesterol and LDL cholesterol were markedly increased. A similar increase in HDL cholesterol concentration was observed in copper-deficient rats in previous studies[80,84] by using precipitation techniques for the determination of HDL cholesterol. The increases in lipid contents of HDL and LDL indicate that there may be corresponding increases in the respective protein as a result of copper deficiency. This is confirmed by the significant increase in total protein content of the HDL and the LDL fractions (Table 1). In addition, the mean concentrations of apo A-I and apo E of the HDL appeared to be elevated in the copper-deficient rats, but the only significant elevation was the 43% increase in apo E (deficient, 17.4 mg/dl; adequate, 12.2 mg/dl).

In order to appreciate the extent of increase in plasma pool size of these lipoproteins, the plasma volume was determined by the use of the Evan's blue dye dilution method.[98] Blood volume was unaffected but plasma volume was increased 55%, indicating an enlarged plasma pool of not only HDL but also VLDL and LDL in copper-deficient rats. Indeed, a twofold increase in plasma HDL cholesterol pool size was recently established when plasma

**TABLE 1**
**Distribution of Protein, Cholesterol, and Triglyceride Among the Various Plasma Lipoprotein Fractions of Control and Copper-Deficient Rats[a]**

| | Lipoprotein fraction | Copper adequate (n = 6) | Copper deficient (n = 5) | *p*-value[b] |
|---|---|---|---|---|
| | | Concentration (mg/100 ml) | | |
| Protein | VLDL | 2.0 ± 0.2 | 1.2 ± 0.3 | NS |
| | LDL | 5.0 ± 0.5 | 7.0 ± 0.6 | <0.05 |
| | HDL | 59.9 ± 2.0 | 77.8 ± 2.3 | <0.001 |
| Cholesterol | VLDL | 5.0 ± 1.4 | 1.6 ± 0.7 | NS |
| | LDL | 23.1 ± 2.6 | 43.0 ± 9.7 | <0.05 |
| | HDL | 45.3 ± 2.6 | 60.5 ± 6.0 | <0.05 |
| Triglyceride | VLDL | 8.4 ± 2.8 | 25.1 ± 11.3 | NS |
| | LDL | 16.2 ± 1.3 | 35.7 ± 9.7 | <0.05 |
| | HDL | 14.4 ± 1.1 | 20.7 ± 5.7 | NS |

[a] Means ± SEM; n = number of observations. Each observation was obtained from lipoproteins isolated from pooled plasma of two or three rats by ultracentrifugation and agarose column chromatography.

[b] *p*-value for treatment comparison of copper effect (df = 1) in analysis of variance; NS, not significant.

From Lei, K. Y., Hassel, C. A., and Allen, D. K., *J. Nutr.*, 113, 2178, 1983. With permission.

volume enlargement was taken into consideration.[93] Furthermore, a recent study indicated that increases in plasma volume as well as plasma cholesterol pool size appeared to precede the actual increase in plasma cholesterol concentration in copper-deficient rats.[99] Thus, in future metabolic studies, not only should plasma concentrations of metabolites be monitored, changes in plasma volume should be determined to establish the absolute plasma pool size.

The cholesterol and protein concentrations of the combined VLDL plus LDL fraction in copper-deficient rats were also examined by Lefevre et al.[88] Although these parameters were not altered, a 13% increase in the proportion of apo B-100 (percent composition of apoproteins) was observed in this fraction of rats fed the copper-deficient diet. Since apo B-100 is synthesized exclusively by the rat liver[100] and is associated mainly with LDL,[101] Lefevre et al.[88] suggested that the higher apo B-100 (percent composition) signifies the beginning of LDL accumulation. Rats in this study were only slightly deficient as judged by their plasma and liver copper levels. If this study was prolonged to increase the severity of the deficiency state, increased LDL protein and cholesterol levels may be evident, as in the study reported by Lei et al.[85]

Consistent increases in HDL cholesterol and protein concentrations, as a result of copper deficiency, have been observed[87-89] in addition to the report by Lei et al.[85] using ultracentrifugation and gel permeation column chromatography (Table 2). These findings have also been repeatedly confirmed in latter studies of Lei and co-workers.[90-93] Data reported for the controls were slightly higher in protein and lower in cholesterol levels of HDL in the studies of Lefevre et al.[87,88] as compared to the studies of Lei et al.[85] and Lee and Koo.[89] Reports of increases for HDL protein and cholesterol ranged from 30 to 110% and 21 to 48%, respectively, in copper-deficient rats. The only exception was in one of the studies by Lefevre et al.;[88] the HDL protein concentration was not significantly increased in rats marginally deficient in copper. Nevertheless, an increase in the plasma concentration of apo A-I, the major apoprotein of HDL, was associated with copper deficiency.[88] This may suggest that HDL particle number is increased. Similarly, the constant ratios of HDL cholesterol to protein or HDL triglyceride to protein observed in the study of Lei et al.[85] suggest that the increase in HDL protein is accompanied by proportional increases in lipids in copper deficiency. These composition data suggest that copper deficiency increased the HDL particle number but the particle size remained relatively constant. Furthermore, the percent composition of lipid and protein components of HDL subfractions reported by Lee and Koo[89] also appeared not to be influenced by copper deficiency.

The apolipoprotein profile of total HDL fraction was examined by the various research groups[86-89] using a common method. Apolipoproteins were separated by dodecyl sulfate-polyacrylamide gel electrophoresis (SDS-PAGE),[102,103] stained with Coomassie blue, scanned densitometrically, and quantitated with an integrator. Differences in chromogenicity of stained apoprotein bands were not corrected. Among the three research groups, the overall apolipoprotein profile appeared to be similar in terms of the direction and magnitude of changes induced by copper deficiency. Increases in apo E and apo A-I concentrations, as a result of copper deficiency, were reported by Lee and Koo[89] as well as in the large factorial experiment of Lefevre et al.[87] A marked increase in apo E concentration was also observed by Croswell and Lei,[86] but the trend for an increase in apo A-I content did not reach significance. When the HDL apolipoprotein data were expressed on the basis of apo E to apo A-I, this ratio appeared to be elevated in the copper-deficient rats used by the three research groups (Table 2). Although both apo E and apo A-I concentrations were increased in HDL, apo E may be elevated to a greater extent than apo A-I. Hence, copper deficiency may increase the population of a HDL subfraction containing apo E, or there may be a selective increase of apo E content in an apo E-rich HDL fraction.

An earlier study by Lei et al.[85] also demonstrated that apo E content of HDL was elevated but to a lesser extent. Data from this study were not included in the comparison with the other studies because the method used was different. In this study HDL was delipidated with 1,1,3,3-tetramethyl-

**TABLE 2**
**Distribution of Cholesterol and Apolipoproteins Among HDL Subfractions**

| | | Plasma concentration (mg/dl) | | | | | | |
|---|---|---|---|---|---|---|---|---|
| | | Cholesterol | Total protein | Apo A-IV | Apo E | Apo A-I | Apo C | Cholesterol/ protein ratio |
| Croswell and Lei (1985)[86] | | | | | | | | |
| HDL | CuD[b] | 71.0 ± 3.6[c] | 142.2 ± 5.6[c] | 4.2 ± 0.7 | 24.0 ± 4.2[c] | 78.7 ± 8.3 | — | 0.50 |
| | CuA | 48.0 ± 3.8[d] | 83.5 ± 11.1[d] | 2.9 ± 0.6 | 8.4 ± 2.2[d] | 57.8 ± 10.9 | — | 0.58 |
| $R_I$ (apo E-poor) | CuD | 30.0 ± 2.1 | 88.9 ± 9.0 | 1.7 ± 0.4 | 0.3 ± 0.2 | 87.0 ± 8.8 | — | 0.34 |
| | CuA | 27.7 ± 5.2 | 62.4 ± 11.8 | 1.3 ± 0.7 | 0.3 ± 0.1 | 60.8 ± 11.3 | — | 0.44 |
| $R_{II}$ (apo-E-rich) | CuD | 37.9 ± 4.9[c] | 37.7 ± 1.7[c] | 0 | 19.1 ± 3.4[c] | 18.6 ± 3.4[c] | — | 1.01 |
| | CuA | 18.0 ± 1.6[d] | 15.1 ± 1.0[d] | 0 | 7.4 ± 1.9[d] | 7.7 ± 2.0[d] | — | 1.19 |
| Lefevre et al. (1986)[88] | | | | | | | | |
| HDL | CuD | 44.0 ± 1.9[c] | 118.2 ± 8.0 | 6.62 | 27.42 | 71.16 | 9.57 | 0.37 |
| | CuA | 36.4 ± 3.1[d] | 111.3 ± 5.6 | 5.78 | 15.58 | 77.13 | 10.13 | 0.33 |
| HDL-AI (apo E-poor) | CuD | — | 93.4 | 8.78 | 2.99 | 66.03 | 11.30 | — |
| | CuA | — | 95.6 | 7.55 | 2.29 | 71.32 | 11.57 | — |
| HDL-E (apo E-rich) | CuD | — | 24.8 | 0.05 | 18.92 | 4.41 | 1.09 | — |
| | CuA | — | 15.4 | 0.06 | 11.27 | 3.42 | 0.65 | — |
| Lee and Koo (1988)[89] | | | | | | | | |
| HDL | CuD | 60.1 ± 4.0[c] | 136.38 | 7.49 | 17.10 | 82.57 | 29.02 | 0.44 |
| | CuA | 42.2 ± 2.8[d] | 65.04 | 3.61 | 4.89 | 42.29 | 14.25 | 0.65 |
| $HDL\text{-}E_O$ (apo E-free) | CuD | 43.8 ± 2.3[c] | 107.95 | 6.73 ± 1.79 | 0 | 73.58 ± 9.58[c] | 27.64 ± 10.36 | 0.41 |
| | CuA | 30.8 ± 3.3[d] | 52.54 | 3.47 ± 0.80 | 0 | 36.07 ± 4.20[d] | 13.00 ± 2.52 | 0.58 |
| $HDL\text{-}E_1$ (apo E-rich) | CuD | 8.0 ± 3.5 | 15.08 | 0.76 ± 0.46 | 5.99 ± 1.04[c] | 7.27 ± 5.47 | 1.06 ± 0.35 | 0.53 |
| | CuA | 5.1 ± 0.9 | 5.44 | 0.14 ± 0.04 | 0.46 ± 0.12[d] | 4.02 ± 0.99 | 0.82 ± 0.33 | 0.94 |
| $HDL\text{-}E_2$ (apo E-very rich) | CuD | 8.3 ± 1.1 | 13.35 | 0 | 11.11 ± 4.74 | 1.72 ± 0.49 | 0.52 ± 0.16 | 0.62 |
| | CuA | 6.1 ± 0.4 | 7.06 | 0 | 4.43 ± 0.45 | 2.20 ± 0.87 | 0.43 ± 0.19 | 0.86 |

[a] Values represent means ± SEM; values without SEM are calculated means.
[b] CuD, copper deficient; CuA, copper adequate.
[c,d] Values in same column with unlike superscripts are significantly different ($p < 0.05$).

Data used with permission.

urea,[104] SDS-PAGE tube gels were used, and apolipoprotein peaks were quantitated manually instead of with a recording integrator. This may have resulted in a generally lower apo E level as compared to the other studies.

### C. Composition of HDL Subfractions

In view of the known heterogenicity of HDL and the role of apo E in the receptor-mediated uptake of lipoproteins,[27] heparin-Sepharose affinity column chromatography was used to partition HDL into apo E-rich and apo E-poor subfractions. The studies of Croswell and Lei[86] and Lee and Koo[89] employed the heparin-affinity method of Weisgraber and Mahley,[105] which involved the use of $MnCl_2$ in sample equilibration and elution of the nonretained fraction consisting mainly of albumin. In contrast, Lefevre et al.[88] used a similar method established by Quarfordt et al.[106] without $MnCl_2$. For compositional studies, the association of $Mn^{++}$ with HDL subfractions is not a major concern, except the apoproteins may precipitate in time during storage. Nevertheless, $Mn^{++}$ should be removed by exhaustive dialysis prior to using these HDL subfractions for chemical analysis or for metabolic studies. Thus HDL subfractions prepared for receptor binding[90,91] and cell culture studies[107,108] were isolated by a modified method of Weisgraber and Mahley[105] without the use of $MnCl_2$. In addition, a narrow cut of the HDL fraction obtained from gel permeation chromatography, using a 6% agarose gel, was carried out to eliminate the presence of albumin in the HDL fraction.

The use of heparin-Sepharose column chromatography by Croswell and Lei[86] to separate HDL resulted in a highly reproducible profile of HDL subfractions. A nonretained fraction, NR (solely albumin), was eluted with 50 m*M* NaCl, 5 m*M* Tris-HCl, and 25 m*M* $MnCl_2$. Two retained fractions, $R_I$ (apo E-poor HDL) and $R_{II}$ (apo E-rich HDL), were eluted with buffers containing 70 and 290 m*M* NaCl, respectively, as well as 5 m*M* Tris-HCl and no $MnCl_2$. In Figure 2, typical elution profiles of HDL subfractions are depicted. Most of the HDL applied were recovered in the retained fractions. Both retained fractions appeared to be higher in the copper-deficient group than in controls as estimated at 280 nm. Protein and lipid compositional analyses of HDL subfractions (Table 2) performed by Croswell and Lei[86] indicate that the concentrations of cholesterol and protein for the entire HDL fraction and the apo E-rich HDL subfraction were significantly higher in copper-deficient rats than in controls. Most of the increase in protein and cholesterol content of the HDL fraction was accounted for by the increase in the apo E-rich HDL subfraction. In addition, there was a trend of elevated protein content in the apo E-poor HDL subfraction, but the trend was not significant. These observations suggest that the enrichment of apo E in the HDL fraction is due mainly to a selective increase in the apo E-rich HDL fraction in the plasma of copper-deficient rats.

Data from the study of Lefevre et al.[88] also confirmed that the apo E-rich HDL fraction was increased by copper deficiency (Table 2). Another

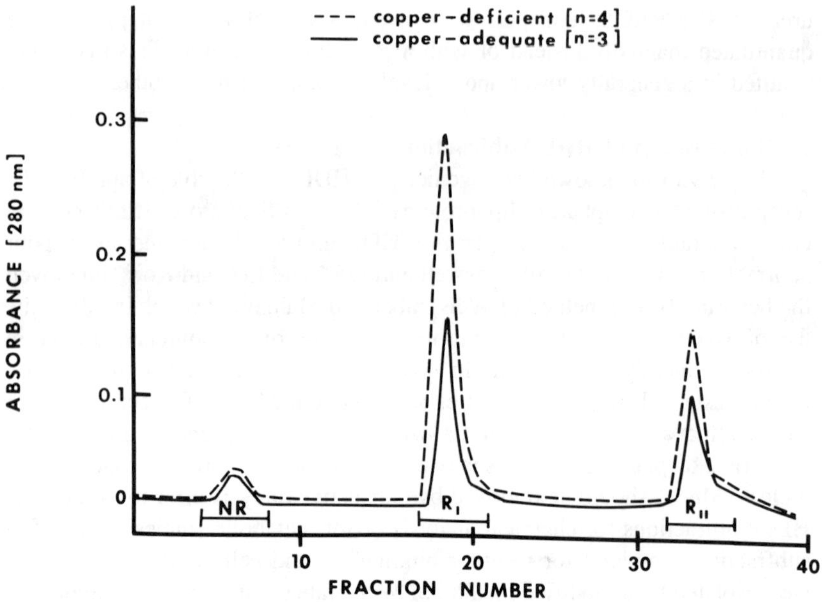


FIGURE 2. Subfractionation of plasma HDL by heparin-Sepharose affinity column chromatography. Approximately 5 mg of HDL protein in 5 m*M* NaCl and 25 m*M* $MnCl_2$ were applied to a 1.0 × 30 cm column. HDL were eluted by an increasing NaCl step gradient at 24 ml/h. A nonretained fraction, NR (solely albumin), was eluted with 50 m*M* NaCl, 5 m*M* Tris-HCl, and 25 m*M* $MnCl_2$. Two retained fractions, $R_I$ (apo E-poor HDL) and $R_{II}$ (apo E-rich HDL), were eluted with buffers containing 70 and 290 m*M* NaCl, respectively, as well as 5 m*M* Tris-HCl and no $MnCl_2$. (From Croswell, S.C., and Lei, K. Y., *J. Nutr.*, 115, 473, 1985. With permission.)

line of evidence, derived from their nondenaturing gradient gel data, indicated that the apo E-rich HDL represented $HDL_1$ and was selectively increased in copper-deficient rats. Previous studies by other investigators have demonstrated that $HDL_1$ is larger in size and enriched with apo E and cholesterol as compared to other HDL subfractions.[109] Although cholesterol values for HDL subfractions were not presented by Lefevre et al.,[88] the cholesterol and protein data from the study of Croswell and Lei[86] demonstrated that this apo E-rich HDL subfraction was indeed enriched with cholesterol.

One major HDL subfraction, HDL-$E_0$ (apo E-free), along with two minor subfractions, HDL-$E_1$ (apo E-rich) and HDL-$E_2$ (apo E-very rich), were isolated by Lee and Koo[89] using heparin-affinity chromatography. All three subfractions appeared to be elevated by copper-deficiency when the overall profiles measured at 280 nm were compared. Unfortunately, the protein content of these subfractions was not provided. They were calculated from the sum of their apolipoprotein concentration data presented by these authors (Table 2). The calculated protein concentration of the HDL-$E_0$, HDL-$E_1$, and HDL-$E_2$ subfractions appeared to be two-, three-, and twofold higher, respectively, in the copper-deficient as compared to the copper-adequate rats. In addition, the cholesterol content reported for apo E-free and apo E-rich HDL subfractions tended to be elevated by copper deficiency but was significant only for apo E-free HDL. Hence, among the various studies, the major difference is in the distribution of cholesterol among these HDL subfractions. In the study of Lee and Koo,[89] copper deficiency increased the cholesterol concentration of apo E-free HDL which accounted for 72% of the increase in cholesterol content of the entire HDL fraction; Croswell and Lei[86] indicated that 87% of the increase in total HDL cholesterol was accounted for by the increase in apo E-rich HDL cholesterol content.

Apolipoprotein content of the HDL subfractions was determined by the same method used by three independent laboratories for the measurement of apolipoprotein profile in the total HDL fraction. In the study of Croswell and Lei,[86] the apo E and apo A-I concentrations of the apo E-rich HDL subfraction were elevated by copper deficiency to about the same extent (2.4- to 2.6-fold increase), contributing to the overall increase in protein content of this fraction. The marked increment of apo E of the apo E-rich fraction also accounted for most of the increase in apo E of the entire HDL fraction. In addition, the apo A-I content of the apo E-poor fraction tended to be elevated by copper deficiency, but the increase was not significant. When the apolipoprotein data from the study of Lefevre et al.[88] were recalculated and expressed as milligrams per deciliter, the apo E concentration of the apo E-rich HDL subfraction appeared to be elevated and subsequently contributed to the increased protein content of this subfraction in copper-deficient rats. The magnitude of the increase in apo E concentration and protein content of the apo E rich subfraction seem to be in close agreement with those reported by Croswell and Lei.[86] However, Lefevre et al.[88] were unable to demonstrate an increase in apo A-I concentration of the apo E-rich HDL subfraction reported previously.[86] A similar elevation in apo E content of the apo E-rich subfraction (HDL-$E_1$) in copper-deficient rats was reported by Lee and Koo.[89] Surprisingly, the apo E content of this fraction was lower in the deficient rats and much lower in the controls when compared to the other studies.[86,88] Apo A-I concentration of this apo E-rich subfraction and the apo E-free HDL subfraction also appeared to be increased by copper deficiency, but the elevation was only significant for the apo E-free HDL subfraction. In contrast, the increase in apo A-I concentration was found to reside in the apo E-rich HDL obtained by Croswell and Lei.[86]

In 1985, Lefevre et al.[87] hypothesized that the enrichment of apo A-I in HDL by copper deficiency was due to a selective increase in $HDL_1$, a larger sized, apo E-rich subpopulation of HDL. The increase in the concentration of apo E-rich fraction, rich in cholesterol and apo E, obtained from plasma of copper-deficient rats by heparin-affinity chromatography[86,88] indicated that $HDL_1$ may be increased relative to $HDL_2$. Supportive data from nondenaturing gel electrophoresis also demonstrated that a larger percentage of particles with diameters greater than 12.5 nm,[88] similar in size (12.5 to 32.5 nm) to $HDL_1$ particles,[109] were found in HDL of copper-deficient rats. In the plasma, HDL without apo E are converted to $HDL_1$ by acquiring apo E from triglyceride-rich lipoproteins.[8] A similar lipoprotein, $HDL_c$, highly enriched in apo E and cholesteryl esters, has been isolated by Mahley[110] in cholesterol-fed animals.

In copper-deficient rats not supplemented with cholesterol, available evidence supports the contention that cholesterol derived from an increased hepatic cholesterol synthesis is packaged into lipoproteins to facilitate an enhanced clearance of nascent cholesteryl ester from the liver to plasma (Chapter 2). An elevated amount of plasma HDL cholesterol is recycled back to the liver to sustain the net efflux of cholesterol from the liver to the plasma (Chapter 4). Hepatic cholesterol derived from *de novo* synthesis and uptake of plasma HDL cholesteryl ester is unavailable to enhance bile acid production and cholesterol elimination. A prolonged net efflux of cholesterol from the liver results in a small depletion of liver cholesterol content, as well as hypercholesterolemia. Thus, the increase in HDL[86,88] may be a compensatory response to aid the reverse cholesterol transport back to the liver to maintain an accelerated lipoprotein production.

Clear distinctions must be made between the copper-deficient animal models used by the three laboratories. The rat model used by Lei and coworkers[85,86] and by Lefevre et al.[87,88] involved no cholesterol supplementation. Hence, the enlarged plasma pool of cholesterol must be derived from an increased endogenous *de novo* synthesis.[111] The capacity of cholesterol synthesis may be a limiting factor for the development of the enlarged plasma cholesterol pool in copper deficiency, whereas the use of the 1% cholesterol supplement by Lee and Koo[89] would undoubtedly suppress hepatic cholesterol synthesis[112] and supply most of the cholesterol needed for the enlarged plasma cholesterol pool. The readily available cholesterol from the diet may enhance HDL production, resulting in markedly elevated HDL. Increased deposition of hepatic cholesterol stores, which normally accompany cholesterol feeding, may not only depress HMG-CoA reductase activity but also reduce the hepatic uptake of cholesterol from HDL or uptake of whole HDL particles. These alterations may contribute to the different cholesterol and apolipoprotein distribution among the various HDL subfractions observed by Lee and Koo.[89] Care must be exercised in the comparison of the available lipoprotein data, since the model used by Lee and Koo[89] is not a pure copper-deficient model but one that is confounded by massive cholesterol feeding.

## IV. HUMAN STUDIES

The first human copper depletion study, which reported increases in total plasma cholesterol and LDL cholesterol levels, was performed by Klevay et al.[113] The study involved only one 29-year-old male subject under constant supervision in a metabolic ward. A 3-d rotating menu of conventional foods was used for the entire duration of the study. The diet provided 0.8 mg of copper, 5.2 mg of zinc (plus 8.5 mg of supplemental zinc), 200 mg of cholesterol, and a linoleate to saturated fatty acid ratio of 0.5, on a daily basis. Prior to depletion, an additional 0.5 mg of copper was supplemented daily in the 43-d control period. No copper supplement was provided during the 15-week depletion period. This was followed by a 39-d repletion period with 2 mg of copper supplemented twice daily. Plasma parameters of copper and cholesterol status were monitored every 2 weeks throughout the study. Cholesterol in plasma lipoproteins was determined by the method of lipid research clinics,[81] but the method for the isolation of LDL was not specified. A net loss of 10.8 mg of copper via urine and feces was attained during depletion. This resulted in marked reductions in plasma copper and cereluplasmin levels as well as erythrocyte superoxide dismutase activity, which indicated that the subject's copper status was compromised. Most of the reduction in plasma copper level was observed at the end of depletion. The lowest plasma copper value was 48 μg/100 ml. A 35% reduction was attained when compared to the lower value of the 95% confidence limits (75 μg Cu/100 ml) established on male staff members of the research center. Thus the subject appeared to be deficient in copper toward the end of depletion. Accompanying the decline in copper status was a 32% increase in plasma total cholesterol level (from 202 to 234 mg/100 ml). An increase in LDL cholesterol level was also reported, but the magnitude of the increment was not presented. When repletion was completed, the subject had retained 15.9 mg more copper than was lost. The parameters measured for copper status promptly reverted toward the normal ranges following repletion. Total cholesterol level also returned to the normal range (198 mg/100 ml), but the fall of LDL cholesterol did not reach the control value. In this study, hematocrit was slightly reduced from 45 to 43 (vol%), but the reduction was not significant. Because of the lack of hematologic change, the authors suggested that lipid metabolism was more sensitive to changes in copper status than is hematopoiesis in this adult subject. Although a single subject was examined in this study, the reversal of lipid parameters and indexes of copper status toward the normal ranges following repletion provided convincing evidence for the hypercholesterolemic effect of copper deficiency in man.

The second human study was carried out by Reiser et al.[114] to examine the influence of copper status on the distribution of plasma cholesterol among lipoproteins in 24 males aged 21 to 57. The study involved an 11-week depletion period with a 7-d rotating menu consisting of conventional foods and purified ingredients which supplied only 1.02 mg of copper and 583 mg of cholesterol per day. This was followed by a 3-week repletion period which provided an additional 3 mg Cu/d in the form of a supplement. During the pretest period, the self-selected diet consumed by the subjects averaged to 1.38 mg of copper and 424 mg of cholesterol per day. A crossover design was originally planned, in which the subjects were to be fed the low-copper diets containing either fructose or cornstarch (7 weeks each) during the depletion period. However, because four instances of heart-related medical complaints were encountered during the first 11 weeks of depletion, the study was terminated after the fourth week of the second dietary carbohydrate period. Since the type of carbohydrate exerted no significant influence on plasma lipid parameters during depletion, all data were pooled to represent a single group of subjects consuming a low level of dietary copper. Plasma lipoproteins were isolated by a combination of precipitation[115,116] and ultracentrifugation[81] methods. HDL was obtained following the precipitation of LDL and VLDL with $MnCl_2$ and heparin. VLDL was isolated by ultracentrifugation.[81] The combined VLDL plus LDL fraction was recovered by ultracentrifugation at density 1.060 g/ml.[81] Cholesterol in the various fractions was determined enzymatically with the Centrifichem 500.[117] LDL cholesterol was derived by subtraction.

The copper status of the subjects appeared to be only slightly depressed since the plasma copper level was mildly but significantly reduced from 122 $\pm$ 4 μg/100 ml at pretest to 102 $\pm$ 4 μg/100 ml after 11 weeks of depletion. In view of the small reduction, and the reduced plasma copper level still within the normal range, the subjects can only be considered to be mildly deficient in copper. Apart from plasma copper, the authors indicated that the erythrocyte superoxide dismutase activity was reduced and reflected the subjects' copper status and the copper density of the diet. Unfortunately, the enzyme activity data were not provided to allow for comparison with other studies. When plasma copper level was used as the index of copper status, the subjects in this study appeared not to be as deficient in copper as the subject reported in the study of Klevay et al.[113] This may explain the inability of this study to demonstrate a significant increase in plasma total cholesterol level in these subjects. Nevertheless, a 14% increase in LDL cholesterol and a 20% reduction in HDL cholesterol were evident at the end of depletion. Copper depletion appeared to have induced a redistribution of cholesterol among lipoproteins resulting in a shift of cholesterol from HDL to the LDL fraction. Upon copper repletion, this adverse distribution of cholesterol among lipoprotein fractions promptly reverted back to the pretest levels.

## V. RABBIT STUDIES

Most studies performed to examine the effect of copper status on lipid metabolism have been conducted with rats since hypercholesterolemia and hyperlipoproteinemia can easily be achieved in copper-deficient rats. However, the rat may not be an ideal animal model for certain aspects of lipoprotein research in view of the known differences in lipoprotein metabolism between rats and humans. In rats, the activity of cholesteryl ester transfer from HDL to VLDL, catalyzed by cholesteryl ester transfer protein, is negligible.[118] Also, rats carry most of the plasma cholesterol in HDL, while humans and rabbits carry a larger fraction in LDL. The matabolism of lipoproteins carrying endogenous and exogenous cholesterol in rabbits is very similar to the pathway in humans.[119] Studies in rabbits indicate that the LDL receptor plays a dual role in cholesterol metabolism. It is involved in the removal of LDL and may also be the main route for the removal of LDL from the circulation.[119,120] In normal human beings, as compared to normal rabbits, very little IDL enter the liver and most are converted to LDL prior to being removed by the LDL receptor.

A study was performed by Hing and Lei[121] to evaluate the suitability of the rabbit as an animal model for the elucidation of mechanisms responsible for hypercholesterolemia induced by copper deficiency. In the past, copper deficiency was induced by feeding rats a diet low in copper. A metal chelator and cupruretic agent, *N*, *N'*-Bis(2 amino ethyl)-1,3-propane diamine HCl or 2,3,2-tetramine HCl (TETA),[122] was used in this study in conjunction with a copper-deficient diet to induce copper deficiency in rabbits. Lipoprotein fractions and apolipoproteins were isolated by ultracentrifugation and gel permeation column chromatography as well as by SDS-PAGE, respectively.[86] A semipurified diet (copper deficient) consisting of 20% soy protein, 54% glucose monohydrate, 5% corn oil, 0.2% DL-methionine, 15% wheat straw (fiber) and mineral and vitamin mixes was used. This basal diet was prepared similar to that designed by Kritchevsky et al.[123] except no copper supplement was included, and the soy protein and wheat straw were washed with 0.25% EDTA to remove any residual copper. The copper-deficient and copper-adequate diets contained 2.3 mg Cu and 13.5 mg Cu, respectively, as well as 80.1 mg Zn/kg of diet. Diets and distilled water were provided *ad libitum*. The use of wheat straw as a fiber source minimized the occurrence of diarrhea often seen in rabbits fed semipurified diets, maintained a reasonable growth response, but resulted in a basal diet only marginally deficient in copper.

In the first experiment,[121] 42 male New Zealand white rabbits, 35 days of age, were randomly divided into three dietary treatments, namely a copper-deficient (2.3 μg/g diet) group, a copper-adequate (13.5 μg/g diet) group, and a chow (21.6 μg/g diet) group. Half of the rabbits from each treatment group were given 100 mg TETA per kilogram body weight per day (orally) for 10 d while the remaining rabbits were untreated. Rabbits fed the copper-deficient diet and treated with TETA demonstrated cardiac hypertrophy and markedly reduced plasma and liver copper levels, which indicated that they were copper deficient right after treatment. Significant elevations (twofold) in LDL protein, cholesterol, triglyceride, and apo B concentrations were observed in TETA-treated rabbits fed the copper-deficient diet. A second experiment[121] was designed to establish that the alterations in LDL are maintained long after TETA treatment. Ten similar rabbits were assigned to three treatments: copper deficient plus TETA ($n = 4$); copper adequate plus TETA ($n = 3$); and copper adequate alone ($n = 3$). The rabbits were fed a TETA dose of 100 mg/d for three 4-d periods over 3 weeks and thereafter maintained on the diets for another 10 weeks. The plasma LDL protein level was found to remain elevated, the plasma copper level was reduced 45%, and the copper level of the heart when expressed as micrograms per gram dry tissue was reduced, 10 weeks post-TETA treatment in rabbits maintained on the copper-deficient diet. However, no significant treatment difference in hepatic copper content was detected.

These studies demonstrated that copper deficiency and hypercholesterolemia were rapidly induced by TETA and were still evident 10 weeks after TETA treatment in rabbits maintained on a copper-deficient diet. In addition, the data suggest that the rabbit, an LDL animal similar to human, may be a suitable animal model to study the influence of dietary copper on lipid metabolism. However, additional research is needed to develop a copper-deficient diet, similar in copper content to that used for rats (about 0.6 μg/g), capable of inducing copper deficiency rapidly and yet maintaining an adequate growth response in rabbits.

## VI. CONCLUSION

In the copper-deficient rat model, most of the changes in lipid and apolipoprotein composition reside mainly in the HDL fraction, whereas in the copper-deficient rabbit model and in human subjects increases in cholesterol and protein content were observed in the LDL fraction. The difference between the rabbits and human subjects as compared to the rats in the distribution of the elevated plasma cholesterol among HDL and LDL may be explained by the apolipoprotein synthetic capacity of the animal models. In Chapter 2, we proposed that the hypercholesterolemia observed in copper-deficient rats is the result of a net efflux of cholesterol from the liver to the plasma, which is sustained by an enhanced hepatic cholesterol synthesis. Cholesterol derived from this enhanced synthesis is packaged into lipoproteins to facilitate an increased clearance of newly synthesized cholesterol from the liver. Thus an increased apolipoprotein synthesis is expected to accompany the elevated cholesterol synthesis in order to maintain an enhanced lipoprotein secretion. Future studies should be performed in the area of apolipoprotein synthesis in order to establish whether the increase in hepatic cholesterol synthesis is in response to the proposed enhanced apolipoprotein synthesis induced by copper deficiency.

In the normal rat liver, the level of apo A-I mRNA is relatively high (1% of total mRNA)[124] and is capable of maintaining a high rate of synthesis and supplying about 50% of plasma apo A-I. Even a small induction of liver apo A-I mRNA by copper deficiency may provide a substantial increase in total plasma apo A-I and HDL levels, whereas the level of apo A-I mRNA in the rabbit liver is very low (0.002% of total mRNA).[125] In the rabbit, plasma apo A-I is mainly of intestinal origin and intestinal apo A-I mRNA is not responsive to dietary manipulation (i.e., cholesterol feeding).[125] Hence, even a large induction of hepatic apo A-I mRNA may produce little or no appreciable increase in total plasma apo A-I and HDL levels in the rabbit.

Although increases in the plasma level of LDL protein[85] and apo B[87] have been observed in copper-deficient rats, the magnitude is small when compared to the changes associated with HDL. Since the capacity of the rat liver to synthesize apo B appears to be low, the relatively small response in LDL protein and apo B levels may have resulted from a small induction of their synthesis by copper deficiency. The increase in LDL cholesterol observed in the human and rabbit studies may also be related to their hepatic capacity to synthesize apo B.

In copper-deficient rats, the findings of enhanced *in vivo* clearance of plasma HDL apolipoproteins, including apo E,[92,93] and increased binding of apo E-rich HDL to hepatic plasma membranes[91] support the contention that the elevated plasma HDL apo E level is *not* the result of decreased catabolism. This conclusion is further supported by the observed increase in hepatic uptake of chylomicron remnants in copper-deficient rats *in vivo*.[126] Thus the elevation of plasma HDL apo E appears to have resulted from an increase in apo E synthesis. Apart from being produced by the liver, apo E has been found to be synthesized by peripheral tissues of humans,[127] nonhuman primates,[128] and rabbits,[125] as well as by mouse peritoneal macrophages.[131] In nonhuman primates as much as 20 to 40% of total body apo E mRNA is estimated to be extrahepatic.[130] Although rat apo E is primarily synthesized in the liver,[60,131] the contribution of peripheral tissues to the overall increase in apo E production should be considered in future research involving the copper-deficient rat model. At present, direct evidence is not yet available to support the proposed induction of apolipoprotein synthesis by copper deficiency. Studies designed to examine apolipoprotein synthesis and lipoprotein secretion are needed to substantiate this contention. In the past, most of the studies performed to study the influence of copper deficiency on lipoprotein metabolism were conducted with rats. More research will have to be performed in other animal models including humans in order to fully understand the mechanism responsible for the hypercholesterolemia associated with copper deficiency.

## REFERENCES


1. **Chapman, M. J.,** Animal lipoproteins: chemistry, structure, and comparative aspects, *J. Lipid Res.,* 21, 789, 1980.
2. **Sata, T., Havel, R. J., and Jones, A. L.,** Characterization of subfractions of triglyceride-rich lipoproteins separated by gel chromatography from blood plasma of normolipemic and hyperlipemic humans, *J. Lipid Res.,* 13, 757, 1972.
3. **Small, D. M. and Shipley, G. G.,** Physical-chemical basis of lipid deposition in atherosclerosis, *Science,* 185, 222, 1974.

4. **Gotto, A. M., Jr., Pownall, H. J., and Havel, R. J.,** Introduction to the plasma lipoproteins, *Methods Enzymol.,* 128, 3, 1986.
5. **Tytgat, G. N., Rubin, C. E., and Saunders, D. R.,** Synthesis and transport of lipoprotein particles by intestinal absorptive cells in man, *J. Clin. Invest.,* 50, 2065, 1971.
6. **Eisenberg, S. and Levy, R. I.,** Lipoprotein metabolism, *Adv. Lipid Res.,* 13, 1, 1975.
7. **Goldstein, J. L. and Brown, M. S.,** Lipoprotein receptors: genetic defense against atherosclerosis, *Clin. Res.,* 30, 417, 1982.
8. **Eisenberg, S.,** High density lipoprotein metabolism, *J. Lipid Res.,* 25, 1017, 1984.
9. **Glomset, J. A.,** The plasma lecithin:cholesterol acyltransferase reaction, *J. Lipid Res.,* 9, 155, 1968.
10. **Miller, G. J. and Miller, N. E.,** Plasma-high-density lipoprotein concentration and development of ischaemic heart-disease, *Lancet,* 1, 16, 1975.
11. **Gordon, T., Castelli, W. P., Hjortland, M. C., Kannel, W. B., and Dawber, T. R.,** High density lipoprotein as a protective factor against coronary heart disease: the Framingham Study, *Am. J. Med.,* 62, 707, 1977.
12. **Segrest, J. P., Jackson, R. L., Morrisett, J. D., and Gotto, A. M., Jr.,** A molecular theory of lipid-protein interactions in the plasma lipoproteins, *FEBS Lett.,* 38, 274, 1974.
13. **Sabesin, S. M. and Frase, S.,** Electron microscopic studies of the assembly, intracellular transport, and secretion of chylomicrons by rat intestine, *J. Lipid Res.,* 18, 496, 1977.
14. **Norum, K. R., Berg, T., Helgerud, P., and Drevon, C. A.,** Transport of cholesterol, *Physiol. Rev.,* 63, 1343, 1983.
15. **Glickman, R. M., Kirsch, K., and Isselbacher, K. J.,** Fat absorption during inhibition of protein synthesis: studies of lymph chylomicrons, *J. Clin. Invest.,* 51, 356, 1972.
16. **Rubin, C. E.,** Electron microscopic studies of triglyceride absorption in man, *Gastroenterology,* 50, 65, 1966.
17. **Casley-Smith, J.R.,** The identification of chylomicra and lipoproteins in tissue sections and their passage into jejunal lacteals, *J. Cell Biol.,* 15, 259, 1962.
18. **Cryer, A.,** Tissue lipoprotein lipase activity and its action in lipoprotein metabolism, *Int. J. Biochem.,* 13, 525, 1981.
19. **Cryer, A. and Jones, H. M.,** The distribution of lipoprotein lipase (clearing-factor lipase) activity in the adiposal, muscular and lung tissues of 10 animal species, *Comp. Biochem. Physiol.,* 63B, 501, 1979.
20. **Robinson, D. S.,** Changes in the lipolytic activity of the guinea pig mammary gland at parturition, *J. Lipid Res.,* 4, 21, 1963.
21. **Olivecrona, T. and Bengtsson, G.,** Molecular basis for the interaction of lipoprotein lipase with triglyceride-rich lipoproteins at the capillary endothelium, *INSERM Symp.,* 87, 149, 1979.
22. **Havel, R. J.,** Origin, metabolic fate, and metabolic function of plasma lipoproteins, in *Hypercholesterolemia and Atherosclerosis,* Steinberg, D. and Olefsky, J. M., Eds., Churchill Livingstone, New York, 1987, 117.
23. **Schaefer, E. J., Jenkins, L. L., and Brewer, H. B.,** Human chylomicron apolipoprotein metabolism, *Biochem. Biophys. Res. Commun.,* 80, 405, 1978.
24. **Cooper, A. D.,** The metabolism of chylomicron remnants by isolated perfused rat liver, *Biochim. Biophys. Acta,* 488, 464, 1977.
25. **Havel, R. J., Goldstein, J. L., and Brown, M. S.,** Lipoproteins and lipid transport, in *Metabolic Control and Disease*, 8th ed., Bondy, P. K. and Rosenberg, L. E., Eds., W. B. Saunders, Philadelphia, 1980, 393.
26. **Cooper, A. D., Nutik, R., and Erickson, S. K.,** Evidence that apo E is the determinant of chylomicron remnant binding to liver plasma membranes, *Clin. Res.,* 28, 518a, 1980.
27. **Hui, D. Y., Innerarity, T. L., and Mahley, R. W.,** Lipoprotein binding to canine hepatic membranes: metabolically distinct apo E and apo B,E receptors, *J. Biol. Chem.,* 256, 5646, 1981.

28. **Jones, A. L., Hradek, G. T., Hornick, C., Renaud, G., Windler, E. E. T., and Havel, R. J.,** Uptake and processing of remnants of chylomicrons and very low density lipoproteins by rat liver, *J. Lipid Res.,* 25, 1151, 1984.
29. **Jost-Vu, E., Hamilton, R. L., Hornick, C. A., Belcher, J. D., and Havel, R. J.,** Fusion between primary lysosomes and multivesicular bodies gives birth to secondary lysosomes, *J. Histochem. Cytochem.,* 34, 120, 1986.
30. **Erickson, S. K. and Cooper, A. D.,** Acyl-coenzyme A:cholesterol acyltransferase in human liver: in vitro detection and some characteristics of the enzyme, *Metabolism,* 29, 991, 1980.
31. **Tall, A. R.,** Plasma lipid transfer proteins, *J. Lipid Res.,* 27, 361, 1986.
32. **Felker, T. E., Fainaru, M., Hamilton, R. L., and Havel, R. J.,** Secretion of the arginine-rich and A-I apolipoproteins by the isolated perfused rat liver, *J. Lipid Res.,* 18, 465, 1977.
33. **Kane, J. P.,** Apolipoprotein B: structural and metabolic heterogeneity, *Annu. Rev. Physiol.,* 45, 637, 1983.
34. **Siuta-Mangano, P., Janero, D. R., and Lane, M. D.,** Association and assembly of triglyceride and phospholipid with glycosylated and unglycosylated apoproteins of very low density lipoprotein in the intact liver cell, *J. Biol. Chem.,* 257, 11463, 1982.
35. **Nilsson-Ehle, P.,** Lipolytic enzymes and plasma lipoprotein metabolism, *Annu. Rev. Biochem.,* 49, 667, 1980.
36. **Eisenberg, S. and Schurr, D.,** Phospholipid removal during degradation of rat plasma very low density lipoprotein in vitro, *J. Lipid Res.,* 17, 578, 1976.
37. **Higgins, J. M. and Fielding, C. J.,** Mechanisms of formation of triglyceride-rich remnant particles from very low density lipoproteins and chylomicrons, *Biochemistry,* 14, 2288, 1975.
38. **Brown, M. S., Kovanen, P. T., and Goldstein, J. L.,** Regulation of plasma cholesterol by lipoprotein receptors, *Science,* 212, 628, 1981.
39. **Pagnan, A., Havel, R. J., Kane, J. P., and Kotite, L.,** Characterization of human very low density lipoproteins containing two electrophoretic populations: double pre-beta lipoproteinemia and primary dysbetalipoproteinemia, *J. Lipid Res.,* 18, 613, 1977.
40. **Havel, R. J.,** The formation of LDL: mechanisms and regulation, *J. Lipid Res.,* 25, 1570, 1984.
41. **Mahley, R. W.,** Apolipoprotein E: cholesterol transport protein with expanding role in cell biology, *Science,* 240, 622, 1988.
42. **Yamamoto, T., Davis, C. G., Brown, M. S., Schneider, W. J., Casey, M. L., Goldstein, J. L., and Russell, D. W.,** The human LDL receptor: a cysteine-rich protein with multiple alu sequences in its mRNA, *Cell,* 39, 27, 1984.
43. **Brown, M. S. and Goldstein, J. L.,** How LDL receptors influence cholesterol and atherosclerosis, *Sci. Am.,* 251, 58, 1984.
44. **Spady, D. K., Turley, S. D., and Dietschy, J. M.,** Receptor-independent low density lipoprotein transport in the rat in vivo: quantitation, characterization, and metabolic consequences, *J. Clin. Invest.,* 76, 1113, 1985.
45. **Spady, D. K., Meddings, J. B., and Dietschy, J. M.,** Kinetic constants for receptor-dependent and receptor-independent low density lipoprotein transport in the tissues of the rat and hamster, *J. Clin. Invest.,* 77, 1474, 1986.
46. **Brown, M. S. and Goldstein, J. L.,** A receptor-mediated pathway for cholesterol homeostasis, *Science,* 232, 34, 1986.
47. **Bilheimer, D. W., Stone, N. J., and Grundy, S. M.,** Metabolic studies in familial hypercholesterolemia: evidence for a gene-dosage effect in vivo, *J. Clin. Invest.,* 64, 524, 1979.
48. **Anderson, R. G. W., Goldstein, J. L., and Brown, M. S.,** Localization of low density lipoprotein receptors on plasma membrane of normal human fibroblasts and their absence in cells from a familial hypercholesterolemia homozygote, *Proc. Natl. Acad. Sci. U.S.A.,* 73, 2434, 1976.

49. **Brown, M. S., Anderson, R. G. W., and Goldstein, J. L.,** Recycling receptors: the round-trip itinerary of migrant membrane proteins, *Cell,* 32, 663, 1983.
50. **Pittman, R. C. and Steinberg, D.,** Sites and mechanisms of uptake and degradation of high density and low density lipoproteins, *J. Lipid Res.,* 25, 1577, 1984.
51. **Rudel, L. L., Parks, J. S., Johnson, F. L., and Babiak, J.,** Low density lipoproteins in atherosclerosis, *J. Lipid Res.,* 27, 465, 1986.
52. **Johnson, F. L., St. Clair, R. W., and Rudel, L. L.,** Studies on the production of low density lipoproteins by perfused livers from nonhuman primates: effects of dietary cholesterol, *J. Clin. Invest.,* 72, 221, 1983.
53. **Johnson, F. L., St. Clair, R. W., and Rudel, L. L.,** Effects of the degree of saturation of dietary fat on the hepatic production of lipoproteins in the African green monkey, *J. Lipid Res.,* 26, 403, 1985.
54. **Gwynne, J. T. and Strauss, J. F., III,** The role of lipoproteins in steroidogenesis and cholesterol metabolism in steroidogenic glands, *Endocrin. Rev.,* 3, 299, 1982.
55. **Reichl, D. and Miller, N. E.,** The anatomy and physiology of reverse cholesterol transport, *Clin. Sci.,* 70, 221, 1986.
56. **Rothblot, G. H., Bamberger, M., and Phillips, M. C.,** Reverse cholesterol transport, *Methods Enzymol.,* 129, 628, 1986.
57. **Roheim, P. S.,** Atherosclerosis and lipoprotein metabolism: role of reverse cholesterol transport, *Am. J. Cardiol.,* 57, 3C, 1986.
58. **Hamilton, R. L., Williams, M. C., Fielding, C. J., and Havel, R. J.,** Discoidal bilayer structure of nascent high density lipoproteins from perfused rat liver, *J. Clin. Invest.,* 58, 667, 1976.
59. **Imaizumi, K., Havel, R. J., Fainaru, M., and Vigne, J-L.,** Origin and transport of the A-I and arginine-rich apolipoproteins in mesenteric lymph of rats, *J. Lipid Res.,* 19, 1038, 1978.
60. **Wu, A-L. and Windmueller, H. G.,** Relative contributions by liver and intestine to individual plasma apolipoproteins in the rat, *J. Biol. Chem.,* 254, 7316, 1979.
61. **Green, P. H. R. and Glickman, R. M.,** Intestinal lipoprotein metabolism, *J. Lipid Res.,* 22, 1153, 1981.
62. **Fielding, C. J.,** Mechanisms of action of lecithin-cholesterol acyltransferase, *Methods Enzymol.,* 129, 783, 1986.
63. **Fielding, C. J., Shore, V. G., and Fielding, P. E.,** A protein cofactor of lecithin:cholesterol acyltransferase, *Biochem. Biophys. Res. Commun.,* 46, 1493, 1972.
64. **Tall, A. R. and Small, D. M.,** Body cholesterol removal: role of plasma high-density lipoproteins, *Adv. Lipid Res.,* 17, 1, 1980.
65. **Biesbroeck, R., Oram, J. F., Albers, J. J., and Bierman, E. L.,** Specific high-affinity binding of high density lipoproteins to cultured human skin fibroblasts and arterial smooth muscle cells, *J. Clin. Invest.,* 71, 525, 1983.
66. **Brinton, E. A., Oram, J. F., Chen, C-H., Albers, J. J., and Bierman, E. L.,** Binding of high density lipoprotein to cultured fibroblasts after chemical alteration of apoprotein amino acid residues, *J. Biol. Chem.,* 261, 495, 1986.
67. **Brinton, E. A., Oram, J. F., and Bierman, E. L.,** The effect of variations in lipid composition of high-density lipoprotein on its interaction with receptors on human fibroblasts, *Biochim. Biophys. Acta,* 920, 68, 1987.
68. **Slotte, J. P., Oram, J. F., and Bierman, E. L.,** Binding of high density lipoproteins to cell receptors promotes translocation of cholesterol from intracellular membranes to the cell surface, *J. Biol. Chem.,* 262, 12904, 1987.
69. **Tabas, I. and Tall, A. R.,** Mechanism of the association of $HDL_3$ with endothelial cells, smooth muscle cells, and fibroblasts: evidence against the role of specific ligand and receptor proteins, *J. Biol. Chem.,* 259, 13897, 1984.

70. **Karlin, J. B., Johnson, W. J., Benedict, C. R., Chacko, G. K., Phillips, M. C., and Rothblat, G. H.,** Cholesterol flux between cells and high density lipoprotein: lack of relationship to specific binding of the lipoprotein to the cell surface, *J. Biol. Chem.*, 262, 12557, 1987.
71. **Johnson, W. J., Mahlberg, F. H., Chacko, G. K., Phillips, M. C., and Rothblat, G. H.,** The influence of cellular and lipoprotein cholesterol contents on the flux of cholesterol between fibroblasts and high density lipoprotein, *J. Biol. Chem.*, 263, 14099, 1988.
72. **Mendel, C. M. and Kunitake, S. T.,** Cell-surface binding sites for high density lipoproteins do not mediate efflux of cholesterol from human fibroblasts in tissue culture, *J. Lipid Res.*, 29, 1171, 1988.
73. **Phillips, M. C. and Rothblat, G. H.,** Cholesterol flux between high-density lipoproteins and cells, *Atherosclerosis Rev.*, 16, 57, 1987.
74. **Graham, D. L. and Oram, J. F.,** Identification and characterization of a high density lipoprotein-binding protein in cell membranes by ligand blotting, *J. Biol. Chem.*, 262, 7439, 1987.
75. **Keso, L., Lukka, M., Ehuholm, C., Baumann, M., Vihko, P., and Olkinuora, M.,** Apolipoprotein A-I-binding protein from human placenta: purification and partial characterization, *FEBS Lett.*, 215, 105, 1987.
76. **Fidge, N., Kagami, A., and O'Connor, M.,** Identification of a high density lipoprotein binding protein from adrenocortical membranes, *Biochem. Biophys. Res. Commun.*, 129, 759, 1985.
77. **Fidge, N. H.,** Partial purification of a high density lipoprotein-binding protein from rat liver and kidney membranes, *FEBS Lett.*, 199, 265, 1986.
78. **Mendel, C. M., Kunitake, S. T., Kane, J. P., and Kempner, E. S.,** Radiation inactivation of binding sites for high density lipoproteins in human fibroblast membranes, *J. Biol. Chem.*, 263, 1314, 1988.
79. **Mendel, C. M., Kunitake, S. T., Hong, K., Erickson, S. K., Kane, J. P., and Kempner, E. S.,** Radiation inactivation of binding sites for high-density lipoproteins in human liver membranes, *Biochim. Biophys. Acta*, 961, 188, 1988.
80. **Allen, K. G. D. and Klevay, L. M.,** Hyperlipoproteinemia in rats due to copper deficiency, *Nutr. Rep. Int.*, 22, 295, 1980.
81. **Anon.,** Lipid and lipoprotein analysis, in *Manual of Laboratory Operations, Lipid Research Clinic Program,* Vol. 1, DHEW Pub. No. NIH 75-628, Washington, D.C., 1974, 56.
82. **Carpenter, K. J., Gotsis, A., and Hegsted, D. M.,** Estimation of total cholesterol in serum by a micro method, *Clin. Chem.*, 3, 233, 1957.
83. **Harvey, P. W. and Allen, K. G. D.,** Lipoproteins and liver lipids in copper-deficient rats, *Nutr. Res.*, 5, 511, 1985.
84. **Lei, K. Y. and Lin, I. M.,** Kinetics of high density lipoprotein cholesterol in copper-deficient rats, *Nutr. Rep. Int.*, 24, 1179, 1981.
85. **Lei, K. Y., Hassel, C. A., and Allen, D. K.,** Alterations in plasma lipid, lipoprotein and apolipoprotein concentrations in copper-deficient rats, *J. Nutr.*, 113, 2178, 1983.
86. **Croswell, S. C. and Lei, K. Y.,** Effect of copper deficiency on the apolipoprotein E-rich high density lipoproteins in rats, *J. Nutr.*, 115, 473, 1985.
87. **Lefevre, M., Keen, C. L., Lönnerdal, B., Hurley, L. S., and Schneeman, B. O.,** Different effects of zinc and copper deficiency on composition of plasma high density lipoproteins in rats, *J. Nutr.*, 115, 359, 1985.
88. **Lefevre, M., Keen, C. L., Lönnerdal, B., Hurley, L. S., and Schneeman, B. O.,** Copper deficiency-induced hypercholesterolemia: effects on HDL subfractions and hepatic lipoprotein receptor activity in the rat, *J. Nutr.*, 116, 1735, 1986.
89. **Lee, C. C. and Koo, S. I.,** Effects of copper deficiency on the composition of three high-density lipoprotein subclasses as separated by heparin-affinity chromatography, *Biochim. Biophys. Acta,* 963, 278, 1988.

90. **Hassel, C. A., Lei, K. Y., Carr, T. P., and Marchello, J. A.,** Lipoprotein receptors in copper-deficient rats: apolipoprotein E-free high density lipoprotein binding to liver membranes, *Metabolism,* 36, 1054, 1987.
91. **Hassel, C. A., Carr, T. P., Marchello, J. A., and Lei, K. Y.,** Apolipoprotein E-rich HDL binding to liver plasma membranes in copper-deficient rats, *Proc. Soc. Exp. Biol. Med.,* 187, 296, 1988.
92. **Carr, T. P. and Lei, K. Y.,** In vivo apoprotein catabolism of high density lipoproteins in copper-deficient, hypercholesterolemic rats, *Proc. Soc. Exp. Biol. Med.,* 191, 370, 1989.
93. **Carr, T. P. and Lei, K. Y.,** High density lipoprotein cholesteryl ester and protein catabolism in hypercholesterolemic rats induced by copper deficiency, *Metabolism,* in press, 1990.
94. **Roschlau, Von P., Bernt, E., and Gruber, W.,** Enzymatic determination of total cholesterol in serum, *Z. Klin. Chem. Klin. Biochem.,* 12, 403, 1974.
95. **Lopez-Virella, M. F., Stone, P., Ellis, S., and Colwell, J. A.,** Cholesterol determination in high density lipoproteins separated by three different methods, *Clin. Chem.,* 23, 882, 1977.
96. **Rudel, L. L., Lee, J. A., Morris, M. D., and Felts, J. M.,** Characterization of plasma lipoproteins separated and purified by agarose-column chromatography, *Biochem. J.,* 139, 89, 1974.
97. **Fainaru, M., Havel, R. J., and Imaizumi, K.,** Apoprotein content of plasma lipoproteins of the rat separated by gel chromatography or ultracentrifugation, *Biochem. Med.,* 17, 347, 1977.
98. **Nour-Eldin, F.,** Iron metabolism and related problems, in *Haematology-Rudimental, Practical and Clinical,* Appleton-Century-Crofts, New York, 1972, 137.
99. **Al-Othman, A. A. and Lei, K. Y.,** Time course development of enlarged plasma volume and cholesterol pool size in copper-deficient rats, *FASEB J.,* 3, A358, 1989.
100. **Elovson, J., Huang, Y. O., Baker, N., and Kannan, R.,** Apolipoprotein B is structurally and metabolically heterogeneous in the rat, *Proc. Natl. Acad. Sci. U.S.A.,* 78, 157, 1981.
101. **Van't Hooft, F. M., Hardman, D. A., Kane, J. P., and Havel, R. J.,** Apolipoprotein B (B-48) of rat chylomicrons is not a precursor of the apolipoprotein of low density lipoproteins, *Proc. Natl. Acad. Sci. U.S.A.,* 79, 179, 1982.
102. **Weber, K. and Osborn, M.,** The reliability of molecular weight determinations by dodecyl sulfate-polyacrylamide gel electrophoresis, *J. Biol. Chem.,* 244, 4406, 1969.
103. **Laemmli, U. K.,** Cleavage of structural protein during the assembly of the head of bacteriophage $T_4$, *Nature (London),* 227, 680, 1970.
104. **Kane, J. P.,** A rapid electrophoretic technique for identification of subunit species of apoproteins in serum lipoproteins, *Anal. Chem.,* 53, 350, 1973.
105. **Weisgraber, K. H. and Mahley, R. W.,** Subfractionation of human high density lipoproteins by heparin-Sepharose affinity chromatography, *J. Lipid Res.,* 21, 316, 1980.
106. **Quarfordt, S. H., Jain, R. S., Jakoi, L., Robinson, S., and Shelburne, F.,** The heterogeneity of rat high density lipoproteins, *Biochem. Biophys. Res. Commun.,* 83, 786, 1978.
107. **Zhang, J., Carr, T. P., and Lei, K. Y.,** Specific binding and uptake of apolipoprotein in E-free high density lipoproteins by cultured liver parenchymal cells of copper-deficient rats, *FASEB J.,* 3, A357, 1989.
108. **Lei, K. Y., Hendriks, H. F. J., Brouwer, A., Bock, I., DeRuiter, C. S. J., and Knook, D. L.,** Influence of copper deficiency on binding and uptake of apolipoprotein E-free high density lipoproteins by isolated rat liver parenchymal and Kupffer cells, *FASEB J.,* 3, A1062, 1989.

109. **Mahley, R. W.,** Alterations in plasma lipoproteins induced by cholesterol feeding in animals including man, in *Disturbances in Lipid and Lipoprotein Metabolism,* Dietschy, J. M., Gotto, A. M., Jr., and Ontka, J. A., Eds., American Physiological Society, Bethesda, MD, 1978, 181.
110. **Mahley, R. W.,** Atherogenic hyperlipoproteinemia: the cellular and molecular biology of plasma lipoproteins altered by dietary fat and cholesterol, *Med. Clin. North Am.,* 66, 375, 1982.
111. **Yount, N. Y., McNamara, D. J., Al-Othman, A. A., and Lei, K. Y.,** The effect of copper deficiency on rat hepatic 3-hydroxy-3-methylglutaryl coenzyme A reductase activity, *J. Nutr. Biochem.,* in press, 1990.
112. **Lei, K. Y.,** Cholesterol metabolism in copper-deficient rats, *Nutr. Rep. Int.,* 15, 597, 1977.
113. **Klevay, L. M., Inman, L., Johnson, L. K., Lawler, M., Mahalko, J. R., Milne, D. B., Lukaski, H. C., Bolonchuk, W., and Sandstead, H. H.,** Increased cholesterol in plasma in a young man during experimental copper depletion, *Metabolism,* 33, 1112, 1984.
114. **Reiser, S., Powell, A., Yang, C-Y., and Canary, J. J.,** Effect of copper intake on blood cholesterol and its lipoprotein distribution in men, *Nutr. Rep. Int.,* 36, 641, 1987.
115. **Bronzert, T. J. and Brewer, H. B., Jr.,** New micro method for measuring cholesterol in plasma lipoprotein fractions, *Clin. Chem.,* 23, 2089, 1977.
116. **Steele, B. W., Koehler, D. F., Azar, M. N., Blaszkowski, T. P., Kuba, K., and Dempsey, M. E.,** Enzymatic determinations of cholesterol in high density lipoprotein fractions prepared by a precipitation technique, *Clin. Chem.,* 22, 98, 1976.
117. **Allain, C. C., Poon, L. S., Chan, C. S. G., Richmond, W., and Fu, P. C.,** Enzymatic determination of total serum cholesterol, *Clin. Chem.,* 20, 470, 1974.
118. **Dolphin, P. J.,** Lipoprotein metabolism and the role of apolipoproteins as metabolic programmers, *Can. J. Biochem. Cell Biol.,* 63, 850, 1985.
119. **Goldstein, J. L., Kita, T., and Brown, M. S.,** Defective lipoprotein receptors and atherosclerosis. Lessons from an animal counterpart of familial hypercholesterolemia, *N. Engl. J. Med.,* 309, 288, 1983.
120. **Kita, T., Brown, M. S., Bilheimer, D. W., and Goldstein, J. L.,** Delayed clearance of very low density and intermediate density lipoproteins with enhanced conversion to low density lipoprotein in WHHL rabbits, *Proc. Natl. Acad. Sci. U.S.A.,* 79, 5693, 1982.
121. **Hing, S. A. O. and Lei, K. Y.,** Copper deficiency and hyperlipoproteinemia induced by N,N′-bis-(2-aminoethyl)-1,3-propane diamine (2,3,2-tetramine) in rabbits, *Fed. Proc. Fed. Am. Soc. Exp. Biol.,* 45, 357, 1986.
122. **Allen, K. G. D., Twedt, D. C., and Hunsker, H. A.,** Tetramine cupruretic agents: comparisons in dogs, *Am. J. Vet. Res.,* 48, 28, 1987.
123. **Kritchevsky, D., Tepper, S. A., Kim, H. K., Moses, D. E., and Story, J. A.,** Experimental atherosclerosis in rabbits fed cholesterol-free diets. IV. Investigation into the source of cholesterolemia, *Exp. Mol. Pathol.,* 22, 11, 1975.
124. **Lin-Su, M-H., Lin-Lee, Y-C., Bradley, W. A., and Chan, L.,** Characterization, cell-free synthesis, and processing of apolipoprotein A-I of rat high-density lipoproteins, *Biochemistry,* 20, 2470, 1981.
125. **Chao, Y-S., Yamin, T-T., Thompson, G. M., and Kroon, P. A.,** Tissue-specific expression of genes encoding apolipoprotein E and apolipoprotein A-I in rabbits, *J. Biol. Chem.,* 259, 5306, 1984.
126. **Koo, S. I., Lee, C. C., and Norvell, J. E.,** Effect of copper deficiency on the lymphatic absorption of cholesterol, plasma chylomicron clearance and postheparin lipase activities, *Proc. Soc. Exp. Biol. Med.,* 188, 410, 1988.
127. **Blue, M. L., Williams, D. L., Zuckler, S., Khan, S. A., and Blum, C. B.,** Apolipoprotein E synthesis in human kidney, adrenal gland, and liver, *Proc. Natl. Acad. Sci. U.S.A.,* 80, 283, 1983.

128. **Williams, D. L., Dawson, P. A., Newman, T. C., and Rudel, L. L.,** Apolipoprotein E synthesis in peripheral tissues of nonhuman primates, *J. Biol. Chem.*, 260, 2444, 1985.
129. **Basu, S. K., Goldstein, J. L., and Brown, M. S.,** Independent pathways for secretion of cholesterol and apolipoprotein E by macrophages, *Science,* 219, 871, 1983.
130. **Newman, T. C., Dawson, P. A., Rudel, L. L., and Williams, D. L.,** Quantitation of apolipoprotein E mRNA in the liver and peripheral tissues of nonhuman primates, *J. Biol. Chem.*, 260, 2452, 1985.
131. **McLean, J. W., Fukazawa, C., and Taylor, J. M.,** Rat apolipoprotein E mRNA, *J. Biol. Chem.*, 258, 8993, 1983.

Chapter 4

# COPPER AND HIGH DENSITY LIPOPROTEIN CATABOLISM *IN VIVO*

**Timothy P. Carr and K. Y. Lei**

## TABLE OF CONTENTS

# I. INTRODUCTION

As pointed out in earlier chapters, dietary copper is essential for the normal maintenance of lipoprotein and cholesterol metabolism. Nutritional copper deficiency often results in elevated plasma cholesterol levels. Much attention has focused on this aspect of copper deficiency because of the high correlation between plasma cholesterol levels and the incidence of coronary heart disease in Western populations.[1,2] While strict copper deficiency in adult humans is unlikely, premature infants[3,4] and young children suffering from general malnutrition[5-8] exhibit typical signs of copper deficiency, such as anemia and osteoporosis, suggesting that diets marginally adequate in copper could contribute to the increased risk of coronary heart disease.

The majority of studies conducted to examine the mechanisms of copper deficiency-induced hypercholesterolemia have involved rats, although the condition has been demonstrated experimentally in mice,[9] monkeys,[10] and humans.[11,12] In the Sprague-Dawley rat, the observed hypercholesterolemia appears to be associated primarily with increased plasma high density lipoproteins (HDL). This observation is not surprising since at least 80% of the total plasma cholesterol in rats is transported in HDL.[13] Unlike humans, rat plasma contains a relatively small amount of low density lipoproteins (LDL) and is devoid of $HDL_3$, but contains $HDL_2$ and large amounts of apolipoprotein (apo) E-rich $HDL_1$.[14,15] Recent evidence strongly suggests that the hypercholesterolemia of copper-deficient rats is attributable to the selective increase in the $HDL_1$ subpopulation relative to $HDL_2$.[16-19] Thus it seems reasonable to assume that changes in the formation and/or catabolism of HDL and possibly the interactions between HDL subpopulations could contribute to the observed hypercholesterolemia of copper-deficient rats.

The present chapter is intended to describe some recent findings in our laboratory which suggest that the catabolism of plasma HDL is indeed altered in the rat as a result of dietary copper deficiency. Such *in vivo* studies have not been reported in copper-deficient animals. It is possible that the mechanisms operating in this hypercholesterolemic model could eventually provide information on the relationship between HDL metabolism and coronary heart disease. A brief review of HDL cholesterol transport in the normolipidemic rat and *in vivo* catabolism is presented. Furthermore, various methodologies available to the investigator are discussed, as well as recommendations for further research.

# II. HDL CHOLESTEROL TRANSPORT

Much literature exists concerning the exogenous and endogenous lipid transport systems in humans and various animal species, whereby cholesterol is eventually delivered to peripheral cells, yet relatively little information is available about the removal of cellular cholesterol and the maintenance of tissue cholesterol levels. As originally described by Glomset,[20] plasma HDL play a pivotal role in cholesterol removal from the periphery and subsequent delivery to the liver for excretion from the body, a process commonly referred to as "reverse cholesterol transport". It should be noted that HDL can also transport cholesterol to steroidogenic organs to provide precursors for steroid synthesis. In peripheral tissues, reverse cholesterol transport is initiated by the efflux of unesterified (free) cholesterol from cells to HDL particles.[21-23] While the mode of interaction between HDL and peripheral cells is not completely understood, it is agreed that $HDL_3$ particles are much more effective than $HDL_2$ in promoting cholesterol removal from cells and reducing cellular cholesterol.[24-26] Newly acquired cholesterol is rapidly esterified in the plasma by lecithin:cholesterol acyltransferase, which appears to further facilitate the transfer of free cholesterol from the cell.[27,28] The newly synthesized cholesteryl ester can be transported directly to the liver in HDL or transferred to other plasma lipoproteins by means of cholesteryl ester transfer protein.[29,30] Cholesteryl ester transferred to very low density lipoproteins (VLDL) and low density lipoproteins (LDL) then have the potential to be delivered to the liver via the LDL receptor.[31]

The rat is considered an "HDL mammal" and possesses the capacity to transport HDL cholesterol from and to peripheral cells. As mentioned above, rat plasma contains large amounts of apo E-rich $HDL_1$.[14,15] Because apo E is a ligand for the LDL receptor,[32] $HDL_1$ can participate in cholesterol delivery to the periphery and in reverse cholesterol transport via the LDL receptor, in addition to nonreceptor and alternative receptor mechanisms. This observation is significant because the rat is deficient in cholesteryl ester transfer activity,[33] a mechanism in other species whereby LDL and VLDL can receive HDL cholesteryl ester. Moreover, the absence of this mechanism in the rat appears to be the basis for the presence of the larger $HDL_1$ particles.[30] Thus it is important to bear in mind that cholesterol transport and lipoprotein metabolism in the rat may be dissimilar to humans in several ways. However, *in vivo* study of HDL metabolism in the copper-deficient rat seems appropriate for three reasons: (1) HDL are the major vehicle for cholesterol transport in the rat and are easily isolated from the plasma; (2) changes in plasma cholesterol levels in the rat are generally resistant to other dietary manipulations, such as cholesterol feeding, but change in response to diets deficient in copper; and (3) the lack of cholesteryl ester transfer activity allows the *in vivo* study of HDL-associated cholesteryl ester without complicating data interpretation by the transfer of this component to other plasma lipoproteins.

## III. HDL CATABOLISM

Because of the central role of the liver in reverse cholesterol transport under normal conditions, hepatic clearance of HDL seems to be a primary mode of catabolism, yet virtually all tissues appear to have the capacity to catabolize, at least partially, HDL particles. To determine the major sites of HDL catabolism, early studies were carried out using iodinated HDL. Results suggested that the liver and possibly the small intestine were the major sites of HDL uptake.[34-36] However, Van Tol et al.[37] found that the fractional catabolic rate of iodinated HDL from the plasma was not altered after partial hepatectomy and further reported that the majority of HDL catabolism occurred in skeletal muscle and not the liver. Since the iodination procedure radiolabels only the protein moiety of HDL particles, further studies were designed to examine the catabolic sites of HDL cholesteryl esters and phospholipids.[38] Results indicated that the liver was indeed the major catabolic site of both HDL cholesteryl esters and phospholipids. It was therefore suggested by Van't Hooft et al.[38] that, at least in the rat, the catabolism of the individual HDL components could occur by separate metabolic pathways depending on the tissue site and metabolic state of the animal.

More recently, nondegradable radioactive ligands have been used to label the various HDL components, thus providing a cumulative marker of tissue uptake and degradation. Stein and co-workers[39] labeled HDL particles with [$^3$H]cholesteryl linoleyl ether to serve as a nondegradable analog of HDL cholesteryl esters. Results confirmed earlier studies that the liver is the major site of HDL cholesteryl ester uptake in the rat, and that uptake in the carcass is also significant. Glass et al.[40] also utilized cholesteryl ether, but, in addition, covalently labeled apo A-I with $^{125}$I-tyramine cellobiose, which also acts as a cumulative marker in cells that degrade apo A-I. These data clearly indicated that the majority of HDL cholesteryl ester was catabolized in the liver. While the liver remained predominant as a catabolic site for apo A-I, substantially less was degraded in this organ as compared to cholesteryl ester. The apparent differential uptake of HDL components was later demonstrated in a variety of tissues;[41] in general, greater uptake of cholesteryl ester occurred in the steroidogenic tissues, whereas the skin and kidneys exhibited greater uptake of apo A-I. Increased uptake of apo A-I observed in the kidney probably reflects degradation of filtered apo A-I not associated with HDL particles.[42] These studies have also demonstrated that HDL cholesteryl ester is cleared from the plasma at a greater rate than HDL protein,[39-41] supporting the concept that a significant fraction of plasma HDL is not degraded as a whole particle.

Differential uptake of HDL protein and cholesteryl ester have also been demonstrated in a variety of cultured cells. HDL cholesteryl ester is preferentially catabolized by primary cultures of rat hepatocytes[41,43,44] and adrenal cells.[41,44,45] Mouse-derived macrophages[46] as well as human Hep G2 hepatoma cells[44,47] and fibroblasts[44,47] also exhibit differential uptake. These studies suggest that HDL can deliver cholesteryl ester to a variety of cells, particularly steroidogenic cells,[48] in addition to removing free cholesterol from them. Furthermore, it appears that preferential delivery is regulated by the cholesterol status of the cell.[47,49] The mechanism of selective cholesteryl ester transfer into the cell apparently does not involve endocytosis of the HDL particle,[44,50] nor is HDL binding to the cell surface required.[47]

Liver perfusion studies confirm the occurrence of a differential uptake of HDL protein and cholesteryl ester components in the liver.[51,52] The uptake of HDL cholesteryl ester was four- to sixfold higher than that of the protein moiety. It appears that a portion of the HDL may bind specifically to the liver, but binding is not necessarily required for the uptake of cholesteryl esters.[52] Furthermore, results indicate that a fraction of the HDL that initially bind are eventually released as smaller, more dense particles, significantly depleted of core cholesteryl ester.[51,52] These data provide further evidence that HDL are not necessarily degraded as whole particles and may be "re-cycled" back to the plasma.

## IV. METHODOLOGICAL CONSIDERATIONS

### A. Kinetic Analysis and Tissue Sites of Uptake

Several attempts have been made at estimating the rates of HDL component removal from the plasma and the tissue sites of uptake. The most commonly used approach is the injection of a single tracer dose of radiolabeled HDL, which represents the catabolism of the entire HDL pool after initial equilibrium. Because the dose can be very small in comparison to the *in vivo* pool, the addition of labeled material does not alter the metabolic system under investigation.[53] Subsequent blood and tissue samples can be taken and determination made of the rate at which radioactivity is cleared from the plasma as well as tissue uptake. An alternative method for the determination of lipoprotein turnover was developed by Dietschy and co-workers,[54,55] which requires a continuous infusion of labeled HDL into the plasma until a constant level of radioactivity is achieved. The clearance rate can then be calculated based on the amount of labeled HDL infused per hour. The latter technique has the advantage of direct *in vivo* determination of clearance rate, rather than deriving values from the mathematical analysis of curves fitted to the plasma disappearance of radiolabeled particles. However, continuous infusion is more cumbersome than single-dose injection and may cause greater trauma to the animals under observation.

The single-dose injection method is used routinely in our laboratory to determine both plasma clearance and subsequent tissue uptake of HDL components. Results are highly reproducible and animals appear to be under minimal stress throughout the procedure. Several methodological criteria, however, must be met in order for accurate interpretation of data. First, the time at which animals are sacrificed after injection should be kept relatively short, preferably 12 h or less, to avoid significant "leakage" of radiolabel from cells in which degradation occurs. Stein et al.[39] demonstrated that loss of radioactive iodine from tissues (e.g., liver, spleen, and adrenal gland) could be detected between 6 and 12 h after injection of iodinated HDL. Preliminary experiments in our laboratory confirm that the recovery of radiolabel with specific tissues could decrease slightly by 12 h after injection

(T. P. Carr and K. Y. Lei, unpublished data). Consequently, inconsistencies can arise if tissue uptake data are expressed as the percent of injected dose recovery with specific tissues. We prefer to express data as the percent (or proportion) of *total tissue* uptake attributable to each tissue, which appears not to be altered by the cellular loss of degraded tracer. The second requirement for validation of this technique is to determine if the radiolabel of HDL redistributes among other plasma components in appreciable amounts. If this occurs, the radioactivity recovered in tissues could represent uptake of other lipoprotein classes from the plasma. Current data[56] suggest that intravascular exchanges of iodinated HDL label are negligible (less than 1% of total HDL radioactivity) within 12 h after injection, adding confidence that tissue-associated radioactive iodine represents the uptake of only HDL protein, at least in the rat. Because the rat lacks cholesteryl ester transfer activity,[33] tissue uptake of labeled cholesteryl ester in this model represents exclusively HDL catabolism. The third criterion for validation is to account for any residual plasma remaining in the tissue, which might yield an overestimation of the tissue-associated radioactivity. Whole-body perfusion[57] is the method of choice in our laboratory, since virtually all residual plasma is flushed from the tissues. This method seems most appropriate since individual tissues exhibit different relative capacities in the space occupied by plasma.[55] The plasma tissue space, then, would have to be determined for each tissue sampled, a procedure which is unnecessary if residual plasma radioactivity is removed. Finally, it should be determined if the radiolabeling procedures alter the HDL particles in such a way (e.g., protein denaturation) as to cause artifactual results. This can be accomplished by injecting radiolabeled HDL which has first been "biologically screened", a process thought to remove radiolabel not associated with intact particles.[34,39,58] This method requires that freshly isolated and labeled HDL first be injected into donor rats and allowed to circulate for a time (approximately 3 to 12 h). Serum is then reisolated and injected directly into recipient animals. While some researchers[34,39] report more consistent results when biologically screened sera are used, others[58] state no difference between data obtained using screened and nonscreened HDL. Results in our laboratory[56] also demonstrate no advantage to biological screening, although methodological variations between laboratories should not be precluded. Thus, the following potential sources of error can be minimized when estimating clearance rates and tissue uptake of rat HDL: (1) loss of degraded radiolabel from cells; (2) tissue uptake of lipoproteins other than HDL; (3) residual plasma radioactivity as contributing to tissue radioactivity; and (4) uptake of altered HDL particles as a result of the radiolabeling procedures.

**B. Radiolabeling of HDL Components**

Several radiolabeling methods exist for HDL depending on the specific chemical moiety to be traced. HDL proteins are generally labeled with radioactive iodine, usually $^{125}I$ or $^{131}I$, according to the original procedure of

McFarlane.[59] Radioactive iodine is attached to the ring structure of tyrosine residues; proteins with a relatively high proportion of tyrosine residues will exhibit greater specific activities compared to proteins with fewer tyrosine residues. Thus it is important to determine the apolipoprotein composition of each HDL preparation, as well as the respective specific activities following iodination. An alternative method for labeling HDL protein was developed by Pittman and co-workers,[57] whereby specific proteins (apo A-I) are labeled with a nondegradable radioactive ligand that becomes trapped within the catabolizing cells as a cumulative marker of cellular uptake. This method has been applied to studies of apo A-I catabolism in the rat.[40-42] HDL phospholipids can be biosynthetically labeled by injecting $H_3{}^{32}PO_4$ into rats and later isolating the HDL from plasma.[60] Plasma clearance rates and tissue uptake of biosynthetically labeled HDL phospholipids have been reported in rats using this method.[38] Finally, HDL cholesteryl ester can be traced by incorporating into the particle core radiolabeled (usually with $^3H$) cholesteryl ester or nondegradable cholesteryl ether. Several procedures of incorporation have been described in the literature, one of which requires lyophilization of HDL, lipid extraction, and reconstitution with the appropriate labeled and unlabeled lipids.[40,41] The second procedure of incorporation utilizes cholesteryl ester transfer protein (CETP), either from human or rabbit plasma. Intralipid® serves as a carrier of cholesteryl ester (or ether) for the CETP reaction in the presence of HDL.[39] A third procedure also exploits CETP, except labeled cholesteryl ester (or ether) is added directly to plasma containing HDL and CETP.[61] The latter procedure of incorporation appears to be the most efficient method of radiolabeling the cholesteryl ester moiety of HDL and is used in our laboratory with a high degree of success.

### C. Nondegradable Ligands

The use of nondegradable ligands has the advantage of minimizing leakage of radioactivity during the assessment of tissue sites of uptake. Pittman and Taylor[57] point out, however, that two conditions must be met for the proper application of this technique. First, the tracer must be metabolically indistinguishable from the molecule in question, at least until it becomes trapped intracellularly. Second, the nondegradable tracer must be adequately trapped within the cell. In practice these molecules are not completely trapped and can leak from tissues at significant rates.[57] Consequently, we prefer to remove and analyze tissues prior to 12 h post-injection regardless of whether or not degradable ligands are used. While trapped label methodology helps to minimize leakage of radioactivity from the cell, it should be mentioned again that interpretation and expression of data as the percentage or proportion of total tissue uptake (not injected dose) can help circumvent this problem irrespective of the labeling technique.

## V. HDL CATABOLISM IN COPPER-DEFICIENT RATS

Experiments in our laboratory have been designed to characterize the catabolism of HDL components in copper-deficient, hypercholesterolemic Sprague-Dawley rats. Copper deficiency is induced by feeding weanling males the AIN-76A diet[62,63] for periods of up to 6 weeks. The clearance rates of HDL apolipoproteins and cholesteryl ester from the plasma have been determined, as well as the tissue sites of catabolism. In some experiments, doubly labeled HDL were used in which the protein was iodinated with $^{125}I$ and the cholesteryl ester moiety was traced by incorporating into the particle core [$^3H$]cholesteryl linoleyl ether. The single dose injection method was used in all experiments, and tissue and blood samples removed at subsequent time intervals for quantitation of radioactivity.

To establish that animals are indeed copper-deficient, commonly used indicators, such as reduced liver copper, cardiac hypertrophy, and reduced hematocrit are routinely employed in our laboratory. Quite notable, however, is the recent observation of increased plasma volume.[16,56] While increased plasma volume may not be an established indicator of copper deficiency, this observation is particularly significant since alterations in plasma volume can potentially affect the pool size of plasma components. For example, if the *concentration* of plasma cholesterol appears to be unaffected by copper deficiency, an increased plasma volume would result in an increased *intravascular pool* of cholesterol. As discussed in Chapter 2, and in contrast to the above example, plasma cholesterol concentration is usually increased by feeding copper-deficient diets. Thus in conjunction with increased plasma volumes, we generally observe increased pools of plasma cholesterol and total protein that are nearly doubled in mass.[56,64] It is apparent, then, that the hypercholesterolemic condition of copper-deficient rats should be judged by the increase in the absolute mass of plasma cholesterol and not just concentration, although the effectiveness of increased plasma volume as an indicator of progressive copper deficiency is uncertain.

In view of these observations, assessment of the clearance rates of HDL protein and cholesteryl ester must necessarily be determined as the absolute mass clearance, since fractional catabolic rate (FCR) considers only the turnover of the plasma pool in question and not the pool size.[53] When expressed as the mass of HDL protein cleared per hour, normalized to 100 g of body weight, copper-deficient rats exhibited an absolute catabolic rate (ACR) that is approximately doubled compared to animals fed diets adequate in copper.[56] This increase may be an obligatory response to the increased plasma pool of HDL protein, which is also approximately doubled in deficient animals. Initial studies[56] suggested the liver to be a major organ of HDL protein catabolism, but could not account for the significant increase in plasma clearance. We speculated that tissue uptake must also be increased in other organs not directly assayed in the copper-deficient animals, particularly the bulk tissues. Subsequent studies[64] now confirm that the skin and muscle probably account for the majority of the increased clearance of HDL protein in copper-deficient rats, whereas uptake in the liver may not increase significantly in response to elevated HDL protein in the plasma.

Plasma clearance of the individual HDL apolipoproteins appears *not* to occur at significantly different rates in copper-deficient rats.[56] Studies of normolipidemic rats[65,66] also indicate that while clearance rates of apolipoproteins fall within a specific range, individual clearance rates are not significantly different. Alternatively, Eisenberg et al.[36] suggested that plasma removal of HDL apolipoproteins may occur at different rates, although their data were not tested statistically and appeared quite similar to the data of the aforementioned studies in which significant differences were not detected. Experimental error could account for slight differences; it is also possible that different rates reflect the heterogeneity of the total HDL population. As no differences have been observed in the plasma clearance of specific HDL apolipoproteins of copper-deficient rats,[56,64] data may be calculated as the catabolism of total HDL protein.

Experiments in our laboratory which utilized doubly labeled HDL particles demonstrated that plasma clearance of HDL cholesteryl ester occurs at a greater rate than HDL protein.[64] Differential clearance of HDL protein and cholesteryl ester moieties has been well documented;[39-41] these data support the conclusion that a significant fraction of HDL are not degraded as whole particles. The magnitude of the disproportionate clearance of cholesteryl ester was also greater in copper-deficient rats.[64] Unlike the increased response of HDL protein uptake to the increased plasma pool, it is possible that the cellular requirement for cholesterol is increased to a greater extent in deficient animals. When calculated as the absolute mass clearance, virtually all of the increased removal of HDL cholesteryl ester observed in copper deficiency was attributable to the liver.[64] Because of the central role of the liver in reverse cholesterol transport,[20] these data imply an increased rate of cholesterol excretion. However, previous studies in our laboratory suggest no increase in bile acid synthesis or fecal cholesterol output in deficient rats, although the capacity to increase excretion clearly exists in response to cholesterol feeding.[67] Thus it appears that, in order to offset the observed increase in plasma clearance of HDL components, the production rate of HDL by the liver must also be increased in copper-deficient rats. Three lines of evidence support this hypothesis: (1) hepatic HMG-CoA reductase activity, the rate-limiting enzyme in cholesterol synthesis, is significantly increased in deficient rats;[68] (2) newly assembled cholesteryl esters (from mevalonate) clear the liver at an accelerated rate in copper deficiency;[69] and (3) total cholesterol concentration in the liver of copper-deficient rats is significantly lower.[70-72] It seems reasonable, then, to suggest that plasma cholesteryl ester delivered to the liver may be ''repackaged'' into HDL at an increased rate. If the production rate of HDL is indeed increased, and in view of the present data, an accelerated flux of HDL components appears to exist in copper-deficient rats. The observed hypercholesterolemia in this model could be the result of an imbalance between influx and efflux of the plasma HDL components, specifically cholesterol and/or cholesteryl ester. Accordingly, inference to a direct cause-and-effect relationship between HDL catabolism and hypercholesterolemia would clearly be inappropriate.

## VI. RECOMMENDATIONS FOR FURTHER RESEARCH

Studies conducted in our laboratory have demonstrated increased *in vivo* catabolism of HDL components in copper-deficient rats. These observations alone do not explain the relationship between hypercholesterolemia and dietary copper deficiency, nor do they directly address the mechanisms contributing to elevated plasma cholesterol in this model. While we have proposed a scheme which seems to explain previous observations of cholesterol and lipoprotein metabolism in copper-deficient rats, further research should most certainly consider intracellular synthesis of HDL components as well as particle assembly and modifications following secretion. Furthermore, the metabolic pathways of *free* cholesterol associated with plasma HDL have yet to be examined *in vivo*. It is hoped that studies of the synthetic events of HDL metabolism will further contribute to an understanding of the mechanisms operating in this hypercholesterolemic model.

## REFERENCES

1. **Lipid Research Clinics Program,** The Lipid Research Clinics Coronary Primary Prevention Trial Results. I. Reduction in incidence of coronary heart disease, *JAMA*, 251, 351, 1984.
2. **Lipid Research Clinics Program,** The Lipid Research Clinics Coronary Primary Prevention Trial Results. II. The relationship of reduction incidence of coronary heart disease to cholesterol lowering, *JAMA*, 251, 365, 1984.
3. **Griscom, N. T., Craig, J. N., and Neuhauser, E. B. D.,** Systemic bone disease developing in small premature infants, *Pediatrics*, 48, 883, 1971.
4. **Ashkenazi, A., Levin, S., Djaldetti, M., Fishel, E., and Benvenisti, D.,** The syndrome of neonatal copper deficiency, *Pediatrics*, 52, 525, 1973.
5. **Graham, G. G. and Cordano, A.,** Copper depletion and deficiency in the malnourished infant, *John Hopkins Med. J.*, 124, 139, 1969.
6. **Al-Rashid, R. A. and Spangler, J.,** Neonatal copper deficiency, *N. Eng. J. Med.*, 285, 841, 1971.
7. **Karpel, J. T. and Peden, V. H.,** Copper deficiency in long-term parenteral nutrition, *J. Pediatr.*, 80, 32, 1972.
8. **Heller, R. M., Kirchner, S. G., O'Neill, J. A., Hough, A. J., Howard, L., Kramer, S. S., and Green, H. L.,** Skeletal changes of copper deficiency in infants receiving prolonged total parenteral nutrition, *J. Pediatr.*, 92, 947, 1978.
9. **Jones, D. G.,** Effects of dietary copper depletion on acute and delayed inflammatory responses in mice, *Res. Vet. Sci.*, 37, 205, 1984.

10. **Milne, D. B., Omaye, S. T., and Amos, W. H.,** Effect of ascorbic acid on copper and cholesterol in adult cynomolgus monkeys fed a diet marginal in copper, *Am. J. Clin. Nutr.,* 34, 2389, 1981.
11. **Klevay, L. M., Imman, L., Johnson, L. K., Lawler, M., Mahalko, J. R., Milne, D. B., Lukaski, H. C., Bolonchuk, W., and Sandstead, H. H.,** Increased cholesterol in plasma in a young man during experimental copper depletion, *Metabolism,* 33, 1112, 1984.
12. **Reiser, S., Powell, A., Yang, C-Y., and Canary, J. J.,** Effect of copper intake on blood cholesterol and its lipoprotein distribution in men, *Nutr. Rep. Int.,* 36, 641, 1987.
13. **Spady, D. K. and Dietschy, J. M.,** Sterol synthesis in vivo in 18 tissues of the squirrel monkey, guinea pig, rabbit, hamster, and rat, *J. Lipid Res.,* 24, 303, 1983.
14. **Lusk, L. T., Walker, L. F., DuBien, L. H., and Getz, G. S.,** Isolation and partial characterization of high-density lipoprotein $HDL_1$ from rat plasma by gradient centrifugation, *Biochem. J.,* 183, 83, 1979.
15. **Oschry, Y. and Eisenberg, S.,** Rat plasma lipoproteins: re-evaluation of a lipoprotein system in an animal devoid of cholesteryl ester transfer activity, *J. Lipid Res.,* 23, 1099, 1982.
16. **Lei, K. Y., Hassel, C. A., and Allen, D. K.,** Alterations in plasma lipid, lipoprotein and apolipoprotein concentrations in copper-deficient rats, *J. Nutr.,* 113, 2178, 1983.
17. **Croswell, S. C. and Lei, K. Y.,** Effect of copper deficiency on the apolipoprotein E-rich high density lipoproteins in rats, *J. Nutr.,* 115, 473, 1985.
18. **Lefevre, M., Keen, C. L., Lönnerdal, B., Hurley, L. S., and Schneeman, B. O.,** Different effects of zinc and copper deficiency on composition of plasma high density lipoproteins in rats, *J. Nutr.,* 115, 359, 1985.
19. **Lefevre, M., Keen, C. L., Lönnerdal, B., Hurley, L. S., and Schneeman, B. O.,** Copper deficiency-induced hypercholesterolemia: effects on HDL subfractions and hepatic lipoprotein receptor activity in the rat, *J. Nutr.,* 116, 1735, 1986.
20. **Glomset, J. A.,** The plasma lecithin:cholesterol acyltransferase reaction, *J. Lipid Res.,* 9, 155, 1968.
21. **Tall, A. R. and Small, D. M.,** Body cholesterol removal: role of plasma high-density lipoproteins, *Adv. Lipid Res.,* 17, 1, 1980.
22. **Reichl, D. and Miller, N. E.,** The anatomy and physiology of reverse cholesterol transport, *Clin. Sci.,* 70, 221, 1986.
23. **Roheim, P. S.,** Atherosclerosis and lipoprotein metabolism: role of reverse cholesterol transport, *Am. J. Cardiol.,* 57, 3C, 1986.
24. **Oram, J. F., Albers, J. J., Cheung, M. C., and Bierman, E. L.,** The effects of subfractions of high density lipoprotein on cholesterol efflux from cultured fibroblasts, *J. Biol. Chem.,* 256, 8348, 1981.
25. **Daniels, R. J., Guertler, L. S., Parker, T. S., and Steinberg, D.,** Studies on the rate of efflux of cholesterol from cultured human skin fibroblasts, *J. Biol. Chem.,* 256, 4978, 1981.
26. **Oram, J. F.,** Effects of high density lipoprotein subfractions on cholesterol homeostasis in human fibroblasts and arterial smooth muscle cells, *Arteriosclerosis,* 3, 420, 1983.
27. **Ray, E., Bellini, F., Stoudt, G., Hemperly, S., and Rothblat, G.,** Influence of lecithin:cholesterol acyltransferase on cholesterol metabolism in hepatoma cells and hepatocytes, *Biochim. Biophys. Acta,* 617, 318, 1980.
28. **Fielding, C. J. and Fielding, P. E.,** Evidence for a lipoprotein carrier in human plasma catalyzing sterol efflux from cultured fibroblasts and its relationship to lecithin:cholesterol acyltransferase, *Proc. Natl. Acad. Sci. U.S.A.,* 78, 3911, 1981.
29. **Zilversmit, D. B.,** Lipid transfer proteins: overview and applications, *Methods Enzymol.,* 98, 565, 1983.
30. **Tall, A. R.,** Plasma lipid transfer proteins, *J. Lipid Res.,* 27, 361, 1986.
31. **Brown, M. S., Kovanen, P. T., and Goldstein, J. L.,** Regulation of plasma cholesterol by lipoprotein receptors, *Science,* 212, 628, 1981.

32. **Mahley, R. W.,** Apolipoprotein E: cholesterol transport protein with expanding role in cell biology, *Science,* 240, 622, 1988.
33. **Ha, Y. C. and Barter, P. J.,** Differences in plasma cholesteryl ester transfer activity in sixteen vertebrate species, *Comp. Biochem. Physiol.,* 71B, 265, 1982.
34. **Roheim, P. S., Rachmilewitz, D., Stein, O., and Stein, Y.,** Metabolism of iodinated high density lipoproteins in the rat. I. Half-life in the circulation and uptake by organs, *Biochim. Biophys. Acta,* 248, 315, 1971.
35. **Rachmilewitz, D., Stein, O., Roheim, P. S., and Stein, Y.,** Metabolism of iodinated high density lipoproteins in the rat. II. Autoradiographic localization in the liver, *Biochim. Biophys. Acta,* 270, 414, 1972.
36. **Eisenberg, S., Windmueller, H. G., and Levy, R. I.,** Metabolic fate of rat and human lipoprotein apoproteins in the rat, *J. Lipid Res.,* 14, 446, 1973.
37. **Van Tol, A., Van Gent, T., Van't Hooft, F. M., and Vlaspolder, F.,** High density lipoprotein catabolism before and after partial hepatectomy, *Atherosclerosis,* 29, 439, 1978.
38. **Van't Hooft, F. M., Van Gent, T., and Van Tol, A.,** Turnover and uptake by organs of radioactive serum high-density lipoprotein cholesteryl esters and phospholipids in the rat in vivo, *Biochem. J.,* 196, 877, 1981.
39. **Stein, Y., Dabach, Y., Hollander, G., Halperin, G., and Stein, O.,** Metabolism of HDL-cholesteryl ester in the rat, studied with a nonhydrolyzable analog, cholesteryl linoleyl ether, *Biochim. Biophys. Acta,* 752, 98, 1983.
40. **Glass, C., Pittman, R. C., Weinstein, D. B., and Steinberg, D.,** Dissociation of tissue uptake of cholesterol ester from that of apoprotein A-I of rat plasma high density lipoprotein: selective delivery of cholesterol ester to liver, adrenal, and gonad, *Proc. Natl. Acad. Sci. U.S.A.,* 80, 5435, 1983.
41. **Glass, C., Pittman, R. C., Civen, M., and Steinberg, D.,** Uptake of high-density lipoprotein-associated apoprotein A-I and cholesterol esters by 16 tissues of the rat in vivo and by adrenal cells and hepatocytes in vitro, *J. Biol. Chem.,* 260, 744, 1985.
42. **Glass, C., Pittman, R. C., Keller, G. A., and Steinberg, D.,** Tissue sites of degradation of apoprotein A-I in the rat, *J. Biol. Chem.,* 257, 7161, 1983.
43. **Leitersdorf, E., Stein, O., Eisenberg, S., and Stein, Y.,** Uptake of rat plasma HDL subfractions labeled with [$^{3}$H]cholesteryl linoleyl ether or with $^{125}$I by cultured rat hepatocytes and adrenal cells, *Biochim. Biophys. Acta,* 796, 72, 1984.
44. **Pittman, R. C., Knecht, T. P., Rosenbaum, M. S., and Taylor, C. A., Jr.,** A nonendocytotic mechanism for the selective uptake of high density lipoprotein-associated cholesterol ester, *J. Biol. Chem.,* 262, 2443, 1987.
45. **Leitersdorf, E., Israeli, A., Stein, O., Eisenberg, S., and Stein, Y.,** The role of apolipoproteins of HDL in the selective uptake of cholesteryl linoleyl ether by cultured rat and bovine adrenal cells, *Biochim. Biophys. Acta,* 878, 320, 1986.
46. **Stein, O., Israeli, A., Leitersdorf, E., Halperin, G., and Stein, Y.,** Preferential uptake of cholesteryl ester-HDL by cultured macrophages, *Atherosclerosis,* 75, 151, 1986.
47. **Rinninger, F. and Pittman, R. C.,** Regulation of the selective uptake of high density lipoprotein-associated cholesteryl esters by human fibroblasts and Hep G2 hepatoma cells, *J. Lipid Res.,* 29, 1179, 1988.
48. **Gwynne, J. T. and Strauss, J. F., III,** The role of lipoproteins in steroidogenesis and cholesterol metabolism in steroidogenic glands, *Endocrin. Rev.,* 3, 299, 1982.
49. **Rinninger, F. and Pittman, R. C.,** Regulation of the selective uptake of high density lipoprotein-associated cholesteryl esters, *J. Lipid Res.,* 28, 1313, 1987.
50. **Pittman, R. C., Glass, C. K., Atkinson, D., and Small, D. M.,** Synthetic high density lipoprotein particles: application to studies of the apoprotein specificity for selective uptake of cholesterol esters, *J. Biol. Chem.,* 262, 2435, 1987.
51. **Mackinnon, M., Savage, J., Wishart, R., and Barter, P.,** Metabolism of high density lipoproteins by the perfused rabbit liver, *J. Biol. Chem.,* 261, 2548, 1986.

52. **Arbeeny, C. M., Rifici, V. A., and Eder, H. A.,** The uptake of the apoprotein and cholesteryl ester of high-density lipoproteins by the perfused rat liver, *Biochim. Biophys. Acta,* 917, 9, 1987.
53. **Shipley, R. A. and Clark, R. E., Eds.,** *Tracer Methods for In Vivo Kinetics,* Academic Press, New York, 1972, chap. 1.
54. **Munford, R. S., Anderson, J. M., and Dietschy, J. M.,** Sites of tissue binding and uptake in vivo of bacterial lipopolysaccharide-high density lipoprotein complexes, *J. Clin. Invest.,* 68, 1503, 1981.
55. **Koelz, H. R., Sherrill, B. C., Turley, S. D., and Dietschy, J. M.,** Correlation of low and high density lipoprotein binding in vivo with rates of lipoprotein degradation in the rat: a comparison of lipoproteins of rat and human origin, *J. Biol. Chem.,* 257, 8061, 1982.
56. **Carr, T. P. and Lei, K. Y.,** In vivo apoprotein catabolism of high density lipoproteins in copper-deficient, hypercholesterolemic rats, *Proc. Soc. Exp. Biol. Med.,* 191, 370, 1989.
57. **Pittman, R. C. and Taylor, C. A., Jr.,** Methods for assessment of tissue sites of lipoprotein degradation, *Methods Enzymol.,* 129, 612, 1986.
58. **Sigurdsson, G., Simon-Pierre, N., and Havel, R. J.,** Quantification of the hepatic contribution to the catabolism of high density lipoproteins in rats, *J. Lipid Res.,* 20, 316, 1979.
59. **McFarlane, A. S.,** Efficient trace-labelling of proteins with iodine, *Nature (London),* 182, 53, 1958.
60. **Rubenstein, B. and Rubenstein, D.,** Interrelationships between rat serum very low density and high density lipoproteins, *J. Lipid Res.,* 13, 317, 1972.
61. **Roberts, D. C. K., Miller, N. E., Price, S. G. L., Crook, D., Cortese, C., LaVille, A., Masana, L., and Lewis, B.,** An alternative procedure for incorporating radiolabelled cholesteryl ester into human plasma lipoproteins in vitro, *Biochem. J.,* 226, 319, 1985.
62. **American Institute of Nutrition,** Report of American Institute of Nutrition ad hoc committee on standards for nutritional studies, *J. Nutr.,* 107, 1340, 1977.
63. **American Institute of Nutrition,** Second Report of the ad hoc committee on standards for nutritional studies, *J. Nutr.,* 110, 1726, 1980.
64. **Carr, T. P. and Lei, K. Y.,** High density lipoprotein cholesteryl ester and protein catabolism in hypercholesterolemic rats induced by copper deficiency, *Metabolism,* in press, 1990.
65. **Roheim, P. S., Hirsch, H., Edelstein, D., and Rachmilewitz, D.,** Metabolism of iodinated high density lipoprotein subunits in the rat. III. Comparison of the removal rate of different subunits from the circulation, *Biochim. Biophys. Acta,* 278, 517, 1972.
66. **Sparks, C. E., Tennenberg, S. D., and Marsh, J. B.,** Catabolism of the apolipoproteins of HDL in control and nephrotic rats, *Biochim. Biophys. Acta,* 665, 8, 1981.
67. **Lei, K. Y.,** Oxidation, excretion, and tissue distribution of [26-$^{14}$C] cholesterol in copper-deficient rats, *J. Nutr.,* 108, 232, 1978.
68. **Yount, N. Y., McNamara, D. J., Al-Othman, A. A., and Lei, K. Y.,** The effect of copper deficiency on rat hepatic 3-hydroxy-3-methylglutaryl coenzyme A reductase activity, *J. Nutr. Biochem.,* in press, 1990.
69. **Shao, M. J. S. and Lei, K. Y.,** Conversion of [2-$^{14}$C]mevalonate into cholesterol, lanosterol and squalene in copper-deficient rats, *J. Nutr.,* 110, 859, 1980.
70. **Lei, K. Y.,** Cholesterol metabolism in copper-deficient rats, *Nutr. Rep. Int.,* 15, 597, 1977.
71. **Hassel, C. A., Lei, K. Y., Carr, T. P., and Marchello, J. A.,** Lipoprotein receptors in copper-deficient rats: apolipoprotein E-free high density lipoprotein binding to liver membranes, *Metabolism,* 36, 1054, 1987.
72. **Hassel, C. A., Carr, T. P., Marchello, J. A., and Lei, K. Y.,** Apolipoprotein E-rich binding to liver plasma membranes in copper-deficient rats, *Proc. Soc. Exp. Biol. Med.,* 187, 296, 1988.

Chapter 5

# LIPOPROTEIN RECEPTORS IN COPPER-DEFICIENT RATS: APPLICATION OF RADIOLIGAND BINDING TECHNIQUES

**Craig A. Hassel and K. Y. Lei**

## TABLE OF CONTENTS

## I. INTRODUCTION

As discussed previously, diet-induced copper deficiency in growing rats frequently results in hypercholesterolemia. The extent of this condition may be influenced by the severity of the imposed deficiency, the type of purified diet, the strain, sex, and age of experimental rats, and other experimental factors. Over the past 10 years, a number of studies have more closely described the metabolic disturbances involved in this condition. Kinetic analyses have demonstrated increases in the size and half-life of the rapidly exchangeable cholesterol pool with copper-deficient (CD) rats.[1] The half-life of free and total plasma high density lipoprotein (HDL) cholesterol was also prolonged.[2] Furthermore, plasma lipoprotein profiles are altered in CD rats.[3,4] Apolipoprotein E-rich HDL ($HDL_1$) is selectively elevated in CD Sprague-Dawley rats, and the increase in this population of particles accounts for most of the observed increases in plasma cholesterol.[5,6] Therefore, it seems reasonable to hypothesize that the hypercholesterolemic condition in CD rats may be facilitated, in part, through impaired receptor-mediated clearance of plasma lipoproteins.

The purpose of this chapter is to discuss the above hypothesis within the framework of rat lipoprotein catabolism and to evaluate the degree to which relevant information can be provided by *in vitro* radioligand binding assays. To properly address this hypothesis, one must maintain an awareness of the species under consideration and the knowledge that multiple and discrete lipoprotein clearance systems operate *in vivo*. In the rat, these systems are not well defined as in other animal models. The utility of radioligand binding studies is ultimately limited to an understanding of how the observed binding is related to operative lipoprotein clearance systems *in vivo*. With this understanding, one may then use lipoprotein binding data directly in a discussion of the hypercholesterolemic condition observed in the CD rat.

## II. REVIEW OF LITERATURE

The importance of lipoprotein receptors as a control mechanism for the regulation of plasma lipoprotein catabolism is now well established. The studies of Goldstein and Brown[7] first demonstrated the role of low density lipoprotein (LDL) receptors in regulating plasma LDL levels. More recently, other lipoprotein receptors have also been defined. These include the apolipoproteins (apo) E receptor and the putative HDL receptor.[8] Each of these receptors appears to possess unique biochemical characteristics. However, their physiological interrelationship and relative importance in contributing to cholesterol homeostasis may vary considerably depending upon the animal species under consideration.

The Sprague-Dawley rat is known as an "HDL mammal", possessing a polydispersed and heterogeneous population of HDL but very little LDL.[9]

Heparin-Sepharose affinity chromatography and rate zonal ultracentrifugation are two methods which have been successfully employed to isolate rat HDL subfractions for further characterization.[10,11] Although these methods differ in their dimension of separation (chemical vs. physical), both yield a large, cholesterol-rich $HDL_1$ particle with a high apo E to apo A-I ratio (often termed apo E-rich HDL), and a smaller, more dense particle with a low apo E to apo A-I ratio. The existence of these discrete HDL subfractions in rat plasma is well established; however, the metabolic distinction between these particles with regard to receptor-mediated catabolism *in vivo* is presently unclear.

Recent advances have allowed investigators to identify the presence of different lipoprotein receptors and measure their expression in tissues and cells through *in vitro* radioligand binding assays. These studies establish the presence of a binding site and provide qualitative and quantitative receptor-ligand information. However, they provide little information about the relative contribution of a given receptor to cholesterol or lipoprotein catabolism *in vivo*. Radioligand binding experiments are designed to examine the binding reaction using criteria which distinguish receptor-mediated from receptor-independent binding. Accepted characteristics of receptor-mediated binding include specificity, saturability (indicating a finite number of receptors or receptor sites), and high affinity. Lipoprotein binding to cell surface receptors should mediate a subsequent physiological event, and in the case of most receptors, the number per cell is not constant and can be regulated by the presence or absence of ligand. In addition to these criteria, other binding characteristics can be used to identify the different lipoprotein receptors described above. For example, the binding of LDL to its receptor requires $Ca^{2+}$, while the binding of HDL to its receptor does not.[12] Sensitivity of the binding to proteolytic enzymes, pH, ionic strength, and chemical modification of the ligand are observations commonly used to identify the lipoprotein receptors responsible for ligand binding. The information derived from radioligand binding assays should be discussed within the context of these criteria, particularly when describing a different experimental system such as the CD rat model.

Numerous *in vitro* binding studies have been performed to assess lipoprotein binding to cells and membranes from various rat tissues. Earlier reports failed to detect significant hepatic LDL binding activity,[13,14] and it is now well established that in the normal adult rat, hepatic LDL receptors are expressed at very low levels. Other studies have established that HDL bind to hepatocytes,[15] smooth muscle cells,[16] intestinal mucosal cells,[17] and to membranes prepared from liver,[18,19] testis,[20] adrenal,[21] ovary,[22] and kidney[23] tissues. The binding observed in these studies was saturable, of high affinity and varying specificity with characteristics distinct from those of the LDL or apo E receptor. Most of these studies were performed using HDL devoid of apo E and apo B (apo E-free HDL). Since this subfraction of HDL lacks the apolipoprotein ligands necessary for interaction with the LDL or apo E receptors, this binding is generally accepted as representing the HDL receptor. This binding site (or receptor) is probably more important physiologically in rat hepatic and extrahepatic tissues than the LDL receptor.

The metabolic disturbances observed in copper deficiency-induced hypercholesterolemia suggest that impaired receptor-mediated clearance of lipoprotein-bound cholesterol may be partly responsible. It is now well established that rat plasma lipoprotein profiles are altered as a result of copper deficiency.[3-6] The apo E content of HDL is increased by copper deficiency, and this observation appears to be the result of an increased concentration of the apo E-rich HDL ($HDL_1$) subfraction. Relatively few studies have investigated the binding of $HDL_1$ in a homologous binding system despite the prominence of this HDL subfraction in rat plasma. Unlike apo E-free HDL, $HDL_1$ possesses the apolipoproteins necessary for interaction with either the LDL or chylomicron remnant receptor.[8] It is quite possible that $HDL_1$ is recognized by more than one lipoprotein receptor in rat liver. If the elevation in $HDL_1$ induced by copper deficiency is the result of impaired interaction with receptors, it is conceivable that any of the lipoprotein receptors may be affected.

To date, three studies have addressed the above hypothesis through application of lipoprotein binding techniques to the CD rat model. The first attempt to evaluate lipoprotein binding in this system was reported by Lefevre et al.[24] in 1986. These investigators hypothesized that the accumulation of $HDL_1$ in CD rats was due to a specific decrease in receptor-dependent uptake of plasma HDL, possibly through a reduction in the number of LDL or chylomicron remnant receptors. A total lipoprotein fraction ($d < 1.21$) was isolated from rats fed a chow diet and selected as the ligand for use in subsequent binding assays. The lipoprotein fraction was radiolabeled and characterized. The preparation had a specific activity of 41 cpm/ng protein and HDL accounted for 89% of the total $^{125}I$-labeled lipoproteins. To test the hypothesis that expression or activity of any receptor recognizing the heterogeneous ligand might be affected by copper deficiency, the authors compared the binding response between crude liver membranes prepared from ten rats of each treatment. The treatment comparison was restricted to one level of radioligand concentration. No mention was made of the concentration of radioligand used in these comparison assays, nor of the rationale for using a one-point comparison experimental design. Results of the comparison yielded no treatment differences when binding activity was expressed on a traditional membrane protein basis. The authors did find that when binding data were expressed on plasma membrane marker enzyme (5′-nucleotidase) activity units, the binding response in thc CD group was significantly lower. Expression of data in this way was justified by the authors because previous studies had shown that lipoprotein receptor binding activity in different membrane preparations correlates highly with 5′-nucleotidase activity. However, the authors provide no information to indicate that the imposed dietary treatment does not affect marker enzyme activity or expression independently of receptor binding. It is possible that copper deficiency may affect the marker enzyme activity directly, or indirectly through treatment induced changes in membrane structure or composition. Such membrane associated alterations may also bring about changes in the recovery of marker enzyme as affected by the membrane isolation procedure. Expression of data on the basis of marker enzyme activity necessarily implies that the assay accurately reflects plasma membrane content. The validity of this assumption must be firmly established before expression of data on this basis can be justified. With this understanding, it is not clear if the data presented by the authors are good evidence of reduced hepatic HDL binding and metabolism, leading to accumulation of $HDL_1$ particles in the plasma of CD rats.

In addition to treatment comparison of binding activity, Lefevre et al.[24] performed studies to characterize the binding of the heterogeneous ligand to hepatic membranes from a control rat. Saturation experiments indicated that the binding consisted primarily of a component which was specific, saturable (maximal equilibrium binding capacity [$B_{max}$] = 5 μg/mg membrane protein), of high affinity and $Ca^{2+}$ insensitive. These characteristics are similar to those of apo E-free binding observed in many other homologous rat systems. Thus, despite the use of a heterogeneous ligand containing apolipoproteins necessary for interaction with the LDL or chylomicron remnant receptor, the majority of the binding appears to be attributable to the putative HDL receptor. The authors indicate that because the hepatic LDL receptor is significantly down-regulated in the normal adult rat, it is unlikely that copper deficiency could increase $HDL_1$ levels by reducing still further the number or activity of LDL receptors. Furthermore, they note that chylomicron remnant receptor activity is resistant to dietary and pharmacological treatments known to affect cholesterol homeostasis. On the basis of this information and the results of their study, the authors indicate that the elevation in $HDL_1$ in CD rats is probably not the result of a down-regulation of either the LDL or chylomicron remnant receptor.

More recently, a study by Hassel et al.[25] examined the binding of apo E-free HDL to crude liver membranes using the CD rat model. In this study, the radioligand was prepared from rats of both treatments by flotation of all lipoproteins ($d < 1.225$) and isolation of HDL by agarose column chromatography. Isolated HDL was then applied to a heparin-Sepharose column, and apo E-free HDL was collected as the nonretained fraction. After iodination, the apo E-free HDL preparations had a specific activity of 300 to 400 cpm/ng protein and no detectable apo E or apo E-associated radioactivity. Crude liver membranes were prepared from rats of each treatment, and the preparations were monitored by 5′-nucleotidase marker enzyme activity. No differences in marker enzyme activity were apparent between treatment preparations at any time during the procedure. Initially, binding of lipoproteins and membranes from CD rats was compared to the binding of lipoproteins and membranes from controls (Figure 1 and Table 1). Saturation experiments

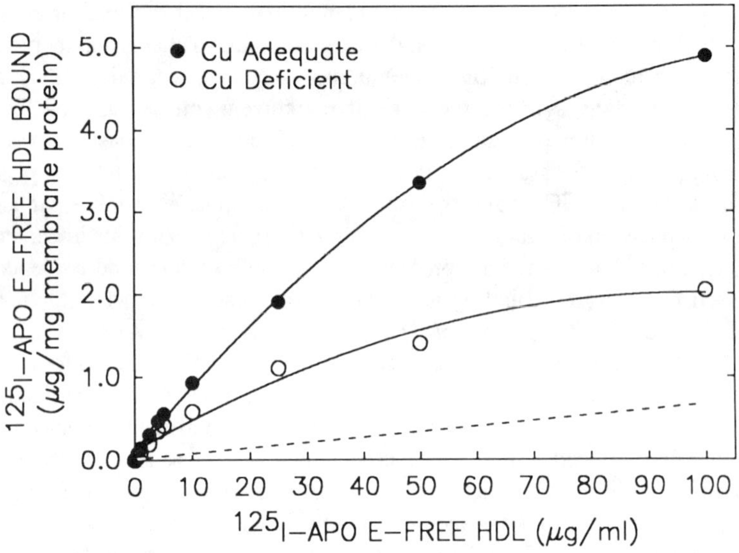


FIGURE 1. Binding of $^{125}$I-apo E-free HDL to liver membranes as a function of HDL concentration. HDL from each treatment group was bound to liver membranes from their respective groups. Solid lines represent total binding as determined in the absence of unlabeled HDL; dashed line represents nonspecific binding determined in the presence of 50-fold excess unlabeled apo E-free HDL and was similar for both treatment groups. (From Hassel, C. A., Lei, K. Y., Carr, T. P., and Marchello, J. A., *Metabolism*, 36, 1054, 1987. With permission.)

**TABLE 1**
**Effect of Dietary Treatment of Apo E-Free HDL Binding Parameters[a]**

| Binding parameter | | Copper status[b] Cu adequate | Cu deficient | *p*-value[c] |
|---|---|---|---|---|
| $K_d$ | $10^{-8}$ *M* | 5.05 ± 0.80 | 3.77 ± 0.26 | NS |
| | μg/ml | 10.1 ± 1.60 | 7.54 ± 0.52 | NS |
| $B_{max}$ | $10^{-8}$ *M* | 10.5 ± 1.90 | 3.66 ± 0.82 | <0.05 |
| | μg/mg[d] | 20.4 ± 2.08 | 8.16 ± 1.49 | <0.05 |

[a] n = 3 binding experiments performed and subjected to computer analysis per treatment.
[b] Cu adequate = membranes and lipoprotein obtained from copper-adequate rats; Cu deficient = membranes and lipoprotein obtained from copper-deficient rats.
[c] NS = not significant.
[d] Per milligram liver membrane protein.

From Hassel, C. A., Lei, K. Y., Carr, T. P., and Marchello, J. A., *Metabolism*, 36, 1054, 1987. With permission.

**TABLE 2**
**Effect of Dietary Treatment on Apo E-Free HDL Maximal Binding ($B_{max}$)[a]**

| | Membrane preparation | |
|---|---|---|
| **Lipoprotein** | **Cu adequate** | **Cu deficient** |
| Cu adequate | 20.4 ± 2.08[b] | 15.4 ± 4.20 |
| | (10.5 ± 1.90)[c] | (8.09 ± 2.53) |
| Cu deficient | 7.19 ± 1.15 | 8.16 ± 1.49 |
| | (3.34 ± 0.55) | (3.66 ± 0.82) |

[a] n = 3 binding experiments performed and subjected to computer analysis per treatment. Treatment comparisons in ANOVA: membrane effect, df = 1, NS; lipoprotein effect, df = 1, $p < 0.05$; interaction, df = 1, NS.

[b] Values expressed as mean ± SEM μg protein bound per milligram membrane protein.

[c] Values in parentheses expressed as $10^{-8}$ mol/l.

From Hassel, C. A., Lei, K. Y., Carr, T. P., and Marchello, J. A., *Metabolism*, 36, 1054, 1987. With permission.

were conducted to observe the binding response over a wide range of radioligand concentration. All quantitative binding data were expressed on a membrane protein basis. Total binding data, specific binding data and computer derived estimates of $B_{max}$ all indicate that less binding was observed when lipoproteins and membranes from CD rats were used in the binding assay compared to controls. The reduced binding implied that some alteration in the interaction between lipoproteins and liver membranes may have occurred *in vitro*. A crossover design was employed in an attempt to determine whether the alteration in binding activity was associated with the lipoprotein binding sites or the lipoprotein particles. Experiments were performed in which lipoproteins isolated from both CD and control rats were bound to liver membranes prepared from CD and control rats. The information from the crossover experiments was combined with the information from the previously described experiments in a 2 × 2 factorial design. No significant differences in the equilibrium dissociation constant ($K_d$) were observed, suggesting the affinity of apo E-free HDL for binding sites was unaffected by copper status. A significant lipoprotein effect was observed for $B_{max}$, indicating that the binding was reduced when lipoproteins derived from deficient animals were used in the assays. Analysis of binding parameters from all experiments demonstrated a significant lipoprotein effect (Table 2), indicating that the reduction in binding is associated with the apo E-free HDL from CD rats and not due to a reduction in binding site numbers. It is difficult to envision how a lipoprotein effect could alter the $B_{max}$ without affecting the $K_d$ of the binding reaction. Perhaps an apolipoprotein conformation is altered by copper deficiency resulting in decreased recognition by binding sites. Alternatively, a change in lipid composition within the lipoprotein particle may in some way alter the surface structure preventing recognition by potential binding sites.

The relevance of the above observations[25] to the hypercholesterolemic CD rat is dependent upon the physiological events which occur subsequent to binding. The observed binding of apo E-free HDL may result in tissue uptake of the entire particle, selective uptake of lipoprotein-bound cholesterol, or efflux of cholesterol from the tissue to the lipoprotein particle. If apo E-free HDL are cleared from plasma by whole particle uptake (receptor-mediated endocytosis), the quantitative contribution of this subfraction to total HDL catabolism *in vivo* is not known. Current hypotheses suggest that small, more dense particles such as apo E-free HDL may collect excess peripheral tissue-cholesterol and evolve into larger apo E-rich $HDL_1$ particles.[26] If the reduced hepatic binding of apo E-free HDL reflects reduced clearance of these particles from plasma, more particles would then be available for conversion into apo E-rich $HDL_1$. This would increase the production rate of apo E-rich $HDL_1$, thereby contributing to an increased pool size of the larger HDL particles.

In the most recent report addressing the role of lipoprotein receptors in the CD rat model, the binding of apo E-rich HDL to hepatic plasma membranes was examined in our laboratory.[27] The apo E-rich $HDL_1$ ligand was prepared from rats of each treatment by ultracentrifugation, agarose column chromatography, and heparin-Sepharose affinity chromatography. The isolated apo E-rich HDL containing both apo B and E were radiolabeled and characterized. The preparations from each treatment were similar in specific activity, apolipoprotein profiles, cholesterol to protein ratio, and in radiolabel distribution among apolipoproteins. Plasma membranes were prepared by applying a crude membrane pellet to a discontinuous sucrose gradient system and centrifuging at high speed in a swinging bucket rotor. The enrichment of the liver plasma membrane preparations was monitored by a 5′-nucleotidase marker enzyme activity (Table 3). The plasma membrane suspensions were enriched 10- to 15-fold over the starting liver homogenates. While the liver homogenates showed no treatment differences, the final plasma membrane preparations showed a highly significant treatment difference. There are two possible explanations for this observation. A procedural artifact unique to this experiment may have occurred. Alternatively, a treatment effect manifested within the membrane structure may have affected the integrity of membrane-associated proteins which, in turn, could result in reduced recovery of enzyme activity during this procedure.

Initial binding characterization experiments were performed to provide qualitative binding information[27] (Table 3). The binding observed in both treatments was insensitive to EDTA, $Cu^{+2}$, and pronase. The binding response was similar between treatments, suggesting a common site is present in both

**TABLE 3**
**Effect of Modification of Incubation Conditions on Binding of $^{125}$I-Apo E-Rich HDL to Hepatic Plasma Membranes[a]**

| Incubation treatment | Cu adequate (ng/mg membrane protein) | Cu deficient (ng/mg membrane protein) |
|---|---|---|
| Standard conditions[b] | 450 ± 12 | 622 ± 21 |
| 30 m*M* EDTA | 442 ± 16 | 641 ± 40 |
| 10 μ*M* $CuCl_2$ | 383 ± 20 | 711 ± 58 |
| Pronase[c] (μg/ml) | | |
| 0 | 380 ± 18 | 602 ± 19 |
| 5 | 390 ± 31 | 600 ± 26 |
| 25 | 354 ± 28 | 663 ± 41 |
| Temperature (°C) | | |
| 0 | 387 ± 21 | 550 ± 31 |
| 37 | 615 ± 18 | 841 ± 25 |
| Time (min) | | |
| 15 | 316 ± 19 | 448 ± 20 |
| 30 | 421 ± 22 | 578 ± 18 |
| 60 | 450 ± 17 | 662 ± 19 |
| 120 | 462 ± 18 | 613 ± 19 |

a All incubations were performed using 10 μg/ml $^{125}$I-apo E-rich HDL and 100 μg membrane protein in 100 μl of 20 m*M* Tis-HCl, 25 m*M* NaCl, 0.5 m*M* $CaCl_2$, 1% BSA at pH 7.5. Values represent the means ± SEM (n = 2).

b Standard conditions = 60 min incubation at 22°C.

c Membranes were treated with the indicated concentration of pronase and incubated for 10 min at 37°C prior to binding. The reaction was stopped by addition of two volumes ice cold 10% BSA.

From Hassel, C. A., Carr, T. P., Marchello, J. A., and Lei, K. Y., *Proc. Soc. Exp. Biol. Med.*, 187, 296, 1988. With permission.

control and copper-deficient plasma membrane preparations. These experiments demonstrate binding properties similar to those reported for apo E-free HDL, supporting the existence of a lipoprotein binding site in rat liver which is distinct from the LDL or chylomicron remnant receptor.

Apo E-rich HDL binding specificity was investigated by binding in the presence of increasing concentrations of unlabeled lipoprotein competitors.[27] Two displacement experiments were performed. The first experiment examined binding site specificity in controls; the second experiment investigated dietary treatment differences. These experiments are visualized in Figure 2. The first of these experiments (Figure 2, Panel A) compared VLDL, LDL, apo E-free HDL, and apo E-rich HDL in their ability to displace labeled apo E-rich HDL from the membranes. Lipoproteins and membranes were obtained from control animals. All lipoproteins studied were capable of competing for the binding site, indicating a relatively nonspecific nature. In another experiment, the effect of dietary treatment was investigated by displacement studies.

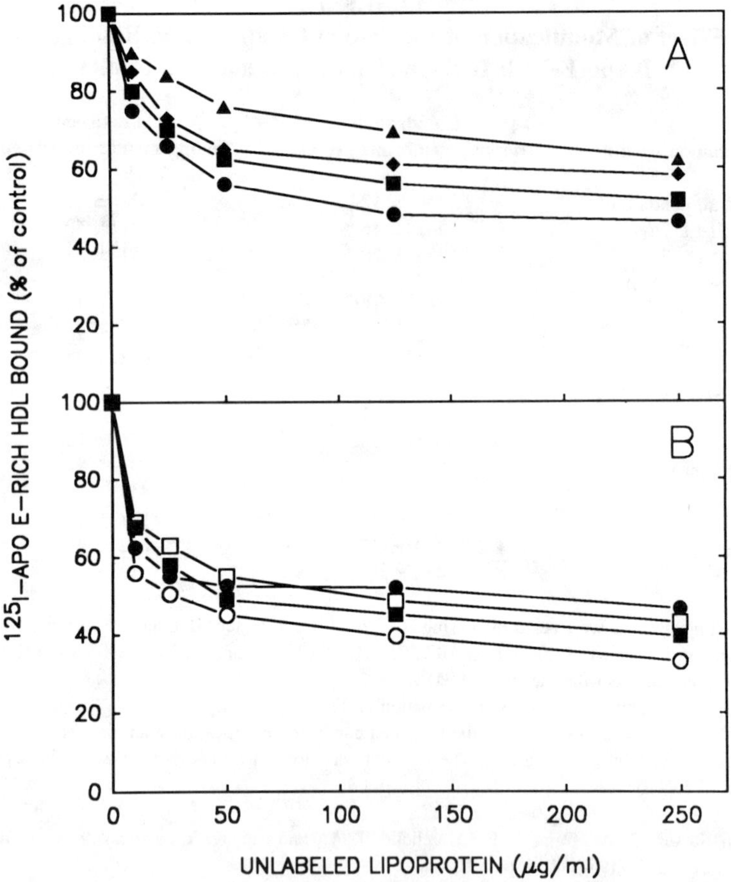


FIGURE 2. Competitive displacement of $^{125}$I-apo E-rich HDL. *Panel A:* Binding site specificity in control assays. Binding of $^{125}$I-apo E-rich HDL in the presence of the indicated concentrations of rat VLDL (▲), LDL (♦), apo E-free HDL (■), and apo E-rich HDL (●). All lipoproteins and plasma membranes were obtained from control animals. *Panel B:* Treatment differences in binding site displacement. Binding of $^{125}$I-apo E-rich HDL from copper-deficient rats to plasma membranes from deficient rats in the presence of the indicated concentrations of apo E-free HDL from copper-adequate (■) and copper-deficient (□) rats, and of apo E-rich HDL from copper-adequate (●) and copper-deficient (○) rats. (From Hassel, C. A., Carr, T. P., Marchello, J. A., and Lei, K. Y., *Proc. Soc. Exp. Biol. Med.*, 187, 296, 1988. With permission.)

Labeled apo E-rich HDL and plasma membranes from copper-deficient rats were incubated with increasing amounts of unlabeled HDL fractions derived from rats of both treatments (Figure 2, Panel B). Apo E-free HDL derived from rats of either treatment group effectively displaced the membrane-bound radioactivity. Thus, in both copper-adequate and copper-deficient assays, apo

**TABLE 4**
**Effect of Dietary Treatment on Specific Binding of $^{125}$I-Apo E-Rich HDL to Liver Plasma Membranes[a]**

| HDL protein concentration (μg/ml) | Dietary treatment: Cu adequate (ng/mg membrane protein) | Dietary treatment: Cu deficient (ng/mg membrane protein) | p-value[b] |
|---|---|---|---|
| 1 | 30 ± 3.3 | 66 ± 5.5 | <0.005 |
| 5 | 153 ± 15 | 269 ± 24 | <0.005 |
| 10 | 290 ± 25 | 530 ± 34 | <0.001 |
| 25 | 653 ± 68 | 1053 ± 101 | <0.01 |
| 50 | 1228 ± 119 | 1892 ± 113 | <0.005 |
| 100 | 2012 ± 202 | 2886 ± 179 | <0.01 |
| 250 | 3131 ± 223 | 4306 ± 273 | <0.01 |

[a] Bound radioactivity was determined by incubation 110 μg membrane protein with the indicated concentration of $^{125}$I-apo E-rich HDL for 60 min at 22°C in 100 μl of 20 m*M* Tris-HCl, 25 m*M* NaCl, 0.5 m*M* $CaCl_2$, 1% BSA at pH 7.5. Values represent means ± SEM (n = 4).
[b] *p*-Values determined by the Student's *t*-test.

From Hassel, C. A., Carr, T. P., Marchello, J. A., and Lei, K. Y., *Proc. Soc. Exp. Biol. Med.*, 187, 296, 1988. With permission.

E did not appear to be a primary requisite for recognition by this binding site. The apo E-rich HDL from deficient rats displaced somewhat more label from the binding site than apo E-rich HDL from controls.

Saturation experiments were conducted over a wide range of label concentrations to provide quantitative treatment comparisons.[27] Total binding data obtained from these experiments were compiled at 16 different label concentrations and then used to construct binding saturation curves. Four curves were constructed for each treatment. Curve comparison by two-way analysis of variance demonstrated that the total binding was greater in the copper-deficient group. Specific binding data, experimentally determined as the difference between total binding and binding observed in the presence of 50-fold excess unlabeled lipoprotein, were compiled at seven different label concentrations. These data are shown in Table 4 as the average specific binding from four pools of membrane preparations for each treatment. The specific binding was significantly higher in the CD group throughout the observed range of ligand concentrations. Thus, both total binding data and experimentally derived specific binding data indicated greater binding when lipoproteins and membranes from CD rats were used in the assays.

The majority of the binding observed in this study[27] appears to be attributable to a high capacity, relatively nonspecific binding site. The higher binding reported here may be due to elevated expression of this nonspecific, high capacity site as a result of copper deficiency. Alternatively, subtle changes in lipoprotein structure induced by copper deficiency could enhance recog-

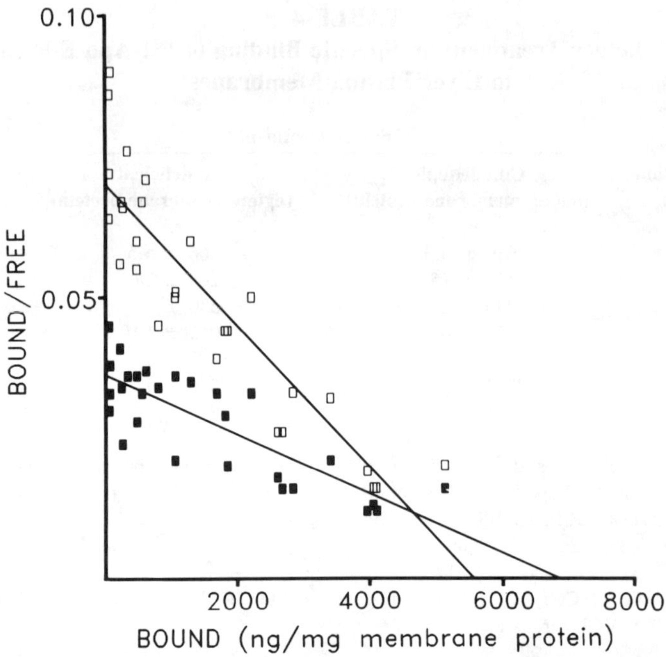


FIGURE 3. Scatchard analysis of experimental derived specific binding data. Specific binding data for all experiments (n = 4 for each treatment) from copper-adequate (■) and copper-deficient (□) assays were subjected to Scatchard analysis. The average $K_d$ and $B_{max}$ for control assays were 169 μg/ml and 6.27 μg bound per milligram membrane protein, respectively. The respective values for copper-deficient assays were 76.1 μg/ml and 5.33 μg bound per milligram membrane protein. Binding incubations were performed at 22°C for 60 min in the presence of 50-fold excess unlabeled apo E-rich HDL. (From Hassel, C. A., Carr, T. P., Marchello, J. A., and Lei, K. Y., *Proc. Soc. Exp. Biol. Med.*, 187, 296, 1988. With permission.)

nition by this binding site. Nevertheless, copper deficiency affects the affinity of the binding reaction within the observed range of label concentrations. When the specific binding data were plotted on Scatchard coordinates (Figure 3) the treatment effect appears largest at lower free ligand concentrations. This observation suggests the possibility of a higher affinity, lower capacity binding site present in the copper-deficient hepatic membranes which may not be present in control membranes. Such a site could conceivably make a contribution to binding while still being masked by the higher capacity site and thereby escape detection as a distinct binding component by kinetic analysis. The higher binding affinity observed in the copper-deficient group may be due to the contribution of high affinity receptors present in the membrane preparations but not detectable as a distinct binding component in the assay. Detection of these receptors as a distinct binding component necessarily entails binding at lower label concentrations. Furthermore, to identify these receptors as such, binding properties should be established at low free ligand concentrations where high affinity sites would be responsible for a larger component of the observed binding.

Despite these possibilities, determination of the *in vivo* significance of these *in vitro* observations remains difficult. Binding of lipoproteins to receptors or binding sites is only the first step in a complex series of metabolic events and can by no means be assumed to represent catabolic activity per se. The mechanisms by which the current observations contribute to the altered cholesterol metabolism observed in copper-deficient rats requires a greater understanding of the metabolic processes subsequent to binding. As with apo E-free HDL, the events subsequent to binding in this system have not been elucidated. With regard to the flow of lipoprotein-bound cholesterol between tissues, three possibilities exist: (1) net influx of lipoprotein-bound cholesterol to hepatic cells; (2) exchange of cholesterol between lipoproteins and hepatic cells with no net unidirectional flux; and (3) net efflux of cholesterol from hepatic cells to lipoproteins. Additionally, there may be at work any of three known mechanisms to facilitate the above possibilities: (1) receptor-mediated endocytosis of intact lipoprotein particles; (2) selective transfer of cholesteryl ester from the particle core to cells; and (3) passive exchange of free cholesterol between the cell surface membrane and the lipoprotein particle surface monolayer. In view of our present understanding of HDL catabolism, apo E-rich HDL probably account for a majority of the total HDL cholesterol removal from plasma[26] either by endocytotic or selective transfer mechanisms. If the binding observed in this study represents the initial step in a sequence of catabolic events, the experimental observations may be interpreted as evidence that receptor-mediated clearance of HDL cholesterol is not impaired in copper deficiency.

From the review presented herein, it is clear that limitations arise when extrapolating information derived from *in vitro* lipoprotein binding studies to the CD rat model. These limitations exist largely because basic knowledge concerning the events of HDL catabolism is incomplete. When discussing lipoprotein binding studies in the CD rat model, one can determine treatment effects on quantitative and qualitative binding properties with little difficulty. The contribution of these sites to *in vivo* clearance of cholesterol from plasma, or the events involved in such a process subsequent to binding, is not well established. Future *in vitro* studies should be performed with intact cells of various rat tissues in which binding, uptake, and degradation of lipoprotein components can be evaluated. Trapped ligand methodology is currently available and will enable the investigator to distinguish receptor-mediated endocytosis from selective transfer processes. Using these procedures, the physiological significance of HDL binding sites will begin to be established. With this information, hypotheses concerning the effect of copper deficiency on receptor-mediated clearance of plasma lipoproteins can be addressed more directly.

# III. METHODS

## A. General Considerations

The application of lipoprotein binding assays to the CD rat system should be approached with several considerations in mind. The binding system should be homologous (lipoproteins and cells/membranes derived from one species) to avoid undue ambiguity in extrapolation of results. Experiments should provide reasonable evidence that a component of the binding has characteristics of receptor-mediated activity. The experiments should include criteria to identify the receptor(s) as described above. If copper deficiency does affect the binding reaction, these observations may be addressed within the framework of several hypotheses. Copper deficiency in rats may alter ligand (lipoprotein) properties, binding site properties, or binding site numbers. Ideally, experiments should be designed to test each of these alternatives. The lipoprotein ligand used for the assay should be selected with appropriate consideration for the experimental objectives. If tissue membranes are to be used, consideration should be given to the method of preparation. Extensive tissue processing which results in a purer preparation necessarily compromises the absolute yield of binding sites. Liver homogenates will yield the best estimate of true binding capacity, whereas purified plasma membranes will provide the most accurate qualitative information. Finally, the binding reaction should be carried out over a wide range of label concentrations to facilitate informative treatment comparisons. The range of ligand concentrations can be conceived as the dimensions of a window through which one observes the binding reaction. An extended range which includes physiological lipoprotein concentrations allows for more accurate binding parameter estimates and greater confidence in interpreting statistical comparisons. For purposes of determining treatment differences, direct comparisons of binding data are preferred over comparison of kinetically derived kinetic parameter estimates.[28]

## B. Iodination of Lipoproteins

HDL subfractions can be easily and routinely radiolabeled with $^{125}I$ using the iodine monochloride method[29] as described by Goldstein et al.[30] Each preparation uses 3 to 5 mg of lipoprotein, equilibrated in buffer "A" (150 m*M* NaCl, 0.24 m*M* EDTA, pH 7.4). The reaction is carried out using 2 mCi Na-$^{125}I$ with 2.64 m*M* freshly prepared iodine monochloride. The stock ICl solution is prepared by injecting 99 mg $NaIO_3$ in 2.0 ml $H_2O$ into 150 mg NaI in 8.0 ml 6 *N* HCl. This mixture is diluted to 40 ml, extracted with $CCl_4$, and immediately before iodination diluted again with 2 *M* NaCl to the final concentration for use in the iodination reaction. The iodination is performed at 0 to 4°C in a fume hood. Lipoproteins are added to 0.5 ml 1 *M* glycine-NaOH buffer and brought to a volume of 1.3 ml. An aliquot of 1 *M* glycine-NaOH buffer (0.4 ml) is added to a V-vial containing the Na-$^{125}I$, and the contents of this vial are subsequently withdrawn and added to a

Vacutainer tube containing the lipoproteins. With the use of a syringe, 0.42 ml of diluted (2.64 m*M*) iodine monochloride is quickly and forcefully injected into the lipoproteins while the Vacutainer tube is being vortexed. The duration of the vortexing should not exceed 2 sec. The mixture is incubated on ice for 5 min, then brought to a final volume of 2.5 ml by addition of 0.28 ml buffer A. The entire volume is applied to a prepacked Sephadex® G-25 column (Pharmacia, Uppsala, Sweden) to separate the iodinated lipoprotein from the free $^{125}$I. The iodinated lipoproteins are then dialyzed against 150 m*M* NaCl and 0.24 m*M* EDTA at pH 7.4 (three changes of 3 l each) for 18 to 24 h. Thereafter, the buffer is changed to 50 m*M* Tris, 25 m*M* NaCl, 1 m*M* $CaCl_2$, pH 7.4 and the dialysis is continued for an additional 24 h. After the final dialysis, the iodinated lipoprotein is sterilized by passage through a 0.45-μm Millipore® filter (Millipore Corp., Bedford, MA) and the solution stored in a plastic tube at 4°C. Iodinated lipoproteins for all experiments should be used within 2 weeks of preparation.

Routinely, each iodinated preparation is characterized to determine the specific activity and the extent of incorporation of $^{125}$I into the HDL lipids. Aliquots of $^{125}$I-lipoprotein are assayed for protein content by the method of Lowry et al.[31] and for radioactivity using an LKB Minigamma 1275 gamma counter (LKB Instruments, Rockville, MD). The $^{125}$I-lipoproteins are not used for binding studies unless they meet the following criteria: (1) 0.5 to 2.0 mg protein per milliliter with a specific activity of 200 to 500 cpm/ng; (2) greater than 98% of the radioactivity is precipitable by trichloroacetic acid; and (3) less than 5% of the radioactivity is associated with lipids as measured by chloroform-methanol (2:1, v/v) extraction.

### C. Crude Liver Membrane Preparation

Freshly obtained 2-g liver samples from each rat are placed in 10 ml ice-cold buffer A (150 m*M* NaCl, 10 m*M* Tris, 1 m*M* phenylmethylsulfonyl fluoride [PMSF], pH 7.5) and liver membrane preparation is initiated within 2 h. All steps in the preparation are carried out at 4°C following the method of Kovanen et al.[32] Liver samples from three rats are pooled and homogenized with two 10-sec pulses of a Polytron homogenizer (Brinkman Instruments, Westbury, NY) at setting no. 10. The whole homogenates are filtered through a nylon net (120 μm) and centrifuged at 500 × *g* for 5 min. The resulting supernatant is centrifuged at 8000 × *g* for 15 min and the 8000 × *g* supernatant is centrifuged at 100,000 × *g* for 60 min. Each 100,000 × *g* pellet is resuspended in 6 ml buffer A by flushing 10 times through a 23-gauge needle, then resedimented at 100,000 × *g* for 60 min. The filtered homogenates, 500 × *g* supernatant, and 100,000 × *g* pellets are assayed for protein content and 5′-nucleotidase activity[33] to monitor enrichment of plasma membranes in the various fractions.

The routine 5′-nucleotidase assay measures the release of inorganic phosphate from AMP and is carried out in a volume of 0.3 ml containing 100 m*M* Tris-HCl (pH 8.5), 10 m*M* $MgCl_2$, and 1 to 2 μg membrane protein. After incubation at 37° C for 20 min, the reaction is stopped by addition of 0.1 ml 10% ascorbic acid and 0.6 ml 0.42% ammonium molybdate in 1 *N* $H_2SO_4$. Samples are incubated at 45°C for 20 min to develop the blue color, which is read at 795 nm. The desired fractions are rapidly frozen and held at −85°C until used for binding. Liver homogenates or 100,000 × *g* pellets can be used in binding assays. On the day of an experiment, the 100,000 × *g* pellets are resuspended in buffer "B" (50 m*M* Tris, 25 m*M* NaCl, 1 m*M* $CaCl_2$, pH 7.5) using a 23-gauge needle as described above. The suspensions are then sonicated in a sonic dismembrator (Fisher Scientific, Pittsburg, PA), model 150, at setting no. 6.

### D. Liver Plasma Membrane Preparation

Under certain conditions, it is advantageous to obtain membrane preparations which are highly enriched in plasma membranes relative to other subcellular organelle membranes. By using a more purified liver plasma membrane preparation in the binding assay, the relative number of cell surface lipoprotein receptors in the membrane aliquot is maximized. This increases the sensitivity of the assay and facilitates identification of binding sites which may be present in low numbers. Furthermore, less interference or background binding is observed, which allows more precise determination of qualitative binding characteristics. The following procedure is a modification of the original method of isolation[34] and is convenient for isolating sufficient amounts of hepatic plasma membranes for use in lipoprotein binding assays. Livers from two or three rats are pooled, minced with scissors and added to 150 ml homogenation media (1 mM $NaHCO_3$, 0.5 m*M* $CaCl_2$, pH 7.5). The mixture is gently homogenized with a Polytron homogenizer, using three 3-sec bursts at setting no. 3. The homogenized suspension is diluted to 500 ml and filtered through two layers of 200-μm nylon mesh. The filtrate is centrifuged at 1500 × *g* for 20 min and the supernatant decanted off the surface of the pink, gelatinous pellets. The pellets are suspended in 40 ml homogenation media and gently mixed in a dounce homogenizer (eight gentle strokes). The mixture is diluted to 150 ml, centrifuged for 15 min at 1500 × *g* and the resultant pellets are again gently homogenized. The mixture is centrifuged a third time at 1500 × *g* for 10 min. The final pellets are taken up with 5.5 volumes of a 70.74% sucrose solution to a density of 1.22 g/ml. This suspension is overlayered using a discontinuous sucrose gradient system,[35] in which sucrose solutions of densities 1.18 and 1.16 g/ml are layered successively over the 1.22 g/ml membrane suspension. The overlayered suspensions are centrifuged for 75 min at 25,000 rpm in a SW 28 rotor. After centrifugation, the plasma membranes are pipetted from the 1.16:1.18 g/ml interface. This modified gradient system rids the preparation of mitochondria found at the 1.18:1.22 g/ml interface and permits the isolation of purer preparations of membrane from rat liver. Aliquots of the plasma membranes are used for protein and

5′-nucleotidase activity determination, and the remainder is quickly frozen at −85°C and used within 2 months.

### E. Binding Studies

Several different types of binding assays should be performed to characterize the events of the binding reaction. Bindability can bc determined by experiments in which very small amounts of $^{125}$I-lipoprotein (3000 cpm/tube) are added to increasing amounts (50 to 500 μg) of membrane protein. The fraction of iodinated lipoprotein capable of binding to the membranes is designated the bindability fraction. Saturation studies are performed by increasing the amount of labeled lipoprotein added to a constant amount of membrane protein. Data from saturation studies are used to estimate parameters (equilibrium dissociation constant, $K_d$, and maximal equilibrium binding capacity, $B_{max}$) of the binding reaction. Displacement studies, in which the increasing amounts of unlabeled lipoprotein competitors are added to constant amounts of membrane protein and labeled lipoprotein, serve to determine the specificity of the binding site. Characterization studies are performed in which time, temperature, and pH of the incubation are adjusted to determine their effect on the binding reaction. In addition, sensitivity of the binding reaction to the presence of proteolytic enzymes, divalent cations, and NaCl concentration can be determined to further characterize the binding reaction.

The incubation temperature is an important consideration and can dramatically influence the quantitative binding response.[23] As with any *in vitro* assay, it is desirable to maintain physiologic conditions as closely as possible. Temperatures ranging from 0 to 37°C have been reported for lipoprotein binding assays. The binding response increases with incubation temperature; however, at 37°C significant membrane-associated proteolytic activity is observed.[23] As a result, radiolabel can dissociate from the lipoprotein particle and/or apolipoprotein so that the binding observed in the assay may not be reflective of actual lipoprotein particle binding. Therefore, physiological temperatures are not recommended for assays with tissue membrane preparations. The investigator should be aware of the compromise between binding response, approximating physiologic conditions, and potential for proteolytic activity in this assay. Frequently, 22°C is the assay temperature of choice.[17-20]

Standard binding incubations are carried out in 20 m*M* Tris-HCl, pH 7.5, 25 m*M* NaCl, 0.5 m*M*$CaCl_2$, and 1% bovine serum albumin-fatty acid free (BSA) in a final volume of 100 μl. Various amounts of membrane protein, labeled lipoprotein, and unlabeled lipoprotein are added as needed, depending on the type of experiments being conducted. After incubation, the reaction mixtures are cooled on ice for 5 min and 80-μl aliquots are transferred to cellulose proprionate centrifuge tubes (Beckman Instruments, Palo Alto, CA). The tubes are centrifuged at 100,000 × *g* for 20 min at 3°C to separate bound from unbound lipoprotein. Supernatant is carefully removed by aspiration,

150 μl of cold incubation buffer is gently layered over the pelleted membranes, and the tubes recentrifuged at 100,000 × *g* for 10 min at 3°C. This step is performed to minimize unbound radioactivity adhering to the surface of the tube. After removal of the supernatant, the tubes are sliced to obtain the pellets and the amount of bound radioactivity is determined in a gamma counter. Aliquots (10 μl) of the original incubation mixture are also counted to assess total radioactivity present in each tube. Tubes containing labeled lipoproteins without membranes were used as assay blanks.

### F. Data Analysis

For purposes of determining treatment differences, direct comparisons of binding data are preferred over comparison of kinetically derived binding parameter estimates.[28] This is due to the problems inherent in deriving reasonable parameter estimates from binding data which frequently show nonlinearity. A proper approach to kinetic analysis of binding data can yield information which is useful to further characterize the binding site(s). To derive the best possible estimates of binding reaction parameters $K_d$ and $B_{max}$, a curve-fitting analysis is highly recommended over the traditional Scatchard analysis. Several curve-fitting programs are now available for the analysis of ligand binding systems. The "Ligand-PC" program[36] has been successfully applied to lipoprotein binding experiments and contains several advantageous features. This method provides optimal estimates of binding parameters using an exact mathematical model that is fitted directly to the data, thereby avoiding possible biases introduced by commonly used approximations. Based on mathematical theory,[37] it describes any number of ligands reacting simultaneously with any number of receptor sites. Furthermore, it uses statistically valid objective measurements for goodness of fit of a given model based on the "extra sum of squares" principle. Nonspecific binding *(n)* can also be estimated simultaneously with other parameters, which provides a more accurate and precise estimate of the binding parameters and their standard errors. Thus, nonspecific binding determined mathematically from total binding provides a better estimate of *n* than experimental determination by addition of excess unlabeled ligand to the binding reaction.

## REFERENCES

1. **Lei, K. Y. and Lin, I. M.,** Kinetics of high density lipoprotein cholesterol in copper-deficient rats, *Nutr. Rep. Int.*, 24, 1179, 1981.
2. **Lin, I. M. and Lei, K. Y.,** Cholesterol kinetic analyses in copper-deficient rats, *J. Nutr.*, 111, 450, 1981.

3. **Lei, K. Y., Hassel, C. A., and Allen, D. K.,** Alterations in plasma lipid, lipoprotein and apolipoprotein concentrations in copper-deficient rats, *J. Nutr.*, 113, 2178, 1983.
4. **Allen, K. G. D. and Klevay, L. M.,** Hyperlipoproteinemia in rats due to copper deficiency, *Nutr. Rep. Int.*, 22, 295, 1980.
5. **Croswell, S. C. and Lei, K. Y.,** Effect of copper deficiency on the apolipoprotein E-rich high density lipoproteins in rats, *J. Nutr.*, 115, 473, 1985.
6. **Lefevre, M., Keen, C. L., Lönnerdal, B., Hurley, L. S., and Schneeman, B. O.,** Different effects of zinc and copper deficiency on composition of plasma high density lipoproteins in rats, *J. Nutr.*, 115, 359, 1985.
7. **Goldstein, J. L. and Brown, M. S.,** The low-density lipoprotein pathway and its relation to atherosclerosis, *Annu. Rev. Biochem.*, 46, 897, 1977.
8. **Mahley, R. W. and Innerarity, T. L.,** Lipoprotein receptors and cholesterol homeostasis, *Biochim. Biophys. Acta,* 737, 197, 1983.
9. **Chapman, M. J.,** Comparative analysis of mammalian plasma lipoproteins, *Methods Enzymol.*, 128, 70, 1986.
10. **Lusk, L. T., Walker, L. F., DuBien, L. H., and Getz, G. S.,** Isolation and partial characterization of high density lipoprotein $HDL_1$ from rat plasma by gradient centrifugation, *Biochem. J.*, 183, 83, 1979.
11. **Oschry, Y. and Eisenberg, S.,** Rat plasma lipoproteins: re-evaluation of a lipoprotein system in an animal devoid of cholesteryl ester transfer activity, *J. Lipid Res.*, 23, 1099, 1982.
12. **Basu, S. K., Goldstein, J. L., and Brown, M. S.,** Characterization of the low density lipoprotein receptor in membranes prepared from human fibroblasts, *J. Biol. Chem.*, 253, 3852, 1978.
13. **Windler, E. E. T., Chao, Y. S., and Havel, R. J.,** Determinants of hepatic uptake of triglyceride-rich lipoproteins and their remnants in the rat, *J. Biol. Chem.*, 255, 5475, 1980.
14. **Windler, E. E. T., Kovanen, P. T., Chao, Y. S., Brown, M. S., Havel, R. J., and Goldstein, J. L.,** The estradiol-stimulated lipoprotein receptor of rat liver. A binding site that mediates the uptake of rat lipoproteins containing apoproteins B and E, *J. Biol. Chem.*, 255, 10464, 1980.
15. **Ose, L., Roken, I., Norum, K. R., Drevon, C. A., and Berg, T.,** The binding of high density lipoproteins to isolated rat hepatocytes, *Scand. J. Clin. Lab. Invest.*, 41, 63, 1981.
16. **Innerarity, T. L., Pitas, R. E., and Mahley, R. W.,** Disparities in the interaction of rat and human lipoproteins with cultured fibroblasts and smooth muscle cells: requirements for homology for receptor binding activity, *J. Biol. Chem.*, 255, 11163, 1980.
17. **Kagami, A., Fidge, N., Suzuki, N., and Nestel, P.,** Characteristics of the binding of high-density $\text{lipoprotein}_3$ by intact cells and membrane preparations of rat intestinal mucosa, *Biochim. Biophys. Acta,* 795, 179, 1984.
18. **Chacko, G. K.,** Human high density lipoprotein ($HDL_3$) binding to rat liver plasma membranes, *Biochim. Biophys. Acta,* 712, 129, 1982.
19. **Mitchel, Y. B., Rifici, V. A., and Eder, H. A.,** Characterization of the specific binding of rat apolipoprotein E-deficient HDL to rat hepatic plasma membranes, *Biochim. Biophys. Acta,* 917, 324, 1987.
20. **Chacko, G. K.,** Characteristics of high density lipoprotein binding sites in rat liver and testis membranes, *Biochim. Biophys. Acta,* 795, 417, 1984.
21. **Gwynne, J. T. and Strauss, J. F., III,** The role of lipoproteins in steroidogenesis and cholesterol metabolism in steroidogenic glands, *Endocrin. Rev.*, 3, 299, 1982.
22. **Ghosh, D. K. and Menon, K. M. J.,** Identification of gonadotropin inducible, high density lipoprotein receptors in the solubilized membranes from rat ovary, *Biochem. Biophys. Res. Commun.*, 134, 1006, 1986.

23. **Van Tol, A., Dallinga-Thie, G. M., Van Gent, T., and Van't Hooft, F. M.,** Specific saturable binding of rat high-density lipoproteins to rat kidney membranes, *Biochim. Biophys. Acta,* 876, 340, 1986.
24. **Lefevre, M., Keen, C. L., Lönnerdal, B., Hurley, L. S., and Schneeman, B. O.,** Copper deficiency-induced hypercholesterolemia: effects on HDL subfractions and hepatic lipoprotein receptor activity in the rat, *J. Nutr.,* 116, 1735, 1986.
25. **Hassel, C. A., Lei, K. Y., Carr, T. P., and Marchello, J. A.,** Lipoprotein receptors in copper-deficient rats: apolipoprotein E-free high density lipoprotein binding to liver membranes, *Metabolism,* 36, 1054, 1987.
26. **Eisenberg, S.,** High density lipoprotein metabolism, *J. Lipid Res.,* 25, 1017, 1984.
27. **Hassel, C. A., Carr, T. P., Marchello, J. A., and Lei, K. Y.,** Apolipoprotein E-rich HDL binding to liver plasma membrane in copper-deficient rats, *Proc. Soc. Exp. Biol. Med.,* 187, 296, 1988.
28. **Munson, P. J. and Rodbard, D.,** Number of receptor sites from Scatchard and Klotz graphs: a constructive technique, *Science,* 220, 979, 1983.
29. **McFarlane, A. S.,** Efficient trace-labelling of proteins with iodine, *Nature (London),* 182, 53, 1958.
30. **Goldstein, J. L., Basu, S. K., and Brown, M. S.,** Receptor-mediated endocytosis of low density lipoprotein in cultured cells, *Methods Enzymol.,* 98, 241, 1983.
31. **Lowry, O. H., Rosebrough, N. J., Farr, A. L., and Randall, R. J.,** Protein measurement with the folin phenol reagent, *J. Biol. Chem.,* 193, 265, 1951.
32. **Kovanen, P. T., Brown, M. S., and Goldstein, J. L.,** Increased binding of low density lipoprotein to liver membranes from rats treated with 17α-ethinyl estradiol, *J. Biol. Chem.,* 254, 11367, 1979.
33. **Widnell, C. C. and Unkeless, J. C.,** Partial purification of a lipoprotein with 5′-nucleotidase activity from membranes of rat liver cells, *Biochemistry,* 61, 1050, 1968.
34. **Neville, D. M., Jr.,** The isolation of a cell membrane fraction from rat liver, *J. Biophys. Biochem. Cytol.,* 8, 413, 1960.
35. **Song, C. S., Rubin, W., Rifkind, A. B., and Kappas, A.,** Plasma membranes of the rat liver: isolation and enzymatic characterization of a fraction rich in bile canaliculi, *J. Cell Biol.,* 41, 124, 1969.
36. **Munson, P. J. and Beveridge, M.,** Ligand PC: A Curve Fitting Data Analysis for Ligand Binding Experiments, National Institutes of Health, Bethesda, Maryland, 1986.
37. **Feldman, H. A., Rodbard, D., and Levine, D.,** Mathematical theory of cross-reactive radioimmunoassay and ligand-binding systems at equilibrium, *Anal. Biochem.,* 45, 530, 1972.

Chapter 6

# EFFECTS OF DIETARY CARBOHYDRATE AND OTHER FACTORS ON BLOOD LIPIDS DURING COPPER DEFICIENCY

**Mary Ann Johnson**

## TABLE OF CONTENTS

# I. INTRODUCTION

The type of dietary carbohydrate strongly determines the pathogical consequences of copper (Cu) deprivation and may influence the dietary requirement for Cu.[1-10] The severity of Cu deficiency is also influenced by strain,[11] age,[11,12] sex,[5] and other dietary factors such as zinc (Zn),[8,14-18] iron (Fe),[9,12] nickel (Ni),[19,20] ascorbic acid,[8,21] and protein.[18] The complex interrelationships among these factors and their impact on human health and Cu status have yet to be determined.

Interactions among dietary carbohydrates and Cu deficiency have been studied primarily in male rats fed semipurified diets that contain 50 to 65% carbohydrate by weight of diet. Male rats fed Cu-deficient diets (<1 mg Cu/kg) with sucrose or fructose rather than starch or glucose have lower growth rates, higher mortality rates, higher heart and liver weights, higher plasma cholesterol and triglyceride levels, are more anemic, and have a higher dietary Cu requirement to maintain certain components of the immune system.[1-10] Some of these changes are evident after only 3 weeks of Cu deprivation.[7-10] After 7 to 9 weeks the mortality rate of Cu-deficient rats fed fructose or sucrose may exceed 60%.[1-3] In contrast, the mortality rates of Cu-deficient rats fed starch or glucose generally range from 0 to 30%.[1-3]

The purpose of this chapter is to review the effects of different carbohydrates and other factors that result in hyperlipidemia during Cu deprivation. The various types of Cu-depletion diets that have been used to produce hyperlipidemia will be discussed. There is little information on the interacting effects of Cu deficiency and dietary carbohydrate on cholesterol and triglyceride levels in animals other than the male rat. Thus the following discussion will focus mainly on information derived from this animal model unless otherwise indicated.

# II. HYPERCHOLESTEROLEMIA IN CU-DEFICIENT RATS

## A. Sucrose

In 1973 Klevay[14] reported that the ingestion of high ratios of Zn to Cu by rats was associated with an increase in plasma cholesterol. In hindsight the diet used for these experiments was nearly ideal for the production of Cu deficiency-induced hypercholesterolemia because it contained 62% sucrose and the Jones-Foster mineral mix which provides ~200 mg Fe/kg diet. Sucrose,[1,2,7,10] Fe,[12] and Zn[10,14-18] are all dietary components that increase the severity of Cu deficiency and thus increase the probability that hypercholesterolemia will occur. In later experiments, the Zn level of the water was adjusted to 10 mg Zn/l for Cu-deficient and Cu-adequate rats.[22,23] Under these experimental conditions the plasma cholesterol levels of Cu-deficient rats were 21 to 38% greater than Cu-adequate rats (2 mg Cu/l drinking water) after 21 to 46 d.[22,23] Similarly, Murthy et al.[15] and Murthy and Petering[16] found that on the average the serum cholesterol levels of Cu-deficient rats were 33% higher than Cu-adequate rats after 60 d.

Allen and Klevay[24-27] examined cholesterol levels of rats fed a Cu-deficient diet (0.6 mg Cu/kg diet) similar to that originally described by Klevay.[14] Two modifications of this diet were a decrease in the sucrose level from 62 to 52% and the addition of 9.5% coconut oil. The plasma cholesterol levels of Cu-deficient rats fed this diet with egg white[24-27] or casein[28] as the protein source were 48 to 129% greater than control rats after 45 to 181 d. Cu-deficient rats did not necessarily become more hypercholesterolemic with time. The addition of saturated fat (coconut oil) may have contributed to the enhanced hypercholesterolemic effect of this diet. Severe cardiovascular abnormalities were observed in Cu-deficient rats fed this diet for only 45 to 61 d.[24]

Copper deficiency has been associated with other changes in cholesterol metabolism such as higher concentrations of plasma free cholesterol, high density lipoprotein (HDL) cholesterol and HDL apolipoproteins E and A-I among rats fed high sucrose diets (59.5% of the diet) for 6 weeks.[29,30] However, Lefevre et al. found that Cu deficiency elevated plasma cholesterol levels in some[30] but not all of these studies.[29] Differences in plasma cholesterol between Cu-deficient and Cu-adequate rats tended to become greater as the Zn contents of the diets were increased.[29] Perhaps longer periods of Cu deprivation or a higher Zn to Cu ratio are needed to consistently elevate cholesterol levels under these experimental conditions.

### B. Glucose

Lei and associates[31-33] demonstrated changes in cholesterol and HDL metabolism in rats fed severely Cu-deficient diets (<1 mg Cu/kg diet) that contained glucose (64 to 65%) as the major source of carbohydrate. These investigators have also used glucose in moderately Cu-deficient diets (2 mg Cu/kg diet vs. 18 mg Cu/kg diet) with high Zn to Cu ratios (~60:1).[34-36] After consumption of these diets for 9 weeks, the Cu-deficient rats had an average of 26 to 40% higher serum cholesterol levels than Cu-adequate rats.[34-36] Several dietary factors were found to influence the hypercholesterolemic effect of Cu-deficiency in these experiments. The differences in serum cholesterol between Cu-deficient and Cu-adequate rats were minimized by high cholesterol intakes (2 vs. 0% of the diet)[34,36] or low Zn intakes (10 vs. 120 mg Zn/kg diet).[35] High fiber intakes (16 vs. 0% of the diet) decreased serum cholesterol levels to a greater extent in Cu-deficient rats compared to Cu-adequate rats.[35]

### C. Starch

Petering et al.[17] demonstrated that hypercholesterolemia occurs in rats fed Cu-deficient diets (0.4 mg Cu/kg) that contain starch (62%) as the major source of carbohydrate. After 109 to 119 d the serum cholesterol levels of

Cu-deficient rats were an average of 52% greater than control rats (16 mg Cu/l drinking water). The greatest differences in serum cholesterol between Cu-deficient and Cu-adequate rats occurred when rats were exposed to 40 mg Zn/l drinking water rather than 2.5 or 10 mg Zn/l drinking water. Unexpectedly, the effect of high Zn was to decrease the serum cholesterol levels of control rats rather than to increase cholesterol among Cu-deficient rats.

### D. Lactose

There is little information on the effects of lactose on cholesterol levels in Cu-deficient rats. A mixture of starch and lactose (~32 and 30%, respectively) increased serum cholesterol levels among Cu-deficient rats after 13 weeks.[18] In contrast, a mixture of sucrose and lactose (3:1) caused a modest but nonsignificant increase in serum cholesterol among Cu-deficient rats after 8 weeks.[37]

Rats fed a mixture of nonfat milk powder (45%) and acid-washed ground corn (39%) had higher serum cholesterol levels when fed Cu-deficient (0.4 to 0.5 mg Cu/kg diet) rather than Cu-adequate diets (5 or 12 mg Cu/kg diet).[20,38] Nielsen et al.[20] found that Cu deprivation elevated plasma cholesterol levels an average of 29% after 8 weeks and 62% after 11 weeks. Nickel was shown to enhance the hypercholesterolemic effect of Cu deficiency.[20]

### E. Carbohydrate Mixtures

No effects of the Zn to Cu ratio (3:1 or 30:1) on serum cholesterol were noted in rats fed a combination of 37.5% glucose and 25% starch for 5, 10, or 15 weeks.[39] The lowest level of dietary Cu was 1.5 mg Cu/kg diet and the highest Zn level was 45 mg/kg diet. Perhaps the Cu content of the diet was not low enough or the Zn to Cu ratio was not high enough to precipitate a severe state of Cu depletion in this study.

Cu deficiency did not increase serum cholesterol levels of female rats fed a mixture of 30% sucrose and 31% starch for 18 weeks.[40] Female rats tend to develop less severe signs of Cu depletion compared to male rats.[5] Thus it is likely that the sex of the animals rather than the type of carbohydrate was the greatest determinant to the outcome of this experiment.[40]

### F. Direct Comparisons of Different Carbohydrates

The type of dietary carbohydrate has been shown to modulate the hypercholesterolemic effect of Cu-deficiency.[1-5] The influence of starch, glucose, sucrose, and fructose on the development of Cu deficiency has been examined primarily in rats. In their intitial experiments Reiser et al.[2] and Fields et al.[1,3] used the Jones-Foster mineral mix that was originally used by Klevay[14] and thus had high dietary Fe levels (~200 mg Fe/kg diet). This level of dietary Fe may have been a factor in the more severe hypercholesterolemia noted in some studies[3] compared to others.[4]

For the purposes of this discussion, Cu-deficient rats will always be

**TABLE 1**
**Plasma Cholesterol Levels (mg/dl) of Male Rats Fed Two Levels of Cu and Four Levels of Fructose and Starch for 4 Weeks**

| % Carbohydrate | | Plasma cholesterol | |
|---|---|---|---|
| Fructose | Starch | Cu deficient | Cu adequate |
| 0 | 60 | 96 ± 7[bc] | 86 ± 6[cd] |
| 20 | 40 | 98 ± 4[bc] | 86 ± 5[cd] |
| 40 | 20 | 105 ± 5[ab] | 79 ± 5[d] |
| 60 | 0 | 118 ± 11[a] | 96 ± 5[bc] |

*Note:* Means ± SEM (n = 6). Means with different superscripts are significantly different at $p < 0.05$ as determined by Fisher's least significant test. Significant effects: Cu, $p < 0.001$; carbohydrate, $p < 0.05$; Cu × carbohydrate, not significant.

compared to their Cu-adequate control fed the identical carbohydrate. Cu-deficient rats fed fructose or sucrose are often, but not always,[2] more hypercholesterolemic than those fed starch. The degree of hypercholesterolemia in Cu-deficient rats fed glucose is less than in those fed fructose[3,4] but may be similar[4] or higher[3] than in those fed starch. These differences in carbohydrate generally are not obvious until after 4 weeks of Cu deprivation. For example, Fields et al.[1] reported that 4 weeks of Cu deprivation increased plasma cholesterol levels by 13% in rats fed sucrose and by 16% in those fed starch. After 8 weeks of Cu deprivation cholesterol levels were increased by 54% in rats fed sucrose but only 26% in those fed starch.[1] Reiser et al[2] found that 7 weeks of Cu deprivation increased plasma cholesterol levels by 61, 36, and 131% in rats fed starch, sucrose, and fructose, respectively. Fields et al. found that plasma cholesterol levels of Cu-deficient rats fed starch, glucose, and fructose were elevated by 8, 81, and 200%, respectively, after 9 weeks[3] and by 27, 27, and 73% after 11 weeks.[4]

The minimal amount of fructose or sucrose required to elevate cholesterol levels in Cu-deficient rats has received little attention. In a short term study conducted in this laboratory, it was observed that fructose tended to be more hypercholesterolemic in rats fed Cu-deficient (<1 mg Cu/kg) compared to Cu-adequate (5 mg Cu/kg) diets (Table 1). However, a marked difference in the hypercholesterolemic effect of fructose was not apparent until fructose comprised 60% of the diet. Since the degree of hypercholesterolemia is usually not severe after 4 weeks of Cu deprivation, longer studies may be needed to determine the effects of incremental levels of dietary fructose.

Male rats have been reported to be more sensitive to the adverse effects of Cu deprivation when fed fructose or sucrose than are female rats.[5] However, the influence of dietary carbohydrate on hypercholesterolemia in male and

**TABLE 2**
**Plasma Cholesterol Levels (mg/dl) of Male and Female Rats Fed Two Levels of Cu and Two Types of Carbohydrate for 4 Weeks**

| | | Plasma cholesterol | |
|---|---|---|---|
| **Sex** | **Carbohydrate** | **Cu deficient** | **Cu adequate** |
| Male | Sucrose | $110 \pm 4^{a}$ | $77 \pm 5^{c}$ |
| | Starch | $82 \pm 2^{c}$ | $81 \pm 3^{c}$ |
| Female | Sucrose | $117 \pm 4^{a}$ | $102 \pm 4^{b}$ |
| | Starch | $110 \pm 6^{ab}$ | $111 \pm 8^{ab}$ |

*Note:* Means ± SEM (n = 6). Means with different superscripts are significantly different at $p < 0.05$ as determined by Fisher's least significance test. Significant effects: Cu, $p < 0.0001$; carbohydrate $p < 0.02$; sex, $p < 0.0001$; Cu × carbohydrate, $p < 0.0001$; Cu × sex, $p < 0.05$; carbohydrate × sex, $p < 0.01$; Cu × carbohydrate × sex, $p < 0.07$.

female Cu-deficient rats is difficult to determine because Cu-adequate rats have not been included in some studies.[5] In this laboratory Cu deficiency (<1 mg Cu/kg diet) is associated with elevated plasma cholesterol levels 4 weeks in both male and female rats fed sucrose, but not starch (Table 2). Cu deficiency increased plasma cholesterol levels by 53% in male rats fed sucrose and by 15% in female rats fed sucrose. However, it should be noted that the cholesterol levels of Cu-deficient male and female rats fed sucrose were identical. Thus, the apparently enhanced hypercholesterolemia in male Cu-deficient rats may actually by an artifact of the lower plasma cholesterol levels in male vs. female rats fed Cu-adequate diets.

### G. Protein

Other macronutrients besides the type of carbohydrate may influence the hypercholesterolemic effect of Cu deficiency. For example, Stemmer et al.[18] reported higher serum cholesterol levels in Cu-deficient rats fed casein or milk powder compared to egg white as the source of protein. These effects of dietary protein were not as evident in Cu-adequate rats. Thus, the type of dietary protein should be noted when evaluating the hypercholesterolemic potential of a Cu-deficient diet. Perhaps the absence of hypercholesterolemia in some Cu-deficient animals is due to the use of egg white rather than casein as the source of protein.[39]

## III. HYPERCHOLESTEROLEMIA IN CU-DEFICIENT MICE

The suitability of mice as a model for studying the hypercholesterolemia of Cu deficiency is uncertain. Cholesterol levels in mice with genetic disorders of Cu metabolism and/or with a dietary deficiency of Cu have been examined. Jones[41] reported that Cu deficiency among female mice fed a diet[42] containing 57% sucrose resulted in 27% increase in plasma cholesterol levels after 7 to 9 weeks. In contrast, Klevay[43] found that mice fed a Cu-deficient diet high in lard (28%) and sucrose (57.5%) developed extensive cardiovascular damage but had no elevations in plasma cholesterol.

Prohaska et al.[11] also reported that hypercholesterolemia did not occur in male mice under a variety of experimental conditions. The semipurified diet used for these experiments was similar to that described by the American Institute of Nutrition[44] and contained 50% sucrose and 15% starch. Pups from dams fed low Cu diets (0.5 mg Cu/kg diet) throughout gestation, lactation, and weaning were not hypercholesterolemic at 12 d of age or at 7 weeks of age. Brindled mice, which have a genetic defect that alters Cu distribution, were also not hypercholesterolemic at 12 d of age. Serum cholesterol levels were not elevated in brindled mice compared to their normal litter mates even when they were born of dams fed a low Cu diet throughout lactation and gestation. In fact, brindled mice born of dams fed Cu-deficient diets tended to have the lowest serum cholesterol levels. Prohaska et al.[11] speculated that mice and rats may differ in their sensitivity to Cu deficiency.

## IV. HYPERCHOLESTEROLEMIA IN CU-DEFICIENT HUMAN AND NONHUMAN PRIMATES

The relationship between Cu deficiency and cholesterol metabolism has been studied most often in rats or mice that are severely depleted of Cu such that they are anemic and have ceruloplasmin activities near or below the detection limit. In extreme cases of experimental Cu depletion the growth rate is impaired and the mortality rate is increased.[1-5] Among humans, severe disturbances in Cu status are most often reported in infants[13] or among individuals with Menkes' disease, an inherited disorder of Cu transport.[45] Unfortunately, little information is available on plasma cholesterol levels in Cu-deficient infants.[13] Blackett et al.[45] reported that three infants with Menkes' disease had plasma cholesterol levels within the normal range despite remarkably low serum Cu and ceruloplasmin in two of the subjects.

From the limited information available, it appears that only modest changes in Cu status among normal adult human and nonhuman primates are associated with significant changes in plasma cholesterol. For example, Milne et al.[21] reported that adult cynomolgus monkeys fed a diet marginal in Cu with 1 mg ascorbic acid per kilogram of body weight per day had significant increases in serum cholesterol after 32 weeks. Serum Cu and ceruloplasmin were not significantly lower than pretest values throughout this depletion period. Similarly, changes in plasma cholesterol levels were observed among adult humans fed low Cu diets that contained either 0.8 or 1.0 mg Cu/d for 15[46] or 11[47,48] weeks, respectively. These levels of Cu intake are well below the recommended safe and adequate intake of 1.5 to 3 mg/d suggested by the National Research Council.[49] In both of these studies the activity of superoxide dismutase in erythrocytes was decreased during Cu depletion.[46,48] However, plasma ceruloplasmin and Cu levels during Cu depletion were either unchanged[47] or showed a gradual decline to somewhat below the normal range.[46] These two studies will be reviewed in some detail.

Klevay et al.[46] were the first investigators to demonstrate that Cu depletion in an adult male (aged 29 years) was associated with increases in total plasma cholesterol and LDL cholesterol. The Cu depletion diet was comprised of conventional foods and contained 0.8 mg Cu/2700 kcal. During the 105 d of Cu depletion the activity of superoxide dismutase in erythrocytes decreased markedly while plasma cholesterol significantly increased from 202 to 234 mg/dl. After 39 d of Cu repletion (4 mg Cu/d) plasma cholesterol levels fell to 198 mg/dl and the activity of superoxide dismutase was restored.

Reiser et al.[47,48] are the only investigators to have examined changes in plasma cholesterol and Cu status in adults fed a low Cu diet with different carbohydrates (fructose or starch, 20% of total calories). In this study[24] male subjects (21 to 57 years) were fed three levels of Cu (mg/day): pretest self-selected (1.38), experimental low Cu (1.02), and experimental repletion (~4). The Zn content of the experimental diets was higher than the self-selected diet (23.5 vs. 13.6 mg Zn/d). Ingestion of fructose rather than starch with the low Cu diet for 11 weeks significantly lowered erythrocyte superoxide dismutase activity but did not alter total plasma cholesterol. Regardless of the type of carbohydrate, subjects fed the low Cu diet had a marked decrease in their HDL cholesterol ratio (HDL/total − HDL) without a change in their total plasma cholesterol. The HDL cholesterol ratios for the three levels of Cu intake (1.38, 1.02, and ~4 mg Cu/d) were 0.35, 0.25, and 0.44, respectively. The change in the HDL cholesterol ratio during the lowest Cu intake was accounted for by a significant increase in LDL and a significant decrease in HDL cholesterol. These changes in the distribution of lipoprotein cholesterol are suspected to increase the risk of heart disease.[50]

## V. HYPERTRIGLYCERIDEMIA IN CU-DEFICIENT ANIMALS

Limited information on the effects of Cu deficiency and hypertriglyceridemia are available in human and nonhuman primates. The effects of Cu depletion on blood triglycerides were not reported in two studies with humans.[46-48] Blackett et al.[45] reported that two out of three subjects with Menkes' disease had elevated plasma triglycerides. In contrast, Milne et al.[21] reported no changes in serum triglyceride levels of adult cynomolgus monkeys fed a diet marginal in Cu for 32 weeks.

Blood triglycerides are not often reported in Cu deficient rodents. Plasma or serum triglycerides may be unchanged[1,40] or increased by as much as 250% in Cu-deficient rats compared to their Cu-adequate controls.[2,4,17,18,20,27] Reiser

**TABLE 3**
**Plasma Triglyceride Levels (mg/dl) of Male Rats Fed Two Levels of Cu and Four Levels of Fructose and Starch for 4 Weeks**

| % Carbohydrate | | Plasma triglyceride | |
|---|---|---|---|
| Fructose | Starch | Cu deficient | Cu adequate |
| 0 | 60 | 128 ± 8[ab] | 86 ± 16[c] |
| 20 | 40 | 149 ± 10[a] | 106 ± 11[bc] |
| 40 | 20 | 138 ± 14[ab] | 117 ± 18[abc] |
| 60 | 0 | 107 ± 12[bc] | 137 ± 18[ab] |

*Note:* Means (n = 6). Means with different superscripts are significantly different at $p < 0.05$ as determined by Fisher's least significance test. Significant effects: Cu, $p < 0.05$; Cu × carbohydrate, $p < 0.05$.

**TABLE 4**
**Plasma Triglyceride Levels (mg/dl) of Male and Female Rats Fed Two Levels of Cu and Two Types of Carbohydrate for 4 Weeks**

| | | Plasma triglyceride | |
|---|---|---|---|
| Sex | Carbohydrate | Cu deficient | Cu adequate |
| Male | Sucrose | 174 ± 14[ab] | 169 ± 8[abc] |
| | Starch | 143 ± 10[cd] | 120 ± 6[d] |
| Female | Sucrose | 155 ± 4[bc] | 190 ± 16[a] |
| | Starch | 146 ± 10[bcd] | 144 ± 10[bcd] |

*Note:* Means (n = 6). Means with different superscripts are significantly different at $p < 0.05$ as determined by Fisher's least significance test. Significant effects: carbohydrate, $p < 0.0001$; Cu × carbohydrate, $p < 0.07$; Cu × sex, $p < 0.05$.

et al.[2] reported that Cu-deficient rats fed starch, sucrose, and fructose had plasma triglyceride levels that were 32, 41, and 88% higher than their Cu-adequate controls. In contrast, Fields et al.[4] reported that after 11 weeks plasma triglyceride levels were doubled among Cu-deficient rats fed fructose, but not those fed starch or glucose.

Recent studies conducted in this laboratory demonstrated only modest changes in plasma triglycerides after male and female rats were fed Cu-deficient diets (<1 vs. 5 mg Cu/kg diet) with various levels of sucrose, fructose, or starch for 4 weeks (Tables 3 and 4). On the average, plasma triglycerides were only 16% higher in rats fed Cu-deficient rather than Cu-adequate diets. In contrast to the reports of other investigators[2,4] it was found that fructose tended to increase plasma triglycerides to a greater extent in Cu-adequate rats compared to Cu-deficient rats (Table 3). It was also found in this laboratory that short-term Cu-deprivation does not consistently result in hypertriglyceridemia in male or female rats fed sucrose or starch (Table 4). It is likely that periods of Cu-depletion longer than 4 weeks are required to consistently produce severe hypertriglyceridemia.

Johnson and Hove[9] reported that a complex relationship among Cu intake, the type of dietary carbohydrate and the level of dietary Fe influenced plasma triglycerides after only 3 weeks of Cu deprivation. Regardless of the dietary Fe level, Cu-deficient rats fed starch did not have marked increases in their plasma triglyceride levels. However, among rats fed sucrose, Cu deficiency elevated plasma triglyceride levels by 52 and 144% when dietary iron levels were 54 and 226 mg Fe/kg diet, respectively. This high level of Fe is similar to that provided by the Jones-Foster mineral mix which has been used in several studies of Cu deficiency.[1-3,14] In another 3-week study conducted in this laboratory no interaction was observed among Cu intake, Zn intake, and carbohydrate (sucrose or starch) in Cu-deficient rats.[10]

## VI. CONCLUSIONS

Hypercholesterolemia has been recognized as a consequence of suboptimal Cu intake in numerous studies with rats and in two studies with humans. In experimental animals, Cu deficiency increases blood cholesterol levels more consistently than it increases blood triglyceride levels. Hyperlipidemia tends to be greatest in Cu-deficient rats fed sucrose or fructose. However, Cu-deficient rats fed starch or glucose may also have higher blood cholesterol and triglyceride levels than their Cu-adequate controls.

For ethical reasons it will be difficult to conduct studies of Cu-depletion in humans. Klevay et al.[46] and Reiser et al.[47,48] reported heart-related abnormalities in their Cu-depleted human subjects. The heart-related medical complaints reported by Reiser et al.[47] prompted these investigators to terminate their study prematurely. Thus our knowledge of the relationship between Cu deficiency and hyperlipidemia will continue to be derived largely from experimental animals. A variety of experimental conditions are available for studies of the hyperlipidemia of Cu deficiency.

# REFERENCES


1. **Fields, M., Ferretti, R. J., Smith, J. C., and Reiser, S.,** Effect of copper deficiency on metabolism and mortality in rats fed sucrose or starch diets, *J. Nutr.,* 113, 1335, 1983.
2. **Reiser, S., Ferretti, R. J., Fields, M., and Smith, J. C.,** Role of dietary fructose in the enhancement of mortality and biochemical changes associated with copper deficiency in rats, *Am. J. Clin. Nutr.,* 38, 214, 1983.
3. **Fields, M., Ferretti, R. J., Smith, J. C., and Reiser, S.,** The interaction of type of dietary carbohydrates with copper deficiency, *Am. J. Clin. Nutr.,* 39, 298, 1984.
4. **Fields, M., Ferretti, R. J., Reiser, S., and Smith, J. C.,** The severity of copper deficiency in rats is determined by the type of dietary carbohydrate, *Proc. Soc. Exp. Biol. Med.,* 175, 530, 1984.
5. **Fields, M., Lewis, C., Scholfield, D. J., Powell, A. S., Rose, A. J., Reiser, S., and Smith, J. C.,** Female rats are protected against the fructose induced mortality of copper deficiency, *Proc. Soc. Exp. Biol. Med.,* 183, 145, 1986.
6. **Failla, M. L., Babu, U., and Seidel, K. E.,** Use of immunoresponsiveness to demonstrate that the dietary requirement for copper in young rats is greater with dietary fructose than dietary starch, *J. Nutr.,* 118, 487, 1988.
7. **Johnson, M. A. and Gratzek, J. M.,** Influence of sucrose and starch on the development of anemia in copper and iron deficient rats, *J. Nutr.,* 116, 2443, 1986.
8. **Johnson, M. A.,** Interaction of dietary carbohydrate, ascorbic acid and copper with the development of copper deficiency in rats, *J. Nutr.,* 116, 802, 1986.
9. **Johnson, M. A. and Hove, S. S.,** Development of anemia in copper-deficient rats fed high levels of dietary iron and sucrose, *J. Nutr.,* 116, 1225, 1986.
10. **Johnson, M. A. and Flagg, E. W.,** Effects of sucrose and cornstarch on the development of copper deficiency in rats fed high levels of zinc, *Nutr. Res.,* 6, 1307, 1986.
11. **Prohaska, J. R., Korte, J. J., and Bailey, W. R.,** Serum cholesterol levels are not elevated in young copper-deficient rats, mice or brindled mice, *J. Nutr.,* 115, 1702, 1985.
12. **Cohen, N. L., Keen, C. L., Hurley, L. S., and Lönnerdal, B.,** Determinants of copper-deficiency anemia in rats, *J. Nutr.,* 115, 710, 1985.
13. **Shaw, J. C. L.,** Copper deficiency and non-accidental injury, *Arch. Dis. Child.,* 63, 448, 1988.
14. **Klevay, L. M.** Hypercholesterolemia in rats produced by an increase in the ratio of zinc to copper ingested, *Am. J. Clin. Nutr.,* 26, 1060, 1973.
15. **Murthy, L., Klevay, L. M., and Petering, H. G.,** Interrelationships of zinc and copper nutriture in the rat, *J. Nutr.,* 104, 1458, 1974.
16. **Murthy, L. and Petering, H. G.,** Effect of dietary zinc and copper interrelationships on blood parameters of the rat, *J. Agric. Food Chem.,* 24, 808, 1976.
17. **Petering, H. G., Murthy, L., and O'Flaherty, E.,** Influence of dietary copper and zinc on rat lipid metabolism, *J. Agric. Food Chem.,* 25, 1105, 1977.
18. **Stemmer, K. L., Petering, H. G., Murthy, L., Finelli, V. N., and Menden, E. E.,** Copper deficiency effects on cardiovascular system and lipid metabolism in the rat; the role of dietary proteins and excessive zinc, *Ann. Nutr. Metab.,* 29, 332, 1985.
19. **Nielson, F. H. and Zimmerman, T. J.,** Interactions among nickel, copper and iron in rats. Growth, blood parameters, and organ wt/body wt/ratios, *Biol. Trace Elem. Res.,* 3, 83, 1981.
20. **Nielson, F. H., Zimmerman, T. J., and Shuler, T. R.,** Interactions among nickel, copper, and iron in rats. Liver and plasma content of lipids and trace elements, *Biol. Trace Elem. Res.,* 4, 125, 1982.

21. **Milne, D. B., Omaye, S. T., and Amos, W. H.,** Effect of ascorbic acid on copper and cholesterol in adult cynomolgus monkeys fed a diet marginal in copper, *Am. J. Clin. Nutr.*, 34, 2389, 1981.
22. **Lau, B. W. C. and Klevay, L. M.,** Postheparin plasma lipoprotein lipase in copper-deficient rats, *J. Nutr.*, 112, 928, 1982.
23. **Lau, B. W. C. and Klevay, L. M.,** Plasma lecithin:cholesterol acyltransferase in copper-deficient rats, *J. Nutr.*, 111, 1698, 1981.
24. **Allen, K. G. D. and Klevay, L. M.,** Cholesterolemia and cardiovascular abnormalities in rats caused by copper deficiency, *Atherosclerosis,* 29, 81, 1978.
25. **Allen, K. G. D. and Klevay, L. M.,** Copper deficiency and cholesterol metabolism in the rat, *Atherosclerosis,* 31, 259, 1978.
26. **Allen, K. G. D. and Klevay, L. M.,** Cholesterol metabolism in copper deficient rats, *Life Sci.*, 22, 1691, 1978.
27. **Allen, K. G. D. and Klevay, L. M.,** Hyperlipoproteinemia in rats due to copper deficiency, *Nutr. Rep. Int.*, 22, 295, 1980.
28. **Harvey, P. W. and Allen, K. G. D.,** Decreased plasma lecithin:cholesterol acyltransferase activity in copper-deficient rats, *J. Nutr.*, 111, 1855, 1981.
29. **Lefevre, M., Keen, C. L., Lönnerdal, B., Hurley, L. S., and Schneeman, B. O.,** Different effects of zinc and copper deficiency on composition of plasma high density lipoproteins in rats, *J. Nutr.*, 115, 359, 1985.
30. **Lefevre, M., Keen, C. L., Lönnerdal, B., Hurley, L. S., and Schneeman, B. O.,** Copper deficiency-induced hypercholesterolemia: effects on HDL subfractions and hepatic lipoprotein receptor activity in the rat, *J. Nutr.*, 116, 1735, 1986.
31. **Lei, K. Y.,** Alterations in plasma lipid, lipoprotein and apolipoprotein concentrations in copper-deficient rats, *J. Nutr.*, 113, 2178, 1983.
32. **Croswell, S. C. and Lei, K. Y.,** Effect of copper deficiency on the apolipoprotein E-rich high density lipoproteins in rats, *J. Nutr.*, 115, 473, 1985.
33. **Hassel, C. A., Carr, T. P., Marchello, J. A., and Lei, K. Y.,** Apolipoprotein E-rich HDL binding to liver plasma membranes in copper-deficient rats, *Proc. Soc. Exp. Biol. Med.*, 187, 296, 1988.
34. **Lei, K. Y.,** Cholesterol metabolism in copper-deficient rats, *Nutr. Rep. Int.*, 15, 597, 1977.
35. **Looney, M. A. and Lei, K. Y.,** Dietary fiber, zinc and copper: effects of serum and liver cholesterol levels in the rat, *Nutr. Rep. Int.*, 17, 329, 1978.
36. **Lei, K. Y.,** Oxidation, excretion, and tissue distribution of [26-$^{14}$C] cholesterol in copper-deficient rats, *J. Nutr.*, 108, 232, 1978.
37. **Eason, S., Carville, D., Strain, J. J., and Hannigan, B. M.,** The influence of dietary carbohydrate on antibody-mediated immunity in copper deficiency, *Biochem. Soc. Trans.*, 16, 1988, 1988.
38. **Nielson, F. H., Shuler, T. R., Zimmerman, T. J., Collings, M. E., and Uthus, E. O.,** Interaction between nickel and iron in the rat, *Biol. Trace Elem. Res.*, 1, 325, 1979.
39. **Fischer, P. W. F., Giroux, A., Belonje, B., and Shah, B. G.,** The effect of dietary copper and zinc on cholesterol metabolism, *Am. J. Clin. Nutr.*, 33, 1019, 1980.
40. **Sherman, A. R.,** Copper and iron deficiencies: effects on serum lipids and tissue minerals, *Nutr. Res.*, 1, 363, 1981.
41. **Jones, D. G.,** Effects of dietary copper depletion on acute and delayed inflammatory responses in mice, *Res. Vet. Sci.*, 37, 205, 1984.
42. **Jones, D. G. and Suttle, N. F.,** The effect of copper deficiency on the resistance of mice to infection with *pasteurella haemolytica, J. Comp. Pathol.*, 93, 143, 1983.
43. **Klevay, L. M.,** Atrial thrombosis, abnormal electrocardiograms and sudden death in mice due to copper deficiency, *Atherosclerosis,* 54, 213, 1985.
44. **Anon.,** Report of the American Institute of Nutrition Ad Hoc Committee on Standards for Nutritional Studies, *J. Nutr.*, 107, 1340, 1977.

45. **Blackett, P. R., Lee, D. M., Donaldson, D. L., Fesmire, J. D., Chan, W. Y., Holcombe, J. H., and Rennert, O. M.,** Studies of lipids, lipoproteins and apolipoproteins in Menkes' Disease, *Pediatr. Res.*, 18, 864, 1984.
46. **Klevay, L. M., Inman, L., Johnson, L. K., Lawler, M., Mahalko, J. R., Milne, D. B., Lukasi, H. C., Bolonchuk, W., and Sandstead, H. H.,** Increased cholesterol in plasma in a young man during experimental copper depletion, *Metabolism,* 33, 1112, 1984.
47. **Reiser, S., Smith, J. C., Mertz, W., Holbrook, J. T., Scholfield, D. J., Powell, A. S., Canfield, W. K., and Canary, J. J.,** Indices of copper status in humans consuming a typical American diet containing either fructose or starch, *Am. J. Clin. Nutr.*, 42, 242, 1985.
48. **Reiser, S., Powell, A., Yang, C-Y., and Canary, J. J.,** Effect of copper intake on blood cholesterol and its lipoprotein distribution in men, *Nutr. Rep. Int.*, 36, 641, 1987.
49. **National Research Council,** Subcommittee on the 10th Edition of the RDAs, *Recommended Dietary Allowances,* National Academy Press, Washington, D.C., 1989.
50. **Miller, N. E., Thelle, D. S., Forde, O. H., and Mjos, O. D.,** The Tromso Heart-Study, High-density lipoprotein and coronary heart-disease: a prospective case-control study, *Lancet,* 1, 965, 1977.

Chapter 7

# COPPER AND THE EXOCRINE PANCREAS: IMPLICATIONS FOR LIPID DIGESTION

**Patsy M. Brannon**

## TABLE OF CONTENTS

Copper deficiency causes a selective atrophy of the exocrine pancreas that impairs its secretion of the major digestive enzymes. The effects of this impaired exocrine function on the digestion of dietary lipid have not been measured directly. However, this atrophy and altered function of the exocrine pancreas have significant implications for the digestion and utilization of dietary lipids. The digestion of dietary lipid and the function of the exocrine pancreas in secreting digestive enzymes will be reviewed prior to the discussion of copper deficiency and the exocrine pancreas.

## I. DIGESTION AND ABSORPTION OF DIETARY LIPID

In the Western diet, 90% of the dietary lipid consists of triglycerides and approximates 40% of the energy intake.[1] Predominantly, this triglyceride consists of long chain fatty acids (LCFA). The remaining 10% of the dietary lipid represents a mixture of phospholipids, cholesterol esters, vitamin esters and other dietary sterols. The major digestive enzymes involved in the hydrolysis of these dietary lipids are presented in Table 1. Later, the characteristics of these enzymes will be described. First, the digestion and absorption of dietary lipids will be summarized. The primary site of digestion of dietary lipid is the small intestinal lumen, although significant hydrolysis of dietary triglyceride also occurs in the stomach. Triglyceride digestion yields fatty acids, monoglycerides, and glycerol that are absorbed primarily in the proximal small intestine.[2]

Triglycerides are initially present in large droplets that form an unstable emulsion with dietary phospholipid. In the stomach, lingual lipase and possibly gastric lipase begin the hydrolysis of this triglyceride, producing diglycerides and fatty acids.[3] Approximately 20% of ingested triglyceride appears as fatty acids in the stomach.[4] Although quantitatively small, these initial hydrolytic products, along with bile salts and dietary phospholipids, may be important in forming smaller stable emulsified triglyceride particles in the small intestine. Short chain fatty acids (SCFA) and medium chain fatty acids (MCFA) may be absorbed directly through the gastric mucosa into the blood.[5]

In the duodenum, chyme containing emulsified triglyceride and its hydrolytic products mixes with pancreatic secretions containing bicarbonate that neutralize the chyme. Fatty acids are then ionized and partition in the interface of the emulsified triglyceride and the aqueous phase. The chyme also mixes with bile containing bile salts in simple micelles. Some bile salts adhere to the surface of the emulsion, stabilizing it. This stabilized emulsion with its larger surface area is important for the subsequent hydrolysis of the remaining triglyceride by pancreatic lipase that acts at lipid-aqueous interfaces.[6] Bile salts also affect pancreatic lipase, both shifting its pH optimum from pH 8—9 to pH 6—7 and inactivating it.[7,8] The proteinaceous cofactor, pancreatic colipase, reverses this inhibition of bile salts and appears to serve as an anchor

## TABLE 1
## Key Enzymes in Dietary Lipid Digestion

| Dietary substrate | Enzyme | Source | Site of action | Product |
|---|---|---|---|---|
| Triglyceride | Lingual lipase | von Ebner's gland | Stomach | Fatty acids; diglycerides |
| | Gastric lipase | Gastric mucosa | Stomach | Fatty acids; glycerides |
| | Pancreatic lipase[a] | Pancreatic acinar cell | Small intestine | Fatty acids; monoglycerides |
| Cholesterol ester; retinyl ester | Carboxyl ester lipase | Pancreatic acinar cell | Small intestine | Fatty acids; alcohols |
| Phospholipid | Phospholipase $A_2$ | Pancreatic acinar cell | Small intestine | Fatty acids; lysophophatides |

[a] Requires the cofactor pancreatic colipase.

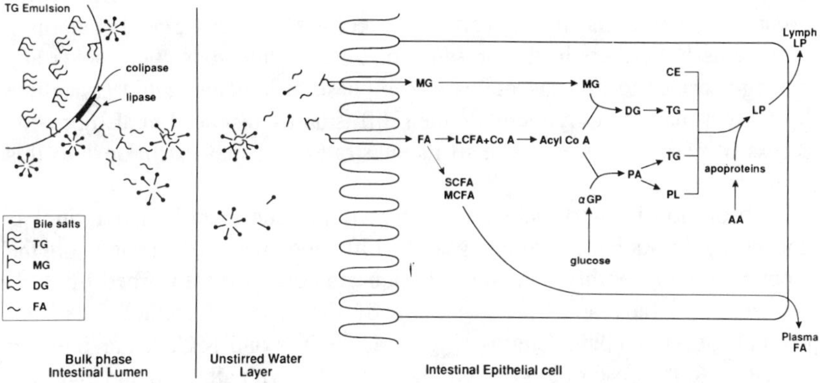


FIGURE 1. Summary of intestinal digestion and absorption of dietary triglycerides. Bile salt-emulsified dietary triglyceride (TG) and its hydrolyzed product diglyceride (DG) are hydrolyzed by lipase with its cofactor colipase. The hydrolyzed products fatty acids (FA) and monoglycerides (MG) form mixed micelles with bile salts, which diffuse through the unstirred water layer. MG and FA are absorbed by diffusion across the microvilli of the intestinal epithelial cell. Long chain fatty acids (LCFA) are reesterified to the absorbed MG or *de novo* synthesized α-glycerol phosphate (α-GP) to TG which is then packaged with reesterified cholesterol ester (CE), phospholipid (PL), and apoproteins (synthesized by the cell from amino acids [AA]) into lipoproteins (LP). These LP are secreted into the lymph through which they are transported to the thoracic duct and enter the circulation. Short chain fatty acids (SCFA) and medium chain fatty acids (MCFA) are absorbed directly into the blood and enter the pool of plasma fatty acids.

for pancreatic lipase at the interface, possibly by displacing bile salts (Figure 1).[9] The action of pancreatic lipase produces monoglycerides and fatty acids, which combine with bile salt micelles to form larger mixed micelles that can diffuse across the unstirred water layer between the luminal bulk phase where digestion occurs and the intestinal epithelial cell where absorption occurs (Figure 1).

Diffusion through this unstirred water layer, an unequilibrated aqueous area 100 to 500 μ thick, appears rate limiting for the absorption of hydrolyzed lipids.[10] Micelles have a slower diffusion rate than fatty acids or monoglycerides alone, but contain high amounts of both. The net effect is a greater transfer of absorbable lipid through the unstirred layer.[11] Bile salts themselves are absorbed in the distal ileum, not in the proximal intestine with hydrolyzed lipids.[12] The dissociation of fatty acids and monoglycerides from mixed micelles is not understood, but may be facilitated by a pH gradient established by the secretion of $H^+$ ions by the intestinal epithelial cells.[13] A low pH facilitates micellar dissociation and protonated fatty acid diffusion.[14]

The digestion of other dietary lipids is less understood than that of dietary triglycerides. Phospholipids are hydrolyzed by phospholipase $A_2$ to fatty acids and lysophosphatides. Cholesterol esters are hydrolyzed to the absorbable products, free cholesterol and fatty acids, by pancreatic carboxyl ester lipase.[15] This enzyme can also hydrolyze retinyl ester to retinol and fatty acids. The digestion of these other dietary lipids may depend on that of dietary triglyceride because, in part, of the affinity carboxyl ester lipase has for surfaces containing fatty acids.[15] Further, in the presence of limiting bile salts, fatty acids and monoglycerides form a viscous isotropic phase of lamellar liquid crystalline lipid[16] that may not only facilitate their diffusion to micelles but also provide access of carboxyl ester lipase to its substrates that may partition into this phase.[15]

Absorption of fatty acids, monoglycerides, cholesterol, and retinol is nonenergy dependent, through passive diffusion that is dependent on the membrane permeability and concentration gradient of the absorbed lipid.[1,17] The absorbed fatty acids are processed by the intestinal epithelial cell depending on their chain lengths (Figure 1). SCFA and MCFA are released directly into the portal blood where they complex with albumin and enter the free fatty acid pool. LCFA are reesterified predominately with absorbed monoglyceride or with α-glycerol phosphate synthesized *de novo* from glucose.[18] This reesterified triglyceride is then packaged with reesterified cholesterol esters, retinyl esters, and other apolar absorbed lipids into chylomicrons and very low density lipoproteins. The apoproteins of these lipoproteins are synthesized *de novo* by the intestinal epithelial cell from amino acids. These lipoproteins are then secreted into the lymph through which they travel to the thoracic duct where they enter the blood system.

### A. Lingual and Gastric Lipase

Lingual lipase is secreted by von Ebner's gland in the proximal oral cavity. This enzyme is active over a broad pH range of 2 to 8 and is stable in the stomach at low pH (2.5 to 3).[19,20] Rat lingual lipase has a molecular weight of 51,000 and is a highly hydrophobic protein.[21] Unlike pancreatic lipase, lingual lipase is not bile salt dependent[5,20] and is only slightly inhibited by high concentrations of bile salts.[5] Lingual lipase selectively hydrolyzes the sn-3 position of triglycerides,[22] with a preference for MCFA (C:8, 10 and 12).[5] Thus, the action of lingual lipase produces long chain diglycerides and MCFA.

Gastric lipase is secreted by the gastric mucosa into the stomach.[23] It preferentially hydrolyzes short chain triglycerides and does not hydrolyze long chain triglycerides.[24,25]

Together lingual and gastric lipases are responsible for the intragastric lipolysis of dietary triglycerides to diglycerides, SCFA and MCFA.[5] However, lingual lipase appears to be the predominant enzyme responsible for intragastric lipolysis. Diversion of oral secretions from the rat stomach results in a 50 to 90% decrease in gastric lipolysis.[26] Further, this diversion of oral secretion also decreases duodenal triglyceride hydrolysis approximately threefold, suggesting that intragastric lipolysis is important for the duodenal hydrolysis by pancreatic lipase. Although the role of intragastric lipolysis in the subsequent hydrolysis of triglycerides is unknown, the hydrolytic products may aid the formation of a stable emulsification of triglyceride that is important for the effective action of pancreatic lipase. These lipolytic products and dietary phospholipids are more effective emulsifiers of triglyceride than bile salts.[27]

### B. Pancreatic Lipase and Colipase

Both human and porcine pancreatic lipases are glycoproteins with molecular weights, respectively, of 48,000 and 52,000.[28,29] As described above, this lipase hydrolyzes long chain and short chain triglycerides at the sn-1 and sn-3 positions. It can also slowly hydrolyze retinyl esters.[3] Its hydrolysis is sequential: triglycerides to diglycerides and diglycerides to monoglycerides.[30] Lipase acts on its triglyceride or diglyceride substrate at lipid-water interfaces. It has little activity against short chain triglycerides at low, soluble concentrations, but has increased activity against these triglycerides at higher, emulsified concentrations.[6]

Pancreatic lipase has a complex interaction with bile salts. Bile salts shift its pH optimum to a range active in the duodenum (6-7) and inhibit its lipolytic activity.[7,8] In fact, bile salts at or above the critical micellar concentration inhibit pancreatic lipase completely. Bile salts may displace lipase from the emulsified triglyceride surface.[9] Such a displacement may explain the mechanism of bile salt inhibition of lipase because the adsorption of lipase to the emulsion surface correlates with its lipolytic activity.[31] Further, its lipolytic activity is dependent on the surface area of the emulsion.

A small pancreatic secretory protein, colipase, reverses the inhibition of pancreatic lipase by bile salts. Colipase has a molecular weight of 10,000.[32] It is secreted as a larger peptide, pro-colipase (up to 111 amino acids) and is activated by N-terminal and C-terminal cleavage by trypsin.[33,34] Colipase complexes both with pancreatic lipase in the presence of bile salts or triglycerides and with bile salts.[8] Thus, colipase may restore the activity of pancreatic lipase by complexing with bile salts on the surface of the triglyceride emulsion and with lipase allowing the enzyme access to its interfacial substrate.

Normally, pancreatic secretions contain plentiful quantities of lipase and colipase. Sufficient lipase and colipase are in 100 ml of duodenal contents to hydrolyze the average daily intake of triglyceride in 1 min.[35]

### C. Other Lipases

Carboxyl ester lipase, also known as cholesterol esterase, hydrolyzes a wide variety of apolar esters to their water-soluble products. Included among its substrates are cholesterol and retinyl esters and also triglycerides.[15] Carboxyl ester lipase can hydrolyze triglycerides at the sn-1, -2 or -3 positions and may have an integrated action with pancreatic lipase in the digestion of triglyceride. It is a large protein with a molecular weight of 65,000 to 70,000 (rat) and 100,000 (human). Further, it requires bile salts for its activity.[36]

Phospholipase $A_2$ is a small peptide of 14,000 molecular weight.[3,37] It hydrolyzes phospholipids from the dietary and biliary secretions. Phospholipase hydrolyzes the sn-2 position of 1,2-diacyl-sn-glycerol-3-phosphocholines producing fatty acids and lysophosphatides. Its activity is bile salt- and calcium-dependent.

## II. THE EXOCRINE PANCREAS

The exocrine pancreas secretes the major enzymes and bicarbonate required for digestion. The acinar cells synthesize and secrete some 20 digestive enzymes, while the ductal cells secrete the bicarbonate-rich fluid. Secretion by both acinar and ductal cells is regulated by various gastrointestinal hormones.

### A. Morphology

The morphology and ultrastructure of the exocrine pancreas is important for its function. Pancreatic lobules contain acinar cells in a three-dimensional array around the ducts throughout the lobule. Acinar cells are the predominant cell type in the pancreas and may comprise 85 to 90% of the organ. Ductal cells line the ducts and are a small percentage of pancreatic cells. The endocrine pancreatic cells in the Islets of Langerhans are dispersed throughout the exocrine pancreas and are also only a small proportion of the pancreatic cells (2 to 3%)

The functional units of enzyme secretion are the acini, which are clusters of acinar cells. Acini have a fivefold greater secretory responsiveness *in vitro* than do single acinar cells.[38] Acinar cells are epithelial and exhibit the typical polarity of epithelium. The basal portion of the acinar cell with its plasma membrane is surrounded by a basement membrane containing collagen type IV, laminin, and proteoglycans.[39] In the basal portion of the cell are its nucleus and rough endoplasmic reticulum. The apical portion of the acinar cell with its luminal membrane faces the lumen of the duct into which secretion occurs. In the apical portion of the cell are zymogen granules and Golgi. This polarization of the acinar cell is related to the unidirectional flux of synthesis and secretion of the digestive enzymes that will be described later. Acinar cells have the tight junctions and gap junctions typical of epithelial cells.[39]

Normally, turnover and proliferation of acinar cells are low. The major period of acinar growth is fetal when the mitotic index is highest at 13 to 14%, postnatal when the mitotic index is 4 to 8%, and weaning when the mitotic index is 1%.[40] In the post-weaning mammal, the acinar cell has a low proliferation rate with a mitotic index of 0.1 to 0.2%.[41] Pancreatectomy or supramaximal stimulation with the secretagogue caerulein increases this low proliferation of acinar cells for a short period of 2 to 5 d.[42,43] The exocrine pancreas, in particular the acinar cells, have a regenerative capacity that is, however, more limited that other highly differentiated organs like the kidney and liver.

## B. Function

### *1. Synthesis and Secretion of Digestive Enzymes*

Considerable work over the last 30 years has elucidated the mechanisms by which the digestive enzymes are synthesized and secreted. (For reviews see References 44 and 45.) This section will summarize these mechanisms, particularly focusing on how the acinar cell segregates secretory proteins from other nascent proteins. Segregation of digestive enzymes and storage of these in membrane-bound vesicles in the cell is one of the ways in which the acinar cell is protected from degradation by the digestive enzymes. Additionally, many of the digestive enzymes are secreted as inactive precursors that are activated by proteolytic cleavage of small peptide fragments following secretion.

Synthesis of the digestive enzymes involves the translation of mRNAs, translocation of the proteins into the lumen of the endoplasmic reticulum, post-translational processing and folding of the enzymes, segregation of the secretory enzymes from lysosomal and membrane proteins, and packaging into membrane-bound vesicles. This process occurs vectorially from the basal to the apical portion of the cell (Figure 2). The mRNAs for the enzymes are transcribed in the nucleus and translocated to the cytoplasm where their translation begins on ribosomes. Each secreted enzyme (as well as lysosomal proteins and membrane proteins) have a "signal peptide" encoded in their N-terminal.[46] Although heterogenous in sequence, these signal peptides are all highly hydrophobic and are required for the translocation of the secretory enzymes into the lumen of the rough endoplasmic reticulum.[46,47] The nascent signal peptide binds to the signal peptide recognition particle (SRP) that is an 11S ribonucleoprotein containing a 7SL RNA and six peptides.[48] This complex (ribosome-mRNA-nascent peptide-SRP) is now targeted to the en-

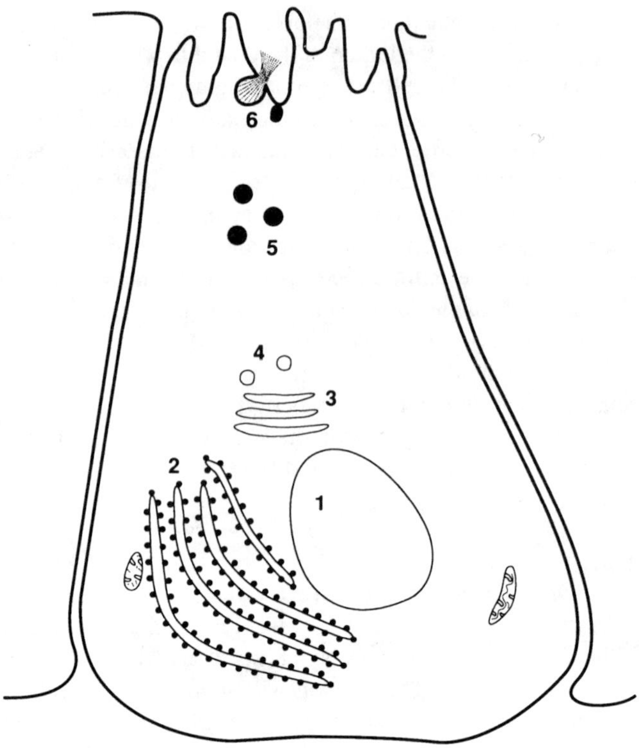


FIGURE 2. Schematic diagram of the synthesis and secretion of digestive enzymes by pancreatic acinar cells. The mRNAs for the digestive enzymes are transcribed and processed in the nucleus (1) and then translocated to the cytoplasm. The mRNAs are translated on the rough endoplasmic reticulum (RER, 2) and enter the lumen of the RER through the interaction of the signal peptides encoded on their N-terminal with the signal recognition particle (SRP) and the docking protein in the RER membrane. Inside the RER, the zymogens are proteolytically cleaved to remove the signal peptide and are further processed. The zymogens move through the Golgi (3) and are packaged into condensing vacuoles on the trans side of the Golgi (4). Mature zymogen granules (ZG, 5) are formed from the condensing vacuoles and move toward the apical membrane of the acinar cell facing the lumen of the ducts. Upon appropriate stimulation, the ZGs fuse with the apical membranes, releasing their contents into the lumen (6).

doplasmic reticulum because the complexed SRP binds to the ''docking protein'', an integral endoplasmic reticular membrane protein. Then, the nascent peptide is translocated through the endoplasmic reticular membrane as its translation proceeds on the rough endoplasmic reticulum. This translocation requires energy produced by nucleotide triphosphate hydrolysis,[47] but is otherwise not well understood. A proposed ''translocon'' may provide a pore through the membrane for the nascent peptide.[48] Inside the lumen, signal peptidases cleave the signal peptide and the nascent proteins undergo additional processing including folding, disulfide bond formation and glycosylation.[45] The proteins move through the endoplasmic reticulum to the Golgi where processing continues. The rate of transport of various digestive enzymes through the endoplasmic reticulum and Golgi vary, perhaps reflecting differences in processing.[49]

In thc Golgi, the secretory proteins are segregated from lysosomal proteins and plasma membrane proteins by a specific recognition mechanism that is not yet understood.[50-52] It is known that lysosomal proteins are similarly segregated by interaction with a mannose-6-phosphate receptor in the trans Golgi.[53] The secretory proteins leave the trans Golgi packaged in condensing vacuoles.[44] The condensing vacuole contents are concentrated to form the mature zymogen granules, which serve as storage vesicles and exocytosis vehicles. The zymogen granules move toward the apical membrane. Upon the appropriate neural or hormonal stimulation, the zymogen granule membrane fuses with the apical membrane and its contents are released into the ductal lumen. Secreted enzymes and fluid are released through the common duct into the duodenum.

### 2. *Regulation of Secretion*

Regulation of acinar secretion occurs through the interaction of secretagogues with two types of plasma membrane receptors differing in their stimulus-coupling mechanisms. One class of receptors acts through a calcium and phosphatidylinositol-mediated pathway while the other class acts through a cyclic AMP (cAMP)-mediated pathway (Figure 3).[54] Four receptors of the former type exist based on their ligand specificity. The cholecystokinin (CCK) receptor binds CCK and the structurally related peptides caerulein and gastrin; the bombesin receptor binds bombesin and litorin; the physalaemin receptor binds physalaemin and substance P; and the muscarinic/cholinergic receptor binds carbamyl choline and various agonists. Two receptors of the latter type exist based again on their ligand specificity. The vasoactive-intestinal peptide (VIP)-preferring receptor binds VIP with high affinity and secretin with 1000-fold lower affinity. In contrast, the secretin-preferring receptor binds secretin with high affinity and VIP with 1000-fold lower affinity. As the occupancy of the calcium-mediated receptors increases to 40%, stimulation of secretion increases progressively; but past 40% occupancy, secretion diminishes in response. However, intracellular calcium increases progressively as occupancy increases to 75%.[54]

Binding of CCK or carbamyl choline to its receptor results in a rapid degradation of plasma membrane phosphatidylinositol by phospholipase C to diacylglycerol and inositol (1,4,5) triphosphate[55] and a large increase of intracellular calcium within seconds of receptor occupancy.[56] This increase in calcium results from the release of calcium by inositol triphosphate from intracellular calcium pools, particularly the rough endoplasmic reticulum.[57]

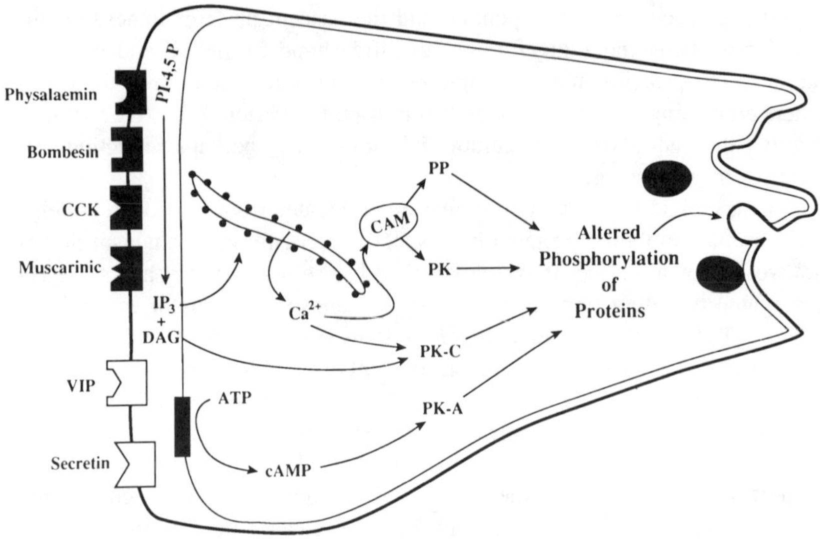


FIGURE 3. Regulation of secretion of digestive enzymes by pancreatic acinar cells. Various agents (secretagogues) stimulate the secretion of the digestive enzymes by interacting with specific receptors on the basolateral membrane of the acinar cells. Interaction of secretagogues with four classes of receptors (cholecystokinin [CCK], muscarinic, bombesin, and physalaemin) results in the hydrolysis of membrane phosphatidylinositol-4,5 bisphosphate (PI-4,5P) by phospholipase C to inositol triphosphate ($IP_3$) and diacylglycerol (DAG). $IP_3$ causes the release of intracellular calcium ($Ca^{2+}$), predominately from the RER. This increases $Ca^{2+}$ can interact directly with phosphokinase C (PK-C) or indirectly with calmodulin (CAM) regulated phosphokinases (PK) and phosphatases (PP) and can, thus, alter the phosphorylation of proteins, resulting in the secretion of the zymogens. DAG can also directly interact with PK-C to affect secretion. Interaction of other secretagogues with two other classes of receptors (vasoactive intestinal peptide [VIP] and secretin) results in the stimulation of secretion by increasing adenylate cyclase activity and cAMP which can stimulate phosphokinase A (PKA) and alter the phosphorylation of proteins.

Both intracellular calcium and diacylglycerol alter the phosphorylation of regulatory and structural proteins in the cell and, thereby, increase the secretion of digestive enzymes. Calcium can activate directly phosphokinase C and indirectly calmodulin-dependent phosphokinases and phosphatases.[56,58] Diacylglycerol also directly activates phosphokinase C.

Binding of secretin or VIP to its receptor results in activation of adenylate cyclase and increased cAMP.[58] This increased cAMP alters the phosphorylation of regulatory and structural proteins by activating protein kinase A.[58,59]

The distal steps in stimulation secretion are not understood. It remains to be determined which acinar proteins are phosphorylated or dephosphorylated and how this altered phosphorylation causes exocytosis of the zymogen granule.

# III. COPPER DEFICIENCY AND THE EXOCRINE PANCREAS

In copper-deficient rats, the acinar cells selectively atrophy. This atrophy, first reported by Muller in 1970,[60] has been confirmed by other investigators.[61-67] Similar pancreatic acinar atrophy has been reported in copper-deficient cattle.[68] These studies of copper deficiency and the exocrine pancreas used different protocols to induce copper deficiency, various initial ages of rats, and variable durations of deficiency (Table 2).

Copper deficiency has been induced by dietary deficiency alone[60,63,66,67] and by copper chelators[61,63-65] in both weanling (50 to 60g), post-weanling (80 to 100 g), and adult (>200 g) rats. The composition of the diets also varies. In the first study reported by Muller,[60] rats were fed nonpurified copper-deficient and copper-supplemented diets consisting of condensed milk with vitamin and mineral supplements. Neither copper content of these diets nor food consumption is reported. The remaining studies all used some type of purified diet. In Folsch and Creutzfeldt's study,[61] the purified diet contained 12% protein by weight (casein and soy protein), 43% sucrose, and 30% cottonseed oil. Again, neither the copper content of either the deficient or supplemented diet nor food consumption is reported. The diet used in the study by Fell and co-workers[62] contained 20% protein (egg albumin) and 60% sucrose. The copper-deficient and copper-supplemented diets had 0.6 and 5 mg Cu/kg diet, respectively. To control for the typically decreased food consumption in copper-deficient rats, supplemented rats were pair-fed to deficient rats. A number of studies[63-65] have used a choline-supplemented, copper-deficient diet containing 20% protein (casein) and 66% sucrose. This copper-deficient diet contained 1.2 mg Cu/kg diet. However, these studies differ in the control diet used. Smith and co-workers[63] used both a nonpurified, closed formula commercial diet (chow) and a copper-supplemented purified diet that contained 20 mg Cu/kg diet. Further, all rats were meal-fed 15 g of diet in the evening to control food intake. Rao and co-workers[64,65] used only a nonpurified, closed formula commercial diet and did not report food consumption. The use of a commercial closed formula diet as the dietary control for a deficient purified diet is inappropriate given the many differences in composition between these two types of diets. Finally, Lewis and co-workers used a purified diet containing 20% protein (egg albumin) and 62% carbohydrate (starch or fructose).[66,67] The deficient and supplemented diets contained 0.6 and 6.0 mg Cu/kg diet, respectively. Food consumption was significantly lower (38%) in copper-deficient fructose-fed rats than all other dietary groups but was not controlled. Despite these differences in dietary composition and the feeding protocols among the studies the types of pancreatic abnormalities — both morphological and functional — are similar (Table 2) and will be discussed in detail below.

Dietary deficiency alone alters copper status in weanling rats.[62,66] After

## TABLE 2
## Effects of Copper Deficiency on the Rat Exocrine Pancreas

| Study | Copper-deficient diet | Chelator | Initial body weight (g) | Pancreatic abnormalities | Onset (weeks) |
|---|---|---|---|---|---|
| Muller[60] | Nonpurified[a] | None | 62 | Acinar atrophy lipomatosis | 7+ |
| Folsch and Creutzfeldt[61] | Purified[b] | Penicillamine[c] | 250—320 | Acinar atrophy and fatty infiltration | 7+ |
| | | | | Complete acinar atrophy | 12 |
| | | | | ↓ Amylase (98%) and trypsin (99%) | |
| | | | | ↓ Basal and stimulated secretion | |
| Fell et al.[62] | Purified[d] | None | (NA)[e] | ↓ Size; ↓ acinar cells; ↓ ZG; ↓ RER | 6 |
| Smith et al.[63] | Purified[f] | None | 100 | ↓ Weight | 6 |
| | | | | ↓ Amylase (50%) | 6 |
| | | Penicillamine[g] | 100 | Acinar atrophy | 5 |
| | | | | ↑ Adiposity | |
| | | | | ↓ Amylase (99%) | 4 |
| | | | | ↓ Basal and stimulated secretion | |
| | | Trien[g,h] | 100 | Acinar atrophy and fatty infiltration | 6 |
| | | | | ↓ Amylase (94%) | 8 |
| | | | | ↓ Basal and stimulated secretion | |
| Rao et al.[64] | Purified[f] | Penicillamine[i] | 220—270 | Acinar atrophy (85—90%) | 8 |
| Rao et al.[65] | Purified[f] | Trien[i] | 80—90 | ↓ Size (61%) | 8 |
| | | | | Acinar atrophy (>90%) | |
| Lewis et al.[66] | Purified[j]—starch | None | 50—60 | ↓ Luminal amylase (40%), lipase (53%), and trypsin (25%) | 5.5 |
| | Purified[j]—fructose | None | 50—60 | ↓ Size (37%) | 5.5 |
| | | | | ↓ Luminal amylase (76%), lipase (53%), and trypsin (25%) | |
| Lewis et al.[67] | Purified[j]—starch | None | 50—60 | ↓ Size (16%) | 5.5 |
| | | | | ↓ Amylase (70%), lipase (69%), and trypsinogen (18%) | |

**TABLE 2 (continued)**
**Effects of Copper Deficiency on the Rat Exocrine Pancreas**

| Study | Copper-deficient diet | Chelator | Initial body weight (g) | Pancreatic abnormalities | Onset (weeks) |
|---|---|---|---|---|---|
| | Purified[j]—fructose | None | 50—60 | ↓ Size (60%)<br>↓ Amylase (70%), lipase (91%), and trypsinogen (64%) | 5.5 |

[a] Condensed milk (7.5% fat, 25.3% dry mass) with vitamin and mineral (no Cu) supplements; Cu content not available.
[b] By weight: 6% casein, 6% soya protein, 3% lactose, 43% sucrose, 4% corn oil, 30% cottonseed oil, 0.1% cystine, 4% fiber, 2% choline, and vitamin and mineral (no Cu) supplements; Cu content no available.
[c] 300 mg/kg body weight daily.
[d] By weight: 20% egg albumin, 60% sucrose, 10% arachis oil, vitamin and mineral (no Cu) supplements; 0.6 mg Cu/kg diet.
[e] Weanling (approximately 40 to 60 g).
[f] By weight: 20% casein, 66% sucrose, 10% peanut oil, 1% vitamin mix and mineral (no Cu) supplements; 1.2 mg Cu/kg diet.
[g] 200 mg/kg body weight daily.
[h] Triethylene tetramine tetrachloride.
[i] 6 g/kg diet.
[j] By weight: 20% egg albumin, 62% carbohydrate (starch or fructose), 9.5% corn oil, 3% cellulose, 3.5% AIN mineral mix (no Cu), 1.0% AIN vitamin mix, 0.002% biotin, 0.3% choline; 0.6 mg Cu/kg diet.

6 weeks, copper-deficient rats had lower hepatic (18% of control) and serum (11%) copper and had anemia (72% of control hemoglobin and 77% of control packed cell volume).[62] After $5^1/_2$ weeks, copper-deficient rats had significantly lower hepatic (42% of control)[66] and pancreatic (60%)[67] copper content. Other studies[60,61,64,65] did not assess copper status.

The chelators used to induce copper deficiency are penicillamine[61,63,64] and triethylene tetramine tetrachloride (Trien).[63,65] These chelators have been used with post-weanling or adult rats whose copper stores are presumably greater than weanling rats. In post-weanling rats, a copper-deficient diet alone after 12 weeks results in no anemia, a decrease in body and pancreatic weight, and a decrease in amylase content.[63] No morphological changes are reported in these copper-deficient rats. These milder effects of only dietary deficiency contrast with the severe atrophy induced much earlier (4 to 6 weeks) with chelator-induced deficiency. Serum copper levels are intermediate in post-weanling dietary deficient rats (4.1 μmol Cu/l) between copper-supplemented rats (18.6 μmol Cu/l) and chelator-induced deficient rats (1.0 μmol Cu/l). These results suggest that it is difficult in post-weanling rats to induce copper deficiency by diet alone and that only a mild, subclinical copper deficiency is produced. No reports of the effects of dietary deficiency exist for adult rats. Of the two chelators used, penicillamine alters copper status and affect pancreatic morphology more rapidly than Trien does.[63] Hypochromic macrocytic anemia is present in the penicillamine-treated rats after 2 to 3 weeks and is pronounced after 4 weeks. However, this anemia is present after 4 weeks in Trien-treated rats and is marked after 6 weeks. Similarly, after 6 weeks, serum copper is over twofold greater in Trien-treated rats (2.5 μmol Cu/l) than penicillamine-treated rats (1.0 μmol Cu/l), even though both are significantly lower than copper-supplemented rats (18.6 μmol Cu/l). These chelators vary, then, in the rapidity of their effects, but both affect copper status and the exocrine pancreas more rapidly than dietary deficiency alone.

The abnormalities in both pancreatic morphology and function are progressive with the duration of copper deficiency.[61] Relatively mild effects are seen prior to 6 or 7 weeks of dietary deficiency and 3 to 4 weeks of chelator-induced deficiency. With more advanced deficiency, however, the severity of abnormalities increases (Table 2). Only with difficulty can the results of early copper deficiency be compared to advanced deficiency among the different studies. Particularly, studies of pancreatic morphology and function prior to 6 weeks of dietary deficiency may be at a stage of deficiency where only mild changes have occurred.

### A. Morphology

Copper deficiency causes a highly selective atrophy of the pancreatic acinar cells, resulting in decreased pancreatic size and weight.[60] This atrophy begins in the periphery of the lobules.[62] Also, this atrophy is more severe in male than female rats.[60] After 7 weeks of copper deficiency, this atrophy is also accompanied by histological changes in the remaining acinar cells which have fewer zymogen granules and pyknotic nuclei.[60,63] Connective tissue and adipocytes increase in the atrophied pancreas. After 12 weeks of copper deficiency, the acinar cells are completely atrophied and replaced by adipose tissue.[61] Acinar atrophy is progressive. After 2 weeks of copper deficiency, acinar cells and lobules are normal, but after 4 weeks, acinar cells decrease in number but still have normal zymogen granules. Acinar atrophy is pronounced after 6 weeks, and acinar cell ultrastructure is abnormal.[62] Complete acinar atrophy occurs much earlier with chelators (5 weeks with penicillamine and 6 weeks with Trien).[63] As acinar atrophy and fatty infiltration progress, the lobular structures are destroyed;[62,63] but the ducts and ductal cells are unaffected by copper deficiency.[61,62] In fact, copper deficiency has been used as an experimental model in which to study the function and regulation of pancreatic ductal cells. Similarly, the Islets of Langerhans and blood vessels are normal in the copper deficient pancreas.[62] Compared to normal pancreas (Figure 4), the loss of acinar cells and disorganized acini are evident in the pancreas of a copper-deficient rat in 6 weeks (Figure 5).

This pancreatic atrophy is noninflammatory and thus differs from pan-

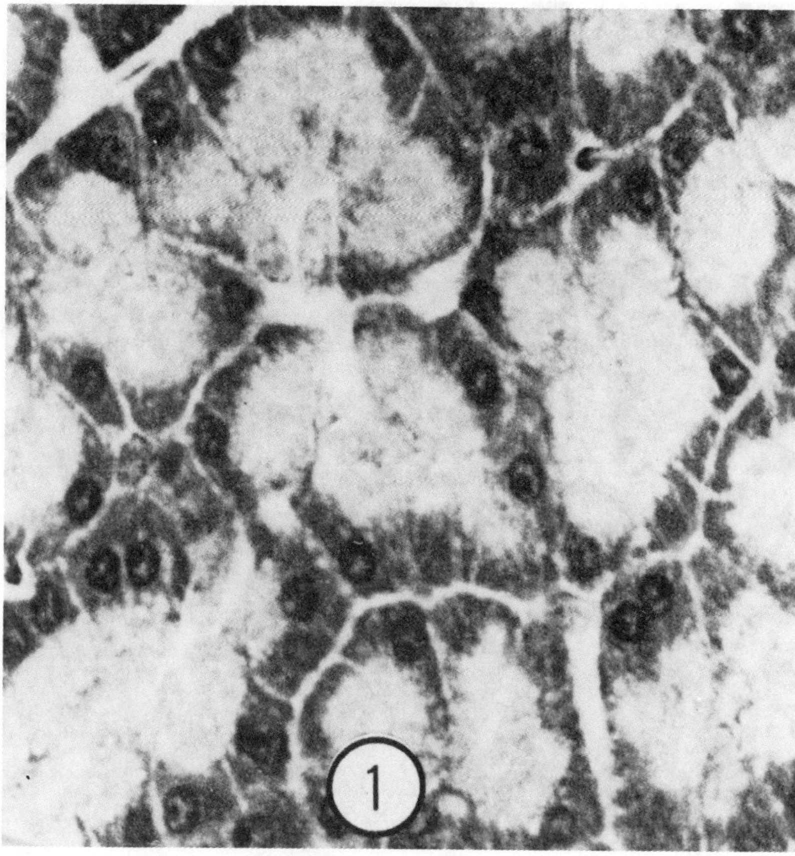


FIGURE 4. Normal appearance of control rat pancreas after pair-feeding with a copper-deficient rat for 6 weeks. Mayer's Haemalum and Basic Fuchsin × 530. (From Fell, B. F. and Davies, N. T., *Histochem. J.*, 14, 665, 1982. With permission.)

creatitis which is associated with inflammatory reactions. Interstitial tissue has increased numbers of histocytes and mast cells,[62] however, and more macrophages surround the acini as well.

A number of ultrastructural changes occur in acinar cells in copper deficiency as well. Mitochondria are hypertrophied, swollen, and distorted.[62] Further, these mitochondria are clustered in groups of two to four and are sometimes stacked within one another. The rough endoplasmic reticulum is decreased in volume and abnormal, exhibiting fragmentation and vessiculation.[62] The Golgi are dilated, but condensing vacuoles exist although they are less numerous and electron-dense than in controls. Zymogen granules also contain less electron-dense material, indicative of decreased secretory protein. The necrotic acinar cells contain autophagic vacuoles and small lipid droplets

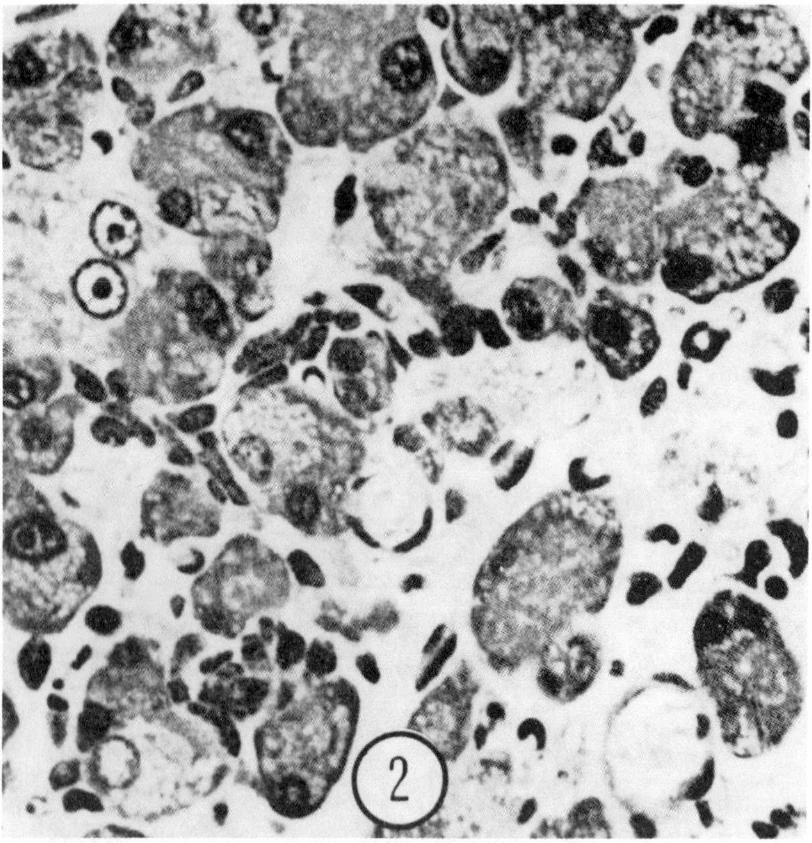


FIGURE 5. Pancreas of rat fed on the copper-deficient diet for 6 weeks. Haemalum and Basic Fuchsin × 530. (From Fell, B. F., King, T. P., and Davies, N. T., *Histochem. J.*, 14, 665, 1982. With permission.)

characteristic of degeneration. In summary, the ultrastructure of the acinar cells degenerates in ways consistent with an impaired capacity to synthesize and package digestive enzymes for secretion.

Histochemically, copper-deficient acinar cells have markedly decreased cytochrome oxidase activity after 6 weeks.[62] In contrast, the monoamine oxidase activity increases. The significance of these changes is unknown. Reduced cytochrome oxidase activity could impair energy production which could affect a number of other cellular functions, particularly protein synthesis which has a high energy demand.

The mechanism of this acinar atrophy and ultrastructure degeneration in copper deficiency is unknown. By implication, copper plays a critical role in the exocrine pancreas, but this role is also unknown. Clearly, future studies need to address the mechanism of copper-deficient acinar atrophy and the role of copper in the normal function of the acinar cell. It is also unclear whether the ultrastructural changes precede or follow the necrosis of the acinar cell. If the ultrastructural changes precede the necrosis and normal synthesis and packaging of digestive enzymes is perturbed, the acinar cell might be exposed to the hydrolytic actions of the digestive enzymes. Presently, it is unknown whether such autodigestion plays a role in the copper-deficient atrophy of acinar cells. The lack of an inflammatory reaction, which is known to occur in other cases of acinar autodigestion, however, suggests that the mechanism of this atrophy may differ from that of autodigestion.

### B. Exocrine Function

Pancreatic enzyme content changes progressively with copper deficiency. When acinar atrophy is complete, pancreatic content of digestive enzymes is severely decreased.[61,63] After 10 weeks of copper deficiency, amylase and trypsin content is 1.5% and 0.9% of control values.[61] Penicillamine-induced copper deficiency decreases amylase earlier (1% at 4 weeks and <6% at 6 weeks).[63]

Early dietary copper deficiency also alters enzyme content. After $5^1/_2$ weeks of deficiency, amylase content is significantly lower in copper-deficient rats (21% of control).[67] However, this early effect of copper deficiency is interactive with the type of dietary carbohydrate. Copper-deficient fructose-fed rats have lower amylase (5% of fructose-fed controls) than copper-deficient starch-fed rats (36% of starch-fed controls).[67] Dietary carbohydrate also has an independent effect on amylase content which is lower (18%) in fructose-fed controls than starch-fed controls. This carbohydrate effect supports the previous report that dietary starch induces a higher amylase content than does dietary fructose.[69] The interaction of dietary carbohydrate and copper deficiency is difficult to interpret, but fructose may act as a mild copper chelator. Although pancreatic copper content is similar in copper deficient and control fructose- and starch-fed rats,[67] hepatic copper content is significantly lower (22%) in fructose-fed controls than starch-fed controls despite the same copper content of the diets.[66] Pancreatic lipase content is significantly lower in copper-deficient rats fructose-fed (9% of control) and starch-fed (31% of control), but carbohydrate has no independent effect on pancreatic lipase content. Early copper deficiency tends to decrease the content of pancreatic chymotrypsinogen (63% of control) and trypsinogen (60% of control). As with amylase and lipase, these decreases tend to be greater in fructose-fed than starch-fed rats. There is a significant carbohydrate × copper interactive effect for both enzymes, but there is no independent effect of either copper or carbohydrate on either enzyme.

An intermediate period (7 weeks) of dietary copper deficiency resulted in profound and significant decreases in both amylase (0.2% of control) and lipase (3.5% of control; Table 3).[70] Copper deficiency was evident in the 52%

**TABLE 3**
**Effects of Copper Deficiency on Digestive Enzymes of the Rat Exocrine Pancreas after 7 Weeks[a]**

| | Control | Copper-deficient | *p* <[b] |
|---|---|---|---|
| N | 6 | 6 | |
| Body weight (g) | 310 ± 3 | 235 ± 6 | 0.002 |
| Total hepatic Cu (μg) | 32.2 ± 2.7 | 16.8 ± 2.9 | 0.001 |
| Pancreas | | | |
| Weight (g) | 1.05 ± 0.35 | 0.17 ± 0.09 | 0.001 |
| Total DNA (mg) | 0.53 ± 0.34 | 0.10 ± 0.03 | 0.02 |
| Total protein (mg) | 53.5 ± 17.0 | 3.2 ± 1.2 | 0.001 |
| Amylase (U/μg DNA) | 11.5 ± 8.4 | 0.03 ± 0.03 | 0.01 |
| Lipase (mU/μg DNA) | 255 ± 95 | 9 ± 3 | 0.001 |

[a] Weanling rats (average 48 g) were fed *ad libitum* a purified diet with (8 mg/kg diet) or without (0.85 mg/kg diet) Cu for 7 weeks and given deionized water. Both diets contained by weight: 20% casein, 0.3% DL-methionine, 65% glucose monohydrate, 5% fiber, 5% corn oil, 3.5% AIN mineral mix, 1.0% AIN vitamin mix, and 0.2% choline bitartrate. Rats were fasted overnight, anesthetized with ether and exsanguinated. Pancreas and liver were removed for analyses. Values represent the mean ± SEM.[71]

[b] Data were analyzed by the Student's *t*-test.

decrease in total hepatic copper content. Pancreatic weight, DNA, and protein were also significantly decreased, suggesting that acinar atrophy had occurred. This diet contained glucose as the major energy source (65% by weight) and is thus comparable to the starch diet fed in the previous study.[67] The difference in the effects of copper deficiency in a glucose- and starch-containing diet between these two studies may reflect the difference in the duration of copper deficiency studied, namely 5+ vs. 7 weeks. If so, these data imply that once the alterations of pancreatic enzyme content begin they proceed quickly.

The mechanisms of alterations of enzyme content are unknown, but have been suggested by Lewis and co-workers[67] to result from acinar atrophy. However, there may be other mechanisms involved especially during early and intermediate copper deficiency. Supporting a role of altered synthesis or secretion are the differential effects of early copper deficiency on pancreatic amylase (21% of control), lipase (20%), trypsinogen (60%) and chymotrypsinogen (60%).[67] If altered enzyme content results from acinar atrophy, the digestive enzymes should all decrease comparably. Also, mild copper deficiency in post-weanling rats results in a moderate (50%) decrease in amylase content with no pronounced pancreatic atrophy.[63] A complicating factor may be changes in gastrointestinal hormones in the copper-deficient rat that in part may result from altered digestion and subsequent hormonal response. Plasma CCK levels are 250% greater in penicillamine-induced copper-deficient rats than in controls.[71] CCK is known to regulate the mRNAs[72] and synthesis[73] of digestive enzymes and stimulates, in particular, the serine proteases. Finally, a number of the ultrastructural abnormalities in the acinar cell in copper deficiency are consistent with an impaired synthetic capacity, but it is unclear whether these ultrastructural changes follow or precede the acinar atrophy.

Secretory responsiveness is also altered in copper deficiency. Basal and stimulated secretion is reduced in advanced penicillamine-induced copper deficiency at 10 weeks.[61] Basal secretion in copper-deficient acini is only 5% of control secretion and secretagogue-stimulated secretion is markedly reduced for both CCK (0.05% of control) and secretin (0.3% of control). Mild dietary copper deficiency for 12 weeks decreases basal (11% of control) and CCK-stimulated (6%) secretion, but not secretin-stimulated secretion (79%).[63] Penicillamine- and Trien-induced copper deficiency results in severe reduction of basal (0.1% of control), CCK-stimulated (<0.006%), and secretin-stimulated (0.02%) secretion.[63] As with other pancreatic abnormalities, copper deficiency progressively alters secretory responsiveness. Trien-induced copper deficiency progressively reduces secretion at 4 (basal — 7%, secretin-stimulated — 7%, and CCK-stimulated — 8.5%) and 6 (basal — 0.05%, secreted-stimulated — 0.02%, and CCK-stimulated — 0.01%) weeks.

Functionally, secretion appears impaired in early copper deficiency as well. The content of digestive enzymes in the small intestinal lumen of copper deficiency after 5.5 weeks is decreased 58% for amylase, 53% for lipase, 25% for trypsin, and 7% for chymotrypsin.[66] These decreases in luminal activities are less than the decreases in pancreatic content reported by these same investigators and described above.[67] This discrepancy is not readily explainable, but may reflect other changes in the secretory process induced by copper deficiency.

### C. Reversibility and Transdifferentiation

Whether these pancreatic abnormalities in copper deficiency can be reversed by copper-repletion is not clear. In one study, morphological changes were reversed in rats treated with penicillamine for 4 weeks and then repleted with copper for 12 weeks.[74] A full recovery of acini following a 1-week period of high proliferation is reported. The fatty infiltration of the pancreas is not reversed; nor is the secretory impairment and decreased amylase content reversed. In contrast, all other studies[63-65] report no morphological or functional recovery following a 5- to 12-week repletion period after penicillamine or Trien treatment for 6 to 10 weeks. The difference among these studies may be that the morphological reverse seen in one study occurred after a 2-week shorter period of deficiency than in the other studies. Given the progressive atrophy that occurs in copper deficiency, there may be early stages when the effects of copper deficiency are reversible and later stages when the effects are irreversible.

An intriguing finding is the growth of hepatocytes during copper repletion of the atrophic pancreas.[64,65] When 85 to 90% of the acinar cells are destroyed, copper repletion results in multiple foci in the pancreas of hepatocytes that stain for albumin and peroxisomal catalase, which are specific to hepatocytes.[64] After 8 weeks of repletion, over 60% of the pancreas of repleted copper-deficient rats is hepatocytes.[65] These cells contain mRNAs for albumin whereas no cells in control pancreas express this liver-specific gene. The mechanism whereby hepatocytes differentiate and grow in the copper deficient-atrophic pancreas during copper repletion is unknown, but represents a unique model of the transdifferentiation of one differentiated cell type to another.

## IV. IMPLICATIONS FOR LIPID DIGESTION

The profound morphological and functional alterations of the exocrine pancreas in copper deficiency could adversely affect the digestion of all dietary lipids. While only pancreatic lipase activity has been measured and shown to decrease over 95% in copper deficiency, the decrease of other digestive enzymes and the severe atrophy of the acinar cells most likely results in decreases of other digestive enzymes including phospholipase $A_2$ and carboxyl ester lipase. Despite the normal excess secretion by the pancreas of lipolytic enzymes,[35] the severe decrease of lipase and the severely impaired secretion in copper deficiency can be expected to decrease duodenal lipolytic activities. Generally, a decrease of 90% of pancreatic enzymes can lead to pancreatic insufficiency and malabsorption. In early copper deficiency, doudenal lipase activity is decreased already by 60%, and this deficit can be expected to decrease considerably more as tissue content decreases 97% by 7 weeks of deficiency. Although no direct measures of lipid digestibility in copper deficiency have been made, the available data support the possibility of severely impaired digestion of triglyceride, cholesterol esters, and vitamin esters in pronounced copper deficiency. No mention of diarrhea in advanced copper deficiency is made in these studies, but only two studies to date[60,61] fed diets with high amounts of dietary lipids (>50% of energy intake). The other studies all used diets low in dietary lipid (20% of energy intake). It is unknown whether salivary function is also altered in copper deficiency. If copper plays a critical role in all secretory epithelial cells, salivary acini could also be impaired, thus eliminating the potential compensatory digestion of triglycerides by lingual lipase.

Paterson and Wormsley[75] propose that Shwachman's syndrome of severe pancreatic insufficiency with steatorrhea may result from infant copper deficiency. Their retrospective studies of Shwachman's syndrome and infant copper deficiency reveal similarities in anemia and skeletal abnormalities in both conditions. If the pancreatic atrophy induced by copper deficiency is irreversible, it could lead to a subsequent chronic pancreatic insufficiency. However, rigorous prospective, clinical and experimental studies need to test this proposal.

Many questions remain unanswered about the role of copper in the pancreatic acinar cell, specifically, and secretory epithelial cell, generally. The mechanism of the selective atrophy of acinar cells needs to be determined. Meanwhile, progressive changes in the availability of pancreatic lipases can be anticipated to impair dietary lipid digestion and utilization. Potentially, this impaired digestion and utilization could complicate advanced copper deficiency with impaired utilization of the fat-soluble vitamins.

## REFERENCES

1. **Friedman, H. I. and Nylund, B.,** Intestinal fat digestion, absorption and transport, *Am. J. Clin. Nutr.*, 33, 1108, 1980.
2. **Gordon, S. G. and Kern, F., Jr.,** The absorption of bile salt and fatty acid by hamster small intestine, *Biochim. Biophys. Acta,* 152, 372, 1968.
3. **Borgstrom, B.,** Luminal digestion of fats, in *The Exocrine Pancreas: Biology, Pathobiology and Diseases,* Go, V. L., Gardner, J. D., Brooks, F. P., Lebenthal, E., DiMagno, E., and Scheele, G. A., Eds., Raven Press, New York, 1986, 361.
4. **Borgstrom, B., Dahlquist, A., Lundh, G., and Sjovall, J.,** Studies on intestinal digestion and absorption in the human, *J. Clin. Invest.*, 36, 1521, 1957.
5. **Hamosh, M.,** Lingual lipase, in *Lipases,* Borgstrom, B. and Brockman, H. L., Eds., Elsevier, Amsterdam, 1984, 50.
6. **Sarda, L. and Desnuelle, P.,** Action de la lipase pancretique sur les ester en emulsion, *Biochim. Biophys. Acta,* 30, 513, 1958.
7. **Morgan, R. G. H. and Hoffman, N. E.,** The interaction of lipase cofactor and bile salts in triglyceride hydrolysis, *Biochim. Biophys. Acta,* 248, 143, 1971.
8. **Borgstrom, B. and Erlanson, C.,** Pancreatic lipase and colipase interactions and effects of bile salts and other detergents, *Eur. J. Biochem.*, 37, 60, 1973.
9. **Hofmann, A. F.,** Lipase, colipase, amphipathic dietary proteins and bile acids: new interactions at an old interface, *Gastroenterology,* 75, 530, 1975.
10. **Wilson, F. A., Sallee, V. L., and Dietschy, J. M.,** Unstirred water layer in the intestine: rate determinant for fatty acid absorption from micellar solutions, *Science,* 174, 1031, 1971.
11. **Westergaard, H. and Dietschy, J. M.,** The mechanism whereby bile acid micelles increase the rate of fatty acid and cholesterol uptake into the intestinal mucosal cell, *J. Clin. Invest.*, 58, 97, 1976.
12. **Borgstrom, B., Lundh, G., and Hofmann, A. F.,** The site of absorption of conjugated bile salts in man, *Gastroenterology,* 45, 229, 1963.
13. **Lucas, M. L., Schneider, W., Haberich, F. J., and Blair, J. A.,** Direct measurement by pH microelectrode of the pH microclimate in the rat proximal jejunum, *Proc. R. Soc. London Ser. B,* 192, 39, 1975.
14. **Shiau, Y. F. and Leine, G. M.,** pH dependence of micellar diffusion and dissociation, *Am. J. Physiol.*, 239, G177, 1980.
15. **Rudd, E. A. and Brockman, H. L.,** Pancreatic carboxyl ester lipase (cholesterol esterase), in *Lipases,* Borgstrom, B. and Brockman, H. L., Eds., Elsevier, Amsterdam, 1984, 185.

16. **Patton, J. S. and Cary, M. C.,** The formation of visible product phases by pancreatic lipase is described, *Science,* 204, 145, 1979.
17. **Johnston, J. M. and Borgstrom, B.,** The intestinal absorption and metabolism of micellar solution on lipids, *Biochim. Biophys. Acta,* 84, 412, 1964.
18. **Shiau, Y. F.,** Mechanisms of intestinal fat absorption, *Am. J. Physiol.,* 240, G1, 1981.
19. **Hamosh, M., Ganot, D., and Hamosh, P.,** Rat lingual lipase: characteristics of enzyme activity, *J. Biol. Chem.,* 254, 12121, 1979.
20. **Roberts, I. M., Montgomery, R. K., and Carey, M. C.,** Rat lingual lipase: partial purification, hydrolytic properties and comparison with pancreatic lipase, *Am. J. Physiol.,* 247, G385, 1984.
21. **Field, R. B. and Scow, R. D.,** Purification and characterization of rat lingual lipase, *J. Biol. Chem.,* 258, 14563, 1983.
22. **Paltauf, R., Esfandi, F., and Holsek, A.,** Stereospecificity of lipases. Enzymatic hydrolysis of enantiomeric alkyl diacylglycerols by lipoprotein lipase, lingual lipase and pancreatic lipase, *FEBS Lett.,* 40, 169, 1974.
23. **Szafran, Z., Szafran, H., Popiela, T., and Trompeter, G.,** Coupled secretion of gastric lipase and pepsin in man following pentagastric stimulation, *Digestion,* 18, 310, 1978.
24. **Barrowman, J. W. and Darnton, S. J.,** The lipase of rat gastric mucosa. A histochemical demonstration of the enzymatic activity against a medium chain triglyceride, *Gastroenterology,* 59, 213, 1978.
25. **Szafran, H., Szafran, Z., Poliela, T., and Wanick, A.,** Electrophoretic properties and specificity of human gastric lipase, *Enzyme,* 23, 187, 1978.
26. **Plucinski, T. M., Hamosh, M., and Hamosh, P.,** Fat digestion in the rat: role of lingual lipase, *Am. J. Physiol.,* 237, E541, 1979.
27. **Linthorst, J. M., Clark, S. B., and Holt, P. R.,** Triglyceride emulsification by amphipaths present in the intestinal lumen during fat digestion, *J. Colloid Interface Sci.,* 60, 1, 1977.
28. **DeCaro, A., Bonicel, J., Pieroni, G., and Gary, O.,** Comparative studies of human and porcine pancreatic lipases: N-terminal sequences, sulfhydryl groups and interfacial activity, *Biochimie,* 63, 799, 1981.
29. **DeCaro, A., Figarella, C., Amie, J., Micel, R., and Gary, O.,** Human pancreatic lipase: a glycoprotein, *Biochim. Biophys. Acta,* 490, 411, 1977.
30. **Verger, R.,** Pancreatic lipases, in *Lipases,* Borgstrom, B. and Brockman, H. L., Eds., Elsevier, Amsterdam, 1984, 84.
31. **Benzonana, G. and Desnuelle, P.,** Etude cinetique de l'action de la pancreatique sur des triglycerides en emulsion. Essais d'une enzymologie en milieu heterogene, *Biochim. Biophys. Acta,* 105, 121, 1965.
32. **Sternby, B. and Borgstrom, B.,** Purification and characterization of human colipase, *Biochim. Biophys. Acta,* 572, 235, 1979.
33. **Borgstrom, B. and Erlanson-Albertsson, C.,** Pancreatic colipase, in *Lipases,* Borgstrom, B. and Brockman, H. L., Eds., Elsevier, Amsterdam, 1984, 152.
34. **Wieloch, T., Borgstrom, B., Pieroni, G., Pattus, F., and Verger, R.,** Porcine pancreatic procolipase and its trypsin activated form: lipid binding and lipase activation on monomolecular films, *FEBS Lett.,* 128, 217, 1981.
35. **Borgstrom, B. and Hildebrand, H.,** Lipase and colipase activities of human small intestinal contents after a liquid test meal, *Scand. J. Gastroenterol.,* 10, 585, 1975.
36. **Mattson, F. H. and Voldenhein, R. A.,** Carboxylic ester hydrolases of rat pancreatic juice, *J. Lipid Res.,* 7, 536, 1966.
37. **Figarella, C., Clemente, F., and Gary, O.,** A zymogen of phospholipase A2 in human pancreatic juice, *Biochim. Biophys. Acta,* 227, 213, 1971.
38. **Peiken, S. R., Rottman, A. J., Batzri, S., and Gardner, J. D.,** Kinetics of amylase release by dispersed acini prepared from guinea pig pancreas, *Am. J. Physiol.,* 235, E743, 1978.

39. **Kern, H. F.,** Fine structure of human exocrine pancreas, in *The Exocrine Pancreas: Biology, Pathobiology and Diseases,* Go, V. L., Gardner, J. W., Brooks, F. P., Lebenthal, E., DiMagno, E., and Scheele, G. A., Eds., Raven Press, New York, 1986, 9.
40. **Werzel, G., Stocker, E., and Heine, W. D.,** Zur zellproliferation in exokrinen pankreas der ratte. Autoradiographische untersuchungen mit 3H-thymidine, *Virchows Arch. B,* 10, 118, 1972.
41. **Lutke, H., Kerne, H. F., and Scheele, G.,** Time course and cellular site of mitotic activity in rat exocrine pancreas during prolonged hormonal stimulation, *Dig. Dis. Sci.,* 29, 985, 1984.
42. **Pearson, K. W., Scott, D., and Torrance, B.,** Effect of partial surgical pancreatectomy in rats. I. Pancreatic regeneration, *Gastroenterology,* 72, 469, 1977.
43. **Elsasser, H. P., Adler, G., and Kern, H. F.,** Time kinetics and cellular source of pancreatic regeneration following acute pancreatitis in the rat., *Dig. Dis. Sci.,* 29, 947, 1984.
44. **Palade, G. E.,** Intracellular aspects of the process of protein synthesis, *Science,* 189, 347, 1975.
45. **Scheele, G.,** Cellular processing of proteins, in *The Exocrine Pancreas: Biology, Pathobiology and Diseases,* Go, V. L., Gardner, J. W., Brooks, F. P., Lebenthal, E., DiMagno, E., and Scheele, G. A., Eds., Raven Press, New York, 1986, 69.
46. **Devillers-Thiery, A., Kindt, T., Scheele, G., and Blobel, G.,** Homology in amino terminal sequences of precursors to pancreatic secretory proteins, *Proc. Natl. Acad. Sci. U.S.A.,* 72, 5016, 1975.
47. **Scheele, G., Dobberstein, B., and Blobel, G.,** Transfer of proteins across membranes. Biosynthesis in vitro of pretrypsinogen and trypsinogen by cell fractions of canine pancreas, *Eur. J. Biochem.,* 82, 593, 1978.
48. **Walter, P. and Lingappa, V. R.,** Mechanism of protein translocation across the endoplasmic reticulum membrane, *Annu. Rev. Cell Biol.,* 2, 499, 1986.
49. **Scheele, G. and Tartakoff, A.,** Exit of nonglycosylated secretory proteins from the rough endoplasmic reticulum is asynchronous in the exocrine pancreas, *J. Biol. Chem.,* 260, 926, 1985.
50. **Pfeffer, S. R. and Rothman, J. E.,** Biosynthetic protein transport and sorting by the endoplasmic reticulum and Golgi, *Annu. Rev. Biochem.,* 56, 829, 1987.
51. **Burgess, T. C. and Kelly, R. B.,** Constitutive and regulated secretion of proteins, *Annu. Rev. Cell Biol.,* 3, 243, 1987.
52. **Wandinger-Ness, A. and Simons, K.,** The polarized transport of surface proteins and lipids in epithelial cells, in *Intracellular Protein Trafficking,* Steere, and Hannover, Eds., Cambridge University Press, Cambridge, 1989, in press.
53. **von Figura, K. and Hasilik, A.,** Lysosomal enzymes and their receptors, *Annu. Rev. Biochem.,* 55, 167, 1986.
54. **Gardner, J. D. and Jensen, R. T.,** Receptors mediating the actions of secretagogues on pancreatic acinar cells, in *The Exocrine Pancreas: Biology, Pathobiology and Diseases,* Go, V. L., Gardner, J. D., Brooks, F. P., Lebenthal, E., DiMagno, E., and Scheele, G. A., Eds., Raven Press, New York, 1986, 109.
55. **Orchard, J. L., Davis, J. S., Larson, R. E., and Farese, R. U.,** Effects of carbachol and pancreozymin (cholecystokinin-octapeptide) on polyphosphoinositide metabolism in the rat pancreas in vitro, *Biochem. J.,* 217, 281, 1984.
56. **Williams, J. A. and Hootman, S. R.,** Stimulation-secretion coupling in pancreatic acinar cells, in *The Exocrine Pancreas: Biology, Pathobiology and Diseases,* Go, V. L., Gardner, J. D., Brooks, F. P., Lebenthal, E., DiMagno, E., and Scheele, G. A., Eds., Raven Press, New York, 1986, 123.
57. **Streb, H., Irvine, R. F., Berridge, M. J., and Schulz, I.,** Release of calcium from a non-mitochondrial intracellular store in pancreatic acinar cells by inositol-1,4,5 triphosphate, *Nature (London),* 306, 67, 1983.

58. **Burnham, P. B. and Williams, J. A.,** Activation of protein kinase activity in pancreatic acini by calcium and cyclic AMP, *Am. J. Physiol.*, 246, G500, 1984.
59. **Jensen, R. T. and Gardner, J. D.,** Cyclic nucleotide-dependent protein kinase activity in acinar cells from guinea pig pancreas, *Gastroenterology*, 75, 806, 1978.
60. **Muller, H. B.,** Der einfluβ kupferarmer kost auf das pankreas, *Virchows Arch. A*, 350, 353, 1970.
61. **Folsch, V. R. and Creutzfeldt, W.,** Pancreatic duct cells in rats: secretory studies in response to secretin, cholecystokinin-pancreozymin and gastrin in vivo, *Gastroenterology*, 73, 1953, 1977.
62. **Fell, B. F., King, T. P., and Davies, N. T.,** Pancreatic atrophy in copper-deficient rats: histochemical and ultrastructural evidence of a selective effect on acinar cells, *Histochem. J.*, 14, 665, 1982.
63. **Smith, P. A., Sunter, J. P., and Case, R. M.,** Progressive atrophy of pancreatic acinar tissue in rats fed copper-deficient diet supplemented with D-penicillamine or triethylene tetramine: morphological and physiological studies, *Digestion*, 23, 16, 1982.
64. **Rao, M. S., Subbarao, V., and Reddy, J. K.,** Induction of hepatocytes in the pancreas of copper-depleted rats following copper repletion, *Cell Differ.*, 18, 109, 1986.
65. **Rao, M. S., Diiedi, R. S., Subbarao, V., Usman, M. I., Scarpelli, D. G., Nemali, M. R., Yeldandi, A., Thangada, S., Kumar, S., and Reddy, J. K.,** Almost total conversion of the pancreas to liver in the adult rat: a reliable model to study transdifferentiation, *Biochem. Biophys. Res. Commun.*, 156, 131, 1988.
66. **Lewis, C. G., Fields, M., Craft, N., Yang, C. Y., and Reiser, S.,** Alteration of pancreatic enzyme activities in small intestine of rats fed a high fructose, low copper diet, *J. Nutr.*, 117, 1447, 1987.
67. **Lewis, C. G., Fields, M., Craft, N., Yang, C. Y., and Reiser, S.,** Changes in pancreatic enzyme specific activities of rats fed a high-fructose, low-copper diet, *J. Am. Coll. Nutr.*, 7, 27, 1988.
68. **Fell, B. F., Farmer, L. J., Farquharson, C., Bremner, I., and Garcia, D. S.,** Observations on the pancreas of cattle deficient in copper, *J. Comp. Pathol.*, 95, 573, 1985.
69. **Howard, F. and Yudkin, J.,** Effect of dietary change upon the amylase and trypsin activities of the rat pancreas, *Br. J. Nutr.*, 17, 281, 1963.
70. **Brannon, P. M., Sloan, D. S., Croswell, S., and Lei, K. Y.,** Differential effects of copper deficiency on the exocrine pancreas and parotid enzymes, *Nutr. Res.*, submitted, 1989.
71. **Folsch, U. R., Schafmayer, A., Ebert, R., Becker, H. D., and Creutzfeldt, W.,** Elevated plasma cholecystokinin concentrations in exocrine pancreatic atrophy in the rat, *Digestion*, 29, 60, 1984.
72. **Renaud, W., Giorgi, D., Iovanna, J., and Dagorn, J. C.,** Regulation of concentrations of mRNA for amylase, trypsinogen I and chymotrypsinogen B in rat pancreas by secretagogues, *Biochem. J.*, 235, 305, 1986.
73. **Schick, J., Kern, H. F., and Scheele, G.,** Hormonal stimulation in the exocrine pancreas results in coordinate and anticoordinate regulation of protein synthesis, *J. Cell Biol.*, 99, 1569, 1984.
74. **Sumii, T., Kimura, T., Ogami, Y., Yamaguchi, H., Kinjo, M., and Ibayashi, H.,** Changes in muscarinic receptors and amylase secretion in regenerating acinar cells of rat pancreas, *Digestion*, 40, 41, 1988.
75. **Paterson, C. R. and Wormsley, K. G.,** Hypothesis: Shwachman's syndrome of exocrine pancreatic insufficiency may be caused by neonatal copper deficiency, *Ann. Nutr. Metab.*, 32, 127, 1988.

Chapter 8

# COPPER AND LONG CHAIN FATTY ACID METABOLISM

**Stephen C. Cunnane**

## TABLE OF CONTENTS



*Abbreviations:* 16:0, palmitic acid; 16:1n-7, palmitoleic acid; 18:0, stearic acid; 18:1n-9, oleic acid; 20:3n-9, eicosatrienoic acid (n-9); 18:2n-9, octadecadienoic acid (n-9); 18:2n-6, linoleic acid; 20:3n-6, dihomo-gamma-linolenic acid; 20:4n-6, arachidonic acid; 18:3n-3, alpha-linolenic acid; 20:5n-3, eicosapentaenoic acid; 22:5n-3, docosapentaenoic acid (n-3); 22:6n-3, docosahexaenoic acid.

# I. INTRODUCTION

In the 1970s the link between copper and long chain fatty acids seemed quite clear; copper supplementation increased the amount of unsaturation in porcine back fat and was associated with enhanced fatty acyl desaturase activity in liver and adipose tissue. Evidence from studies of the copper-deficient rat showed substantial agreement with those done earlier in the pig. Since 1985, however, the picture has become more complicated. Some studies using the rat and evaluating fatty acid composition (particularly the ratio of stearic acid (18:0) to oleic acid (18:1n-9)) of whole organs have confirmed the earlier data; other studies of apparently equal merit but studying primarily subcellular fractions of organs such as liver have not given comparable results. In spite of diverse effects on the profiles of individual fatty acids, the degree of unsaturation of membrane fatty acids, especially those of 22 carbons, appears more or less consistently increased in copper deficiency. The mechanism of the effect of copper on the profile of long chain fatty acids is probably multifactorial including effects on fatty acid desaturation, esterification, and mobilization from triglyceride, as well as synthesis of phospholipids. There seems little doubt that the hypomyelination associated with genetic copper deficiency is partly a result of impaired synthesis and/or myelin incorporation of very long chain (>22 carbons).

# II. COPPER SUPPLEMENTATION

## A. Pig

Several studies have shown that copper supplementation at 250 mg/kg of diet results in softer back fat in pigs.[1-3] The increased softness of the back fat was attributed in part to increased Δ6 desaturation of oleoyl-CoA to linoleoyl-CoA (18:2n-9) and increased Δ9 desaturation of palmitoyl-CoA (16:0) to palmitoleoyl-CoA (16:1n-7) in hepatic and adipose tissue microsomes[3,4] (see Table 1). It was recognized that fatty acids within the back fat triglyceride may also have been redistributed thereby altering their melting point and hence the softness of the triglyceride in the fat.[1] It has been pointed out that the back fat triglycerides of the copper-supplemented pigs contained more of a fatty acid with 18 carbons and 2 double bonds which would have contributed to the increased softness of the back fat.[3] Desaturated derivatives of oleic acid, e.g., octadecadienoic acid (18:2n-9) as well as linoleic acid (18:2n-6), could both have contributed to the total "18:2" measured (the isomers were not identified). The 18:2n-9 could be synthesized by the pig in the presence of supplemental copper[3] but increased 18:2n-6 in the back fat triglycerides would, as an essential fatty acid which cannot be synthesized by mammals, have to have been consumed or mobilized from another body site.

**TABLE 1**
**Specific Activities (nmol/min/mg Microsomal Protein) of Long Chain Fatty Acyl Desaturases in Control and Copper-Supplemented Pigs[3] (Mean ± SD, n = 6)**

| | Control | Copper (+) |
|---|---|---|
| 18:1n-9 to 18:2n-9 | | |
| Liver | 0.31 ± 0.05 | 0.36 ± 0.07[a] |
| Fat | 0.41 ± 0.09 | 0.58 ± 0.12[a] |
| 16:0 to 16:1n-7 | | |
| Liver | 0.035 ± 0.009 | 0.058 ± 0.01[a] |
| Fat | 0.96 ± 0.17 | 1.15 ± 0.22[b] |

[a] $p < 0.01$.
[b] $p < 0.05$.

## B. Rat

The tissue fatty acid composition of copper-supplemented rats has been reported.[5] Addition of copper at 250 mg/kg to a semi-purified rat diet for 12 weeks resulted in changes in plasma and tissue fatty acid composition similar to those reported for the copper-supplemented pig, e.g., higher 16:1n-7 and 18:1n-9, and generally lower levels of polyunsaturated fatty acids in both phospholipids and triglycerides (Table 2). The increase in 16:1n-7 and 18:1n-9 was assumed to arise from Δ9 desaturation of their saturated precursors (16:0 and 18:0, respectively).

An important similarity between the fatty acid data from the copper-supplemented pig and rat is the presence of metabolites of 18:1n-9 in the tissue lipids in both species. In the pig, the presence of 18:2n-9 was surmised and its hepatic microsomal synthesis from radiolabeled 18:1n-9 was demonstrated *in vitro*.[3] In the copper-supplemented rat, 18:2n-9 was not detected but the longer chain derivative eicosatrienoic acid (20:3n-9) was detected at 0.2 to 0.5% of tissue and plasma lipids.[5]

The presence of 20:3n-9 has traditionally been associated with dietary deficiency of 18:2n-6 which allows the further metabolism of 18:1n-9 to 20:3n-9 (Figure 1). The metabolism of 18:1n-9 is usually prevented by the presence of adequate amounts of 18:2n-6 and, to a lesser extent by alpha-linolenic acid (18:3n-3). However, the stimulation by copper of the synthesis of 18:1n-9 appears to alter the competitive balance between 18:1n-9 and 18:2n-6 toward 18:1n-9 allowing the synthesis of small amounts of 20:3n-9 which is then detectable in the tissue lipids of the copper-supplemented rats (Table 2).

In the pig, copper supplementation at 250 mg/kg increases rate of body weight gain but this level of copper supplementation in the rat (100 times requirement) impairs growth rate.[5] Although it was not established, the slower

**TABLE 2**
**Profiles of 16:0, 16:1n-7, 18:0, 18:1n-9, and 20:3n-9 (mg%) in Plasma and Tissue Lipids of Control Rats (Ctl) or Rats Supplemented for 10 Weeks With 250 mg/kg Dietary Copper (Cu+)[5]**

**Phospholipids**

| | Plasma | | Liver | |
|---|---|---|---|---|
| | Ctl | Cu+ | Ctl | Cu+ |
| 16:0 | 27.9 | 27.1 | 19.0 | 22.1[a] |
| 16:1n-7 | 0.4 | 2.0[a] | 0.7 | 1.3[a] |
| 18:0 | 13.2 | 15.8[a] | 18.3 | 17.5 |
| 18:1n-9 | 6.6 | 14.2[a] | 7.8 | 8.8[a] |
| 20:3n-9 | ND[b] | 0.4[a] | ND[b] | 0.4[a] |

**Triglycerides**

| | Fat | | Liver | |
|---|---|---|---|---|
| | Ctl | Cu+ | Ctl | Cu+ |
| 16:0 | 17.5 | 25.6[a] | 26.2 | 32.6[a] |
| 16:1n-7 | 3.5 | 7.9[a] | 2.3 | 3.6[a] |
| 18:0 | 2.0 | 3.5[a] | 2.4 | 2.0 |
| 18:1n-9 | 30.2 | 38.3[a] | 29.7 | 39.7[a] |
| 20:3n-9 | ND[b] | ND[b] | ND[b] | 0.5[a] |

[a] $p < 0.05$ compared to Ctl.
[b] ND — trace (< 0.1 mg%).

growth rate in the copper-supplemented rats may have been due to an excessively imbalanced copper to zinc ratio which probably induced secondary zinc deficiency. Zinc deficiency affects tissue fatty acid composition in a manner analogous to that of copper supplementation (see Section VI).

## C. Brown Algae

The effect of increased copper on proportional fatty acid composition of membrane and storage lipids is not limited to mammals. Two members of the family of brown algae (*Fucus* spp.) have been shown to have markedly different rates of 14-C acetate incorporation into lipids as well as different endogenous fatty acid composition depending on whether they were growing in coastal waters of low (normal) copper content or in waters contaminated primarily with copper.[6] Both *F. vesiculosis* and *F. serratus* normally incorporate 14-C acetate primarily into 16:0 (55 to 60%) but in water in which they were exposed to 4 to 46 times as much copper due to waste from a local copper mine, they incorporated 30 to 50% less 14-C acetate into 16:0 and 30

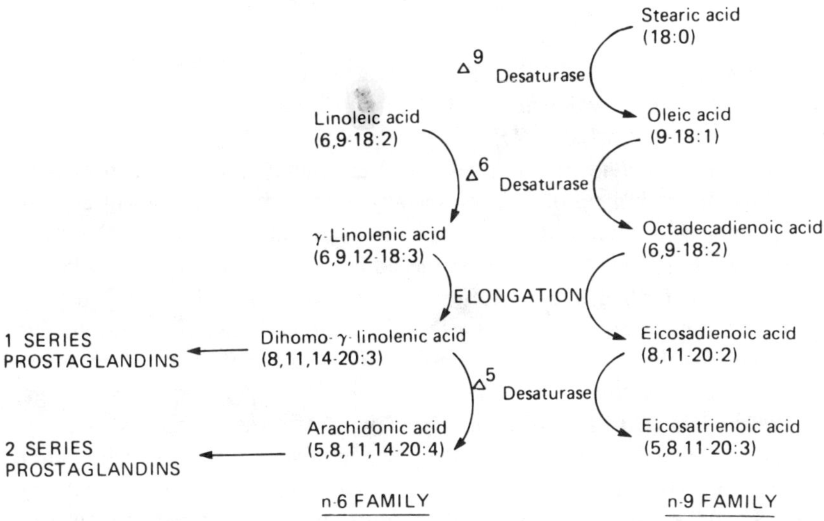


FIGURE 1. Effect of copper on long chain fatty acid metabolism. Fatty acids of the n-6 (linoleic) and n-9 (oleic) families are shown. Linoleic acid is metabolized to arachidonic acid through the Δ6 desaturase, elongase, and Δ5 desaturase. Oleic acid competes with linoleic acid for access to the same enzyme system. In the presence of adequate linoleic acid, oleic acid metabolism is limited; when linoleic acid is not present in adequate amounts, oleic acid is metabolized to eicosatrienoic acid (20:3n-9). Copper stimulates the conversion of stearic acid to oleic acid (Δ9 desaturation) as well as the further Δ6 and Δ5 desaturation of oleic acid to eicosatrienoic acid (n-9). Hence, octadecadienoic acid (n-9) and eicosatrienoic acid (n-9) are observed in copper-supplemented animals. With copper-enhanced metabolism of stearic and oleic acids there is a concomittant decrease in linoleic acid metabolism to arachidonic acid. The reverse (low oleic acid and higher 20 and 22 carbon fatty acids) is usually seen in copper deficiency but not necessarily in all cell types or under conditions of maximal copper depletion.

to 60% more 14-C acetate into 18:0 and 18:1n-9. The endogenous lipids of these brown algae contained significantly less 16:0 and n-3 polyunsaturated fatty acids and significantly more 18:1 and frequently more 18:2n-6 when they were obtained from copper-polluted water.[6] It cannot be stated with certainty that these effects on lipid composition and 14-C acetate incorporation in copper-polluted algae were uniquely due to copper but the overwhelming difference in their environment was a raised copper concentration in the water.

In view of the distinct similarity of the observations made by unrelated researchers studying effects of excess copper on lipid metabolism in the rat, pig, and brown algae, it seems clear that copper supplementation definitely affects the synthesis or metabolism of long chain fatty acids with much of the effect centering on synthesis or metabolism of 18:1n-9. Nevertheless, it is also clear that high levels of copper supplementation have been present in all these studies. The threshold at which copper affects 18:1n-9 metabolism has not been established.

## III. COPPER DEFICIENCY

### A. Animals

#### *1. Dietary Copper Deficiency*

##### *a. Whole Tissue Analysis—Rat*

Wahle and Davies[7] have investigated the relation between copper deficiency and metabolism of long chain fatty acids. Their analyses of adipose tissue fatty acid composition in copper-deficient rats showed that the effect of copper deficiency depended on the adipose tissue site studied; whereas subcutaneous fat had a 30 to 40% reduction in the ratio of 18:1n-9 to 18:0, this ratio in perirenal fat was not affected by copper intake. Hence, although their data for the subcutaneous fat site of the copper-deficient rats was consistent with the *increased* 18:1n-9 to 18:0 ratio in copper-*supplemented* pigs, it was evident that the effects of copper were affected by the location and possibly metabolic role of the triglycerides involved. Since 18:1n-9 is more readily oxidized than 18:0,[8] it is possible that copper deficiency selectively alters the rate of mobilization of 18:1n-9 from adipose tissue, thereby affecting its ratio to 18:0 in the fat.

Subsequent to the study of Wahle and Davies,[7] the effects of moderate copper deficiency on fatty acid in other organs besides fat and on the polyunsaturated fatty acids was published.[5] This report supported the previous research indicating that desaturation of 18:0 appeared impaired in copper deficiency. It also demonstrated that in liver and plasma phospholipids and in liver triglycerides, the fatty acid unsaturation index (roughly analogous to the percent composition of polyunsaturated fatty acids) was increased in copper deficiency. In the phospholipids of the heart, however, the increased unsaturation was limited to the n-3 fatty acids. Hence, copper deficiency appeared to allow increased synthesis of the fatty acids derived from 18:2n-6 and 18:3n-3 while suppressing the synthesis of the n-9 fatty acids (primarily 18:1n-9). That this was more or less specific for copper and not altered food intake (or weight gain) was suggested by the fact that copper supplementation (which suppressed weight gain in the rat) had exactly the opposite effect to moderate copper deficiency (no effect on weight gain) on the composition of the polyunsaturated fatty acids.[5]

At the same time, another paper reported the effects of copper deficiency in the rat on fatty acid composition of liver *total* lipids which were at variance with the previous data by suggesting that higher levels of 18:1n-9 were associated with copper deficiency.[9] Two differences with the study of Cunnane et al.[5] may have contributed to the different results. First, the level of copper depletion in the study of Singh and Medeiros[9] was greater (lower copper intake over a longer period of time) which resulted in decreased weight gain compared to the controls. Second, and of lesser importance, liver total lipids were used to determine the fatty acid profiles without separation of the lipid classes. While the rat and pig studies previously cited make it clear that the

### TABLE 3
### Profiles of the Essential Fatty Acids (mg/100 g) in Tissue Triglycerides of Control (Ctl) and Copper-Deficient (Cu−) Rats[10]

| | Liver | | Heart | | Fat (mg/g) | |
|---|---|---|---|---|---|---|
| | Ctl | Cu− | Ctl | Cu− | Ctl | Cu− |
| 18:2n-6 | 351 | 578[a] | 206 | 333 | 175 | 174 |
| 20:3n-6 | 6 | 7 | 2 | 4[a] | 0.4 | 0.5[a] |
| 20:4n-6 | 708 | 648 | 185 | 263[a] | 3.8 | 5.2[a] |
| 22:5n-6 | 170 | 247[a] | 81 | 99 | 1.4 | 2.2[a] |
| 18:3n-3 | 6 | 9 | 2 | 5 | 4.3 | 4.2 |
| 22:5n-3 | 4 | 8[a] | 1 | 3[a] | 0.1 | 0.1 |
| 22:6n-3 | 79 | 103[a] | 22 | 28[a] | 0.2 | 0.4 |

[a] $p < 0.05$.

effects of copper deficiency or supplementation on composition of 18:0 or 18:1n-9 occur similarly in liver phospholipids as triglycerides, severe copper deficiency causing reduced weight gain may not have the same effect on fatty acid composition as copper deficiency alone.

Medeiros' group[10] has elaborated further on their initial research on the effects of copper deficiency on fatty acid profiles and has shown that in comparison with pair-fed or *ad libitum*-fed controls, weanling rats made maximally copper deficient over 8 weeks have quantitatively higher amounts of 20 and 22 carbon fatty acids in tissue lipids, especially triglycerides of heart, liver, and adipose tissue[10] (Table 3). These results substantiate the initial work suggesting that long chain polyunsaturated fatty acids are increased in copper-deficient rats.[5,11] Therefore, copper appears to affect both the *metabolism* of the polyunsaturated fatty acids and their net esterification into triglycerides. In the adipose tissue from the copper-deficient rats, among fatty acids with <20 carbons, only 18:0 was significantly increased.[10] This is consistent with both its slow rate of oxidation during normal copper intake[8] as well as slower metabolism to 18:1n-9 caused by copper deficiency. This is corroborated by the observation that 18:0 is only slowly mobilized from adipose tissue into serum free fatty acids in comparison with other long chain fatty acids and this difference is enhanced by copper.[2]

*b. Whole Tissue Analyses—Mouse*

The weight gain of mice which have been fed a diet maximally depleted of copper for 7 weeks is substantially unaffected by the copper depletion in spite of a 20% reduction in hematocrit and 99% reduction in plasma ceruloplasmin activity.[12] The fatty acid composition of the phospholipids and triglycerides of several organs was measured and showed, with one exception, no consistent effect of copper deficiency either between the organs studied or between the mouse compared to the rat. The fatty acid unsaturation index was increased in the liver but not in the kidney, heart, or brain. The 18:0 to 18:1n-9 ratio was higher in phospholipids of liver and heart but not in kidney or brain.[12] The one change in fatty acid composition caused by dietary copper deficiency that was consistent between the rat and mouse in studies which have reported the levels of dihomo-gamma linolenic acid (20:3n-6) was a decrease in the amount of this fatty acid in all organs studied.[5,12,13]

Korte and Prohaska[13] have reported further on the fatty acid composition of the copper-depleted mouse. As in the previous report,[12] they showed that copper deficiency appeared to have no consistent effect on fatty acid composition of plasma membranes from different cell types. Membranes of erythrocyte ghosts and splenocytes from the copper-deficient mice contained an increased proportion of 18:1n-9 but thymocyte membranes contained lower 18:1n-9. Splenocyte phospholipids contained higher arachidonic acid (20:4n-6) and a higher index of fatty acid unsaturation but the opposite was true in erythrocytes and thymocytes. These results have been obtained from a well-defined model of dietary copper depletion from which purified plasma membranes of selected cell types have been studied yet have shown no consistent effect of copper depletion on the fatty acid composition or ratios among related fatty acids.[13]

In attempting to resolve the substantially different effects of copper deficiency observed by different researchers, the issue of *degree* of copper depletion becomes apparent as a major difference between the study of Cunnane et al.[5] on the one hand and those of Singh and Medeiros,[9] Cunnane et al.,[12] and Korte and Prohaska[13] on the other. Cunnane et al.[5] studied rats which were made moderately copper deficient but which were not growth-retarded over the 10 weeks of the study. They obtained results which supported a role of copper in the synthesis and metabolism of 18:1n-9 and which were compatible with those obtained by others studying copper supplementation in both pig and the rat. Compatibility between the effects of copper deficiency and supplementation is important in evaluating whether copper is directly or indirectly linked to the metabolism of long chain fatty acids. When the effects of copper deficiency and supplementation are not compatible, it seems likely that other factors, e.g., food intake or qualitative aspects of the diet, ultimately determine the fatty acid composition of the membranes or tissues studied.

Singh and Medeiros[9] and Korte and Prohaska[13] have studied relatively severe dietary copper deficiency in which copper depletion was quite evident from biochemical indices of copper status as well as body weights. Their studies produced results suggesting no obvious connection between copper and 18:1n-9 metabolism. However, food intake and normal body weight gain are important determinants of the activity of the Δ9 desaturase which controls the synthesis of 18:1n-9 (see Section IV).

*c. Subcellular Analyses*

Gallagher and Reeve[11] analyzed the fatty acid composition of liver mi-

**TABLE 4**
**Essential Fatty Acid Profiles (mg%) of Mitochondrial Lipids from Control (Ctl) and Copper-Deficient (Cu−) Rats; A Comparison of Three Studies**

| | Gallagher and Reeve[11] | | Balevska et al.[14,a] | | Lawrence et al.[15] | |
|---|---|---|---|---|---|---|
| | Ctl | Cu− | Ctl | Cu− | Ctl | Cu− |
| 18:2n-6 | 8.3 | 11.7[b] | 11.8 | 15.2[b] | 14.2 | 13.3 |
| 20:4n-6 | 11.0 | 12.6 | 18.0 | 15.0 | 25.5 | 23.4[b] |
| 22:4n-6 | NR[c] | | NR | NR | 8.7 | 11.3 |
| 22:5[d] | NR | | 0.8 | 0.6 | 2.4 | 3.3[b] |
| 22:6n-3 | 11.6 | 13.7[b] | 6.9 | 9.8[b] | 3.5 | 3.1 |

[a] Mitochondrial phospholipids.
[b] $p < 0.05$ compared to Ctl.
[c] NR, not reported.
[d] n-6 and n-3 isomers not distinguished.

tochondria from copper-deficient rats fed a milk-based diet and showed that levels of 18:1n-9 were significantly lower but that total polyunsaturated fatty acids were higher in the mitochondria from the copper-deficient rats (see Table 4 (Gallagher and Reeve[11])). They attributed the inability of liver mitochondria from copper-deficient rats to appropriately bind and transport ATP to the altered fatty acid composition. Balevska et al.[14] also used a milk-based diet to induce copper deficiency in rats over a 4-week period. They reported that the fatty acid composition of liver microsomal and mitochondrial membranes had higher 16:1n-7 and docosahexaenoic acid (22:6n-3), lower 18:2n-6 and 20:4n-6, but no change in 18:0 or 18:1n-9 (Table 4 (Balevska et al.[14])). The only effect on mitochondrial fatty acids which was consistent with that of Gallagher and Reeve[11] (who also used rats fed a milk-based diet but whose rats were more copper deficient) was the increased 22:6n-3.

Liver mitochondrial composition and function in copper-deficient rats has also been assessed by Lawrence et al.[15] Copper deficiency (8 weeks) was shown to cause a reduction in the total amount and activity of cytochrome c oxidase. Swelling properties of isolated liver mitochondria from control or copper-deficient rats did not differ and this was considered to be related to the fact that the mitochondrial fatty acid composition varied only marginally between the two groups (Table 4 (Lawrence et al.[15])). Hence, three of four studies of rat liver mitochondria have shown that the level of 22:6n-3 is elevated in the copper-deficient group. This effect is consistent with the whole organ data from copper-deficient rats.[5,8] However, neither the 18:0 to 18:1n-9 ratio nor the total fatty acid unsaturation index appears to have been consistently affected by copper depletion in any of these studies.

**TABLE 5**
**Fatty Acid Profiles (mg%) of Liver Triglycerides of Genetically Copper-Deficient Mice (Brindled)**

| | Brindled | Control |
|---|---|---|
| 16:0 | 29.4 ± 4.2 | 36.3 ± 1.7[a] |
| 18:0 | 5.5 ± 2.6 | 3.1 ± 0.8[a] |
| 18:1n-9 | 26.9 ± 5.3 | 32.4 ± 2.0[a] |
| 18:2n-6 | 17.7 ± 4.7 | 17.0 ± 2.0 |
| 20:4n-6 | 2.6 ± 0.7 | 1.7 ± 0.8 |
| 22:5n-3 | 11.6 ± 5.5 | 6.1 ± 2.1[a] |
| 22:6n-3 | 1.2 ± 0.8 | 0.9 ± 0.3 |

[a] $p < 0.01$.

The fatty acid unsaturation index (double bond index) is an antiquated term of debatable value as an index of composition because it refers to long chain fatty acids from 18 to 22 carbons in chain length and from 1 to 6 double bonds in unsaturation. These fatty acids are synthesized by different enzymes and presumably have different functions in membranes. If one generality can be made, it is that those of 22 carbon chain length probably have more in common and can replace each other functionally in membranes more readily than can those of 18 carbons (18:0, 18:1n-9, and 18:2n-6). The 20 carbon fatty acids (primarily 20:3n-6 and 20:4n-6) are intermediary both in function and, presumably, replaceability. Hence, if use is to be made of the fatty acid unsaturation index, perhaps the 22 carbon fatty acids could be referred to specifically. In the context of copper deficiency, this would help in the search for similarities between studies which otherwise appear totally inconsistent. In fact, most studies whether in the rat or mouse, or studying the fatty acid composition of whole organs or subcellular organelles, have shown that the unsaturation index of 22 carbon fatty acids is more or less consistently increased in copper deficiency (Tables 3 through 6).

The fatty acid composition of hepatic plasma membranes from copper-deficient rats has also been studied. The most notable difference in the phospholipid fatty acid composition of the membranes from copper-deficient rats was a 27% decrease in 20:4n-6.[16] 22:6n-3 was increased 2.5 times (not significant) and the unsaturation index of 22 carbon fatty acids was significantly increased in the copper-deficient rats. The increase in the 22 carbon fatty acids balanced the decrease in 20:4n-6 with respect to the change in total fatty acid unsaturation but the fluoresence of a membrane probe indicated decreased membrane fluidity in the copper-deficient group. Whether 20:4n-6 content of plasma membranes is a more important determinant of membrane fluidity than the 22 carbon fatty acids, especially in copper deficiency, remains to be established.

**TABLE 6**
**Fatty Acid Profiles (mg%) in White Matter of Brain in Menke's Disease (MD) Compared to Healthy Controls (Ctl)[19]**

| | PE | | PC | | PS | | SM | |
|---|---|---|---|---|---|---|---|---|
| | MD | Ctl | MD | Ctl | MD | Ctl | MD | Ctl |
| 16:0 | 8 | 5 | 45 | 34 | 1 | 5 | 5 | 3 |
| 16:1n-7 | 1 | 1 | 3 | 3 | tr | tr | tr | tr |
| 18:0 | 35 | 12 | 14 | 15 | 48 | 45 | 63 | 35 |
| 18:1n-9 | 14 | 30 | 26 | 40 | 21 | 31 | | |
| 23:0 to 26:0 + 23:1 to 26:1 | | | | | | | 29 | 59 |
| 20:4n-6 | 14 | 12 | 7 | 2 | 2 | 2 | | |
| 22:4n-6 | 8 | 19 | 1 | 1 | 5 | 4 | | |
| 22:5[a] | 3 | 3 | | | 4 | 2 | | |
| 22:6n-3 | 14 | 8 | 1 | 1 | 14 | 4 | | |

[a] Isomer not specified.

### 2. *Genetic Copper Deficiency*

One reason why the dietary models of copper deficiency have yielded inconsistent data regarding changes in fatty acid profiles may be because that they are not homogeneous models, i.e., many factors may alter the eventual effects of dietary copper depletion. It was therefore considered that a genetic model of copper depletion might provide valuable data or at least a further perspective on how long chain fatty acid metabolism is affected by copper depletion. One such study has been conducted using the "brindled" mouse as the model of genetic copper depletion. Unless cross-fostered to a healthy lactating dam, this mutant does not survive weaning. It expresses copper deficiency in the form of cardiac hypertrophy, low liver copper, and low serum ceruloplasmin levels. The fatty acid composition of phospholipids of liver, heart, and kidney is only different from that of healthy controls to a minor extent. As with the dietary model of murine copper deficiency, the fatty acid profiles of tissue phospholipids appear to be relatively unaffected in the brindled mouse.[5] However, the liver triglycerides have significantly lower 16:0 and 18:1n-9, and higher 18:0, 20:4n-6, and docosapentaenoic acid (22:5n-3) (Table 5). This data is in agreement with the generalized increase in proportion of 22 carbon fatty acids in copper deficiency. It also supports a probable inhibitory effect of copper deficiency on the metabolism of 18:0 to 18:1n-9.

## B. Humans

The defect in copper metabolism in Menkes' Disease is related to both its absorption and transport into various cell types.[17] Except for a single report of low 22:6n-3 in brain phospholipids,[18] no detailed analyses of lipids or fatty acids have been reported for Menkes' Disease until recently. Hara and

Taketomi[19] have shown that myelin lipids were significantly lower in the brain in Menkes' Disease than in controls. Long chain fatty acids in phospholipids of grey matter were more or less normal but in white matter, marked differences with controls were described (Table 6), including higher 18:0 and 22 carbon fatty acids. Oleic acid and long chain monounsaturated fatty acids were significantly lower in Menkes' patients. These differences in brain fatty acid profiles in Menkes' Disease are consistent with those seen in subtotal copper depletion in animals. The 60 to 70% lower amounts of 24:0, 24:1, 25:0, 25:1, 26:0, and 26:1 in sphingomyelin (all derived from 18:0 and 18:1n-9) strongly suggests that copper deficiency impairs the metabolism of the latter fatty acids which, in turn, contributes to impaired myelin synthesis in the brain in Menkes' Disease.[19]

The effects of experimental copper deficiency on serum fatty acid profiles in humans recently have been investigated.[20] Over the course of 11 weeks during which daily copper intake was reduced from 1.4 to 1.0 mg/d, serum phospholipids were shown to contain 27% less 18:1n-9, 38% more 20:4n-6, and 94% more 22:6n-3. These effects, particularly the increased 22:6n-3, are consistent with the majority of fatty acid data from animals experimentally depleted of copper. The fact that higher serum levels of 22:6n-3 are present in animals in which arrhythmias have been experimentally induced[21] and in humans experiencing angina or myocardial infarction[22] suggests that the changes in 22:6n-3 which occur in copper deficiency may be of relevance to the occurrence of arrhythmias in humans[23] and cardiac arrest in rats.[24]

## IV. COPPER AND THE FATTY ACYL DESATURASE SYSTEM

Copper supplementation has been established as increasing the proportion of 18:1n-9 and 18:2 (mixed isomers) in subcutaneous fat of the pig.[3,4] These and earlier studies of the effects of copper on the softness of the back fat of pigs provided the basis on which the effects of copper on synthesis and metabolism of 18:1n-9 have been investigated. Ho and Elliot[3] have shown that, *in vitro*, the synthesis of 16:1n-7 from 16:0 and 18:2n-9 from 18:1n-9 are both increased in pigs fed supplemental copper. Thompson et al.[4] have clearly demonstrated that copper modulates the hepatic desaturation of 18:0 and that in addition to affecting the initial velocity of the enzyme, it may also affect its stability. These reactions occur in the "fatty acyl desaturase system" which is present in the microsomal fraction of many cell types. This system consists of electron transport from a donor (primarily NADH but also NADPH to a lesser extent) through flavoproteins and cytochrome b5 to the terminal desaturase protein itself which is also known as the cyanide-sensitive factor.[25,26]

Among divalent cations, iron has a central role in the activity of the terminal desaturase.[27] In *Mycobacterium phlei* grown in metal-deficient media, 0.2 to 0.3 m*M* iron is required to attain full stearoyl (Δ9) desaturase activity. Nevertheless, 25 to 100 μ*M* iron is inhibitory when added to the growth medium *in vitro*.[26] Copper has a similar dual effect on the activity of the stearoyl desaturase; in *M. phlei* grown in a cation-deficient medium, 1 to 10 μ*M* copper increases desaturase activity but at higher concentrations (<100 μ*M*) copper no longer increases the desaturase activity.[26] Similarly, in microsomal fractions from copper-deficient rats, stearoyl desaturase activity is 65% less than in *ad libitum*-fed controls and 46% less than in pair-fed controls. Copper added to the diet or injected into copper-deficient rats prior to assaying the Δ9 desaturase activity caused a significant increase in desaturase activity.

Nevertheless, copper added *in vitro* has no effect on (at 30 nmol/mg protein) or inhibited (>200 nmol/mg protein) rat liver microsomal desaturase activity.[7] Copper when added to an *in vitro* chick liver microsomal system at 10 μ*M* to 1 m*M* has been shown to inhibit stearoyl-CoA desaturase activity by 20 to 80% but effects of lower copper concentrations were not reported.[28] This inhibitory effect on desaturase activity by copper could not be prevented by EDTA. However, the copper had to be present in the *in vitro* microsomal system to have the inhibitory effect; its removal prior to initiating the desaturation sequence prevented its inhibitory effect. These results are consistent with those of Wahle and Davies[7] and Kashiwabara et al.[26] who have also shown the inhibitory effect of copper on microsomal desaturation when it is added *in vitro*.

It is therefore evident that the terminal desaturase protein itself is sensitive to the presence of copper; in the absence of copper the desaturation process is inhibited as it is in the presence of excess copper. Iron but not copper is present in stoichiometric amounts in the desaturase protein.[27] Copper's effect therefore appears to be one of modulating access of either electrons or molecular oxygen to the desaturase protein. Copper may also affect the synthesis or the stability of the desaturase protein which appears to be the main mechanism by which its activity is regulated.[29] Whether in fact the mechanism by which the stimulation by copper and inhibition by copper deficiency is the same awaits further study. Also awaiting further study is the evaluation of the activity of other desaturases besides the Δ9 desaturase. The synthesis of 20:4n-6 and 22:6n-3 depend on the Δ6, Δ5, and Δ4 desaturases and, at present, we have no information concerning the effect of copper on their activity.

## V. COPPER AND PHOSPHOLIPIDS

The presence of a fatty acid profile in liver total phospholipids which is altered, for example, by decreased copper intake generally can be attributed to one of two causes. The first and most common explanation is that the metabolism of the fatty acids is altered such that the amount of a product fatty acid may be increased or decreased to the extent that the change in flux through that pathway affects the acylation of that fatty acid into membrane phospholipids resulting in a change in the fatty acid profile of the phospholipids of the tissue. The second major cause of a change in the fatty acid profile of total phospholipids is a change in the relative contribution of an individual phospholipid class to the total. Since individual phospholipid classes have markedly different fatty acid profiles, this, too, can be a determinant of the fatty acid profile of the total phospholipid pool. This is relevant to understanding how copper may alter fatty acid profiles in tissues because both mechanisms probably apply; 18:1n-9 may be increased by the presence of a higher copper intake through increased desaturation of 18:0. 22:6n-3 may, in turn, be increased in microsomal phospholipids in copper deficiency through a change in proportional phospholipid composition. We should not expect one mechanism to explain all the changes which may be induced by a change in copper intake.

Gallagher and Reeve[30] have shown that copper deficiency impairs the synthesis of phospholipid in liver mitochondria but not in liver microsomes. This difference was shown to be due to the inability of copper-deficient mitochondria to adequately bind ATP which is required for phospholipid synthesis. They showed that although total phospholipid content of whole liver or whole brain was not significantly altered by copper deficiency, liver mitochondrial phospholipid content was 22% lower than in control mitochondria.[30] In their study, the proportional distribution of phospholipid classes was not affected by copper deficiency. Subsequently, however, it has been shown that the brains of 25-day-old mice from dams which had been fed for 9 weeks on a copper-deficient diet had altered brain lipid composition including lower cholesterol, sulfatides and galactocerebrosides, and 39% higher phosphatidylcholine.[31] These findings occurred in conjunction with constant tremor and abnormal myelination. Zimmerman et al.[32] have shown that after raising three consecutive generations of rats on a copper-deficient diet, the third generation had lower cholesterol but significantly higher phosphatidylserine plus phosphatidylinositol in the myelin of the whole brain. Total brain myelin yield in the copper-deficient rats was only 34% of that in the controls. In genetically copper-deficient mice, total lipid/brain has been shown not to be different from controls but based on the dry weight of each brain, total phospholipid was significantly higher in the copper-deficient mice. This increase was accounted for by higher amounts of sphingomyelin and phosphatidylserine.[33]

In none of these studies reporting different proportional composition of phospholipids in copper deficiency has phospholipid fatty acid composition been described. Nevertheless, phosphatidylserine is a fraction enriched in 18:0 and 22:6n-3. Hence, a significant increase in phosphatidylserine could alter the contribution of 18:0 and 22:6n-3 to the fatty acid profiles of total phospholipids. This is borne out by the fact that in Menke's Disease, total brain phospholipids are decreased but this decrease is due mainly to lower phosphatidylcholine.[19] Since phosphatidylethanolamine and phosphatidylserine contain high amounts of 18:0 and 22:6n-3 and sphingomyelin also contains high 18:0, the increased proportion of these two fatty acids in fatty acid profiles of *total* phospholipids may be, in part, due to lower phosphatidylcholine in copper deficiency.

## VI. COPPER-ZINC INTERACTIONS OF RELEVANCE TO LONG CHAIN FATTY ACIDS

Based on the research which showed an effect of copper on the metabolism of 18:0 and 18:1n-9 and that which showed that zinc interacted with the metabolism of n-6 essential fatty acids,[34] it was hypothesized that the balance of zinc to copper intake would be an important determinant of the degree to which these competing families of fatty acids would be metabolized[35,36] (Figure 1). Studies involving feeding rats diets partially depleted of copper or zinc or supplemented with zinc have shown the remarkable similarity in plasma and liver fatty acid profiles of the depletion of copper and excess of zinc or depletion of zinc and excess of copper.[5,36,37]

These observations may have a direct application in Wilson's disease which involves a genetic defect in copper metabolism resulting in copper accumulation in the liver, liver necrosis and, if untreated, neurological abnormalities and death. This disease has been classically treated with chelators to remove the excess copper and to minimize copper absorption. Over the past 10 years, two groups have demonstrated that oral zinc supplementation can be the treatment of choice because of lower toxicity and equal or greater efficacy compared to chelators.[39,40] It is known that copper tends to destabilize hepatocyte membranes and that zinc competitively inhibits this effect.[35] The toxic effects of excess copper on hepatocytes and their inhibition by zinc may also involve a perturbation of the phospholipid fatty acid profiles of hepatocyte membranes but this remains to be established.

## VII. CONCLUSION

Research on copper's effect on the composition of long chain fatty acids shows a number of intriguing effects. First, it appears quite clear from experimental and clinical data and from studies of both copper supplementation and deficiency that copper modulates the activity of the $\Delta 9$ desaturase although all studies reporting fatty acid profiles (especially of membranes) do not fully agree on this point. Second, although the fatty acid unsaturation index is usually higher in organs of copper-deficient rats and mice, there is as yet no other direct or indirect evidence that copper affects the desaturases controlling essential fatty acid metabolism. That some of its effects on fatty acid ratios and desaturation may occur through metabolic competition with zinc appears to be a distinct possibility. Third, the increased proportion of 20 and 22 carbon fatty acids is most consistently seen in the triglyceride fraction of tissue lipids.

This may occur as a result of greater mobilization of 16 and 18 carbon fatty acids from triglycerides in copper deficiency or may be the result of selective synthesis of triglycerides with specific esterification of the 20 and 22 carbon fatty acids. Fourth, whether the differential effect of copper on composition of free fatty acids in serum is of physiological significance awaits further research.

Changes in fatty acid and lipid metabolism induced by copper are of immediate relevance to cardiovascular disease particularly in the context of cholesterol metabolism, atherosclerosis, and cardiac arrhythmias. Menke's disease has provided some of the clearest evidence of the vital role of copper in conditions involving dys- or hypomyelinization. The importance of adequate copper for myelinization and acquisition of long chain fatty acids by the brain during early development should also be an area of fruitful research.

## ACKNOWLEDGMENT

The Natural Sciences and Engineering Research Council of Canada is thanked for financial support of the author's research.

## REFERENCES

1. **Christie, W. W. and Moore, J. H.,** The effect of dietary copper on the structure and physical properties of adipose tissue triglycerides in pigs, *Lipids,* 4, 345, 1969.
2. **Myres, A. W. and Bowland, J. P.,** Effects of environmental temperature and dietary copper on growth and lipid metabolism in pigs. III. Concentration and composition of plasma free fatty acids, *Can. J. Anim. Sci.*, 53, 127, 1973.
3. **Ho, S. K. and Elliot, J. I.,** Fatty acid composition of porcine depot fat as related to the effect of supplemental dietary copper on the specific activities of fatty acyl desaturase systems, *Can. J. Anim. Sci.*, 54, 23, 1974.
4. **Thompson, E. H., Allen, C. E., and Meade, R. J.,** Influence of copper on stearic acid desaturation and fatty acid composition in the pig, *J. Anim. Sci.*, 36, 868, 1973.
5. **Cunnane, S. C., Horrobin, D. F., and Manku, M. S.,** Contrasting effects of low or high copper intake on rat tissue lipid essential fatty acid composition, *Ann. Nutr. Metab.*, 29, 103, 1985.
6. **Smith, K. L., Bryan, G. W., and Harwood, J. L.,** Changes in endogenous fatty acids and lipid synthesis associated with copper pollution in *Fucus* spp., *J. Exp. Bot.*, 36, 663, 1985.
7. **Wahle, K. W. J. and Davies, N. T.,** Effects of dietary copper deficiency in the rat on fatty acid composition of adipose tissue and desaturase activity of liver microsomes, *Br. J. Nutr.*, 34, 105, 1975.
8. **Leyton, J., Drury, P. J., and Crawford, M. A.,** Differential oxidation of essential and non-essential fatty acids in vivo in the rat, *Br. J. Nutr.*, 57, 383, 1987.
9. **Singh, N. P. and Medeiros, D. M.,** Effect of copper deficiency and sodium intake upon liver lipid and mineral composition in the rat, *Biol. Trace Elem. Res.*, 6, 423, 1984.

10. **Ovecka, G. D., Miller, G., and Medeiros, D. M.,** Fatty acids of liver, cardiac and adipose tissues from copper deficient rats, *J. Nutr.*, 118, 480, 1988.
11. **Gallagher, C. H. and Reeve, V. E.,** Copper deficiency in the rat. Effect on liver and brain lipids, *Aust. J. Exp. Biol. Med. Sci.*, 49, 453, 1971.
12. **Cunnane, S. C., McAdoo, K. R., and Prohaska, J. R.,** Lipid and fatty acid composition of organs from copper deficient mice, *J. Nutr.*, 116, 1248, 1986.
13. **Korte, J. J. and Prohaska, J. R.,** Dietary copper deficiency alters protein and lipid composition of murine lymphocyte plasma membranes, *J. Nutr.*, 117, 1076, 1987.
14. **Balevska, P. S., Russanov, E. M., and Stanchev, P. I.,** Copper deficiency-induced changes in the fatty acid composition of mitochondrial and microsomal membranes of rat liver, *Biol. Trace Elem. Res.*, 8, 211, 1985.
15. **Lawrence, C. B., Davies, N. T., Mills, C. F., and Nicol, F.,** Studies on the effects of copper deficiency on rat liver mitochondria. I. Changes in mitochondrial composition, *Biochim. Biophys. Acta,* 809, 351, 1985.
16. **Lei, K. Y., Rosenstein, F., Shi, F., Hassel, C. A., Carr, T. P., and Zhang, J.,** Alterations in lipid composition and fluidity of liver plasma membranes in copper deficient rats, *Proc. Soc. Exp. Biol. Med.*, 188, 335, 1988.
17. **Danks, D. M., Campbell, P. E., Stevens, B. J., Mayne, V., and Cartwright, E.,** Menkes' kinky hair syndrome. An inherited defect in copper absorption with widespread effects, *Pediatrics,* 50, 188, 1972.
18. **O'Brien, J. S. and Sampson, E. L.,** Kinky Hair Disease. II. Biochemical studies, *J. Neuropathol. Exp. Neurol.*, 25, 523, 1966.
19. **Hara, A. and Taketomi, T.,** Cerebral lipid and protein abnormalities in Menke's Steely Hair Disease, *Jpn. J. Exp. Med.*, 56, 277, 1986.
20. **Cunnane, S. C.,** The profile of long chain fatty acids in serum phospholipids. A possible indicator of copper status in humans, *Am. J. Clin. Nutr.*, 48, 1475, 1988.
21. **Gudbjarnason, S., Doell, B., and Oskarsdottir, G.,** Docosahexaenoic acid in cardiac metabolism and function, *Acta Biol. Med. Ger.*, 37, 777, 1978.
22. **Skuladottir, G., Hardarson, T., Sigfusson, N., Oddsson, G., and Gudbjarnason, S.,** Arachidonic acid levels in serum phospholipids of patients with angina pectoris or fatal myocardial infarction, *Acta Med. Scand.*, 218, 55, 1985.
23. **Reiser, S., Smith, J. C., Jr., Mertz, W., Holbrook, J. T., Scholfield, D. J., Powell, A. S., Canfield, W. K., and Canary, J. J.,** Indices of copper status in humans consuming a typical American diet containing either fructose or starch, *Am. J. Clin. Nutr.*, 42, 242, 1985.
24. **Klevay, L. M.,** Coronary heart disease. The zinc/copper hypothesis, *Am. J. Clin. Nutr.*, 28, 764, 1975.
25. **Oshino, N. and Sato, R.,** The dietary control of the microsomal stearoyl-CoA desaturation enzyme system in rat liver, *Arch. Biochem. Biophys.*, 149, 369, 1972.
26. **Kashiwabara, Y., Nakagawa, H., Matsuki, G., and Sato, R.,** Effects of metal ions in the culture medium on the stearoyl-coenzyme A desaturase activity of *Mycobacterium phlei, J. Biochem.*, 78, 803, 1975.
27. **Strittmatter, P., Spatz, L., Corcoran, D., Rogers, M. J., Setlow, B., and Redline, R.,** Purification and properties of rat liver microsomal stearoyl coenzyme A desaturase, *Proc. Natl. Acad. Sci. U.S.A.*, 71, 4565, 1974.
28. **Shreekrishna, K. and Joshi, V. C.,** Inhibition of the microsomal stearoyl coenzyme A desaturation by divalent copper and its chelates, *Biochim. Biophys. Acta,* 619, 267, 1980.
29. **Jeffcoat, R.,** The biosynthesis of unsaturated fatty acids and its control in mammalian liver, *Essay Biochem.*, 15, 1, 1979.
30. **Gallagher, C. H. and Reeve, V. E.,** Copper deficiency in the rat. Effect on synthesis of phospholipids, *Aust. J. Exp. Biol. Med. Sci.*, 49, 21, 1971.

31. **DiPaolo, R. V., Kanfer, J. N., and Newberne, P. M.,** Copper deficiency and the central nervous system, *J. Neuropathol. Exp. Neurol.*, 33, 226, 1974.
32. **Zimmerman, A. W., Matthieu, J-M., Quarles, R. H., Brady, P. O., and Hsu, J. M.,** Hypomyelination in copper-deficient rats, *Arch. Neurol.*, 33, 111, 1976.
33. **Cunnane, S. C., Prohaska, J. R., and Bourre, J-M.,** Tissue fatty acid and lipid composition in genetically copper deficient mice, *FASEB J.*, 2, A1205, 1988.
34. **Cunnane, S. C., Manku, M. S., and Horrobin, D. F.,** Essential fatty acids in tissue phospholipids and triglycerides of the zinc deficient rat, *Proc. Soc. Exp. Biol. Med.*, 177, 441, 1984.
35. **Cunnane, S. C.,** Differential regulation of essential fatty acid metabolism to the prostaglandins: possible basis for the interaction of zinc and copper in biological systems, *Prog. Lipid Res.*, 21, 73, 1982.
36. **Cunnane, S. C.,** Contrasting effects of zinc and copper on the composition of long chain fatty acids in tissue lipids of the rat, in *Trace Elements in Man and Animals,* Mills, C. F., Bremner, I., and Chesters, J. K., Eds., Commonwealth Agricultural Bureaux, Slough, U.K., 1985, 70.
37. **Cunnane, S. C.,** Essential fatty acid-mineral interactions with reference to the pig, in *Fats in Animal Nutrition,* Wiseman, J., Ed., Butterworths, London, 1984.
38. **Cunnane, S. C.,** Adverse effects of zinc deficiency on essential fatty acids are independent of food intake in adult rats, *Br. J. Nutr.*, 57, 273, 1988.
39. **Brewer, G. J.,** Oral zinc therapy for Wilson's Disease, *Ann. Intern. Med.*, 99, 314, 1983.
40. **Hoogenraad, T. U.,** Effective treatment of Wilson's Disease with oral zinc, *Arch. Neurol.*, 39, 672, 1982.

Chapter 9

# COPPER AND PROSTAGLANDIN METABOLISM

**Kenneth G. D. Allen and Melvin M. Mathias**

## TABLE OF CONTENTS

# I. INTRODUCTION

Prostaglandins (PG) and thromboxanes (TX) are eicosanoids synthesized from phospholipid derived arachidonic acid by prostaglandin H (PGH) synthase (EC 1.14.99.1). These autocoids are synthesized by virtually all tissues and act as local (paracrine) effectors of cellular metabolism and function. Considerable attention has been directed to $TXA_2$ synthesis by platelets and prostacyclin ($PGI_2$) synthesis by vascular endothelium in view of their importance in cardiovascular disease.

PGH synthase, which has both cyclooxygenase and peroxidase activity, catalyzes the rate controlling step of PG and TX synthesis. The arachidonic acid substrate for PGH synthase is derived from the C-2 position of tissue phospholipid pools. In view of the possible influence of precursor pools of arachidonate on subsequent eicosanoid production, it is important to consider the events leading to arachidonate release from phospholipid pools. Evidence points to the importance of both specific phosphoinositide phospholipase C and nonspecific phospholipid phospholipase $A_2$ in arachidonate release. Early arachidonic acid release in all tissues that synthesize eicosanoids is associated with phosphoinositide turnover.[1] The events leading to the release of substrate arachidonate occur in response to cellular stimuli, usually agonist cell surface receptor interactions, which stimulates phospholipase C activity. In part, arachidonic acid substrate for PGH synthase is derived from this released 1,2-diacylglycerol by subsequent diacylglycerol and monoacylglycerol lipase activity,[1] but the exact contribution due to this pathway is not clear and apparently varies with tissue sources. The polyphosphoinositol formed by phospholipase C activity serves to amplify arachidonic acid release from other phospholipids by $Ca^{2+}$ mobilization from endogenous pools, particularly the endoplasmic reticulum.[1] Phospholipase $A_2$ activity has a high Ca requirement, and this Ca mobilization stimulates phospholipase $A_2$ activity and additional arachidonate release from other phospholipids such as phosphatidylcholine and phosphatidylethanolamine, as well as additional phosphatidylinositol. In platelets it has been estimated that the arachidonic acid released in the early times after stimulation of platelets is derived from 1,2-diacylglycerol mainly from phosphatidylinositol.[1] Phosphatidylcholine provides most of the arachidonate liberated when platelets are maximally challenged.[1] The sequence of events leading to arachidonate release emphasizes the potential importance of specific phospholipid pools and their arachidonate content on subsequent PG and TX production.

PGH synthase, also referred to as PG endoperoxide synthetase, catalyzes the initial and rate controlling step common to all PG and TX synthesis.[2] PGH synthase is a microsomal glycoprotein, molecular weight 70,000, which has both cyclooxygenase and peroxidase activity.[3] The cyclooxygenase activity of PGH synthase catalyzed the bisdioxygenation of arachidonic acid to form the 15-hydroxy-9,11-endoperoxy-5,13-dienoic derivative of arachidonic acid, $PGG_2$. The peroxidase activity of PGH synthase, which is apparently a unique peroxidase distinct from other peroxidases, reduces the 15-hydroperoxy group to a hydroxyl group forming $PGH_2$. Both cyclooxygenase and peroxidase activities of PGH synthase require heme.[3]

The cyclooxygenase activity of PGH synthase is a free radical-initiated process.[3,4] The currently accepted mechanism of action points to PGH synthase ferriheme binding of lipid hydroperoxides (free fatty acid hydroperoxides). Ferriheme reaction with enzyme-bound free fatty acid hydroperoxides (ROOH) produces the lipid hydroperoxy radical, ROO·;

$$\text{Heme } Fe^{3+} + \text{ROOH} \rightarrow \text{Heme } Fe^{2+} + \text{ROO}\cdot + H^+$$

Enzyme bound ROO· interacts with enzyme bound arachidonic acid to abstract H at arachidonate C-13 forming an arachidonate radical (A·);

$$\text{AH} + \text{ROO}\cdot \rightarrow \text{A}\cdot + \text{ROOH}$$

Subsequent radical rearrangement and bisdioxygenation produces $PGH_2$. The peroxidase activity of PGH synthesis reduces the 15-hydroperoxy group of $PGG_2$ to form $PGH_2$. $PGH_2$ is the terminal common metabolite of subsequent tissue-specific PG and TX synthesis. Evidence in favor of this free radical-initiated process and the physiological role of lipid hydroperoxides has been summarized.[3] Using purified PGH synthase and monitoring oxygen consumption has been employed to assess the influence of peroxides on PGH synthase activity. Lipid hydroperoxides at approximately 10 n*M* begin to activate the enzyme. In the lipid hydroperoxide concentration range 0.01 to 1.0 μ*M*, PGH synthase activity is apparent. Above 1.0 μ*M* irreversible inactivation of the synthase occurs.[3] This irreversible inactivation probably involves enzyme destruction, and is apparently due to intermediates which arise during catalysis. The chemical identity of the oxygen metabolite released by the peroxidase activity of PGH synthase is not clear, but is believed to be involved in this irreversible inactivation, and appears to be an oxygen centered radical described as $[O_x]$ which has some of the properties of OH·.[2] The unique feature of autoactivation and irreversible inactivation of PGH synthase by lipid hydroperoxides provide for control of tissue PG and TX synthesis. Hydrogen peroxide also stimulates PGH synthase at 20 μ*M* concentration presumably by a similar mechanism; however, evidence favors lipid hydroperoxides as the physiologically relevant control mechanism.[3] Studies involving oxygen radical scavengers, radical-producing treatments such as addition of $H_2O_2$ and transition metal ions, and synthetic antioxidants, all point to the importance of oxygen-derived radicals in the control of PGH synthase. Furthermore, nutrients involved in oxygen radical metabolism pathways, such as selenium for glutathione peroxidase (SeGSP) activity and vitamin E, have been shown to influence PG and TX synthesis in a manner consistent with the role of lipid hydroperoxides, and resultant radicals, in PGH synthase activity.[5-18]

Recent work has suggested that SeGSP is a major constraint on PG and TX synthesis and functions by controlling tissue ROOH concentrations and hence activator concentrations of lipid hydroperoxide.[19] Autoinactivation of PGH synthase limits the extent of PG production once stimulation has occurred.

Tissue-specific metabolism of $PGH_2$ results in effector PG and TX production. For example, platelets produce $TXA_2$ from $PGH_2$ by the action of TX synthase. In the vascular endothelium prostacyclin (PGI) synthase produces $PGI_2$. $PGF_{2\alpha}$ arises from $PGH_2$ either by nonenzymatic reduction or by the enzyme endoperoxide F reductase. Isomerization of $PGH_2$ by endoperoxide isomerase produces $PGE_2$ which by subsequent 9-keto reductase activity can also form $PGF_{2\alpha}$.

The autoacceleration and irreversible inactivation of PGH synthase by lipid hydroperoxides would appear to control all tissue PG and TX synthesis since $PGH_2$ production is common to all subsequent tissue-specific PG and TX synthesis. However, lipid hydroperoxides also have exclusively inhibitory effects on PGIs.[20] Other tissue-specific reactions using PGH as substrate, such as $TXA_2$, $PGE_2$, and $PGF_{2\alpha}$ synthesis, are unaffected by lipid hydroperoxides. PGI synthase is inhibited by a variety of lipid hydroperoxides, although not in an analogous manner to PGH synthase. The inhibition of PGI synthase appears to be enzyme destruction due to an oxidant $[O_x]$, released by the peroxidase activity of PGH synthase. In particular, 15-hydroperoxyeicosatetraenoic acid (15-HPETE) inhibits PGI synthase with a 50% inhibition concentration ($IC_{50}$) of 1.2 $\mu M$. In general endogenous lipid hydroperoxides, such as $PGG_2$, are especially potent in irreversible inactivation of PGH synthase whereas other lipid hydroperoxides that are not involved in the enzymatic pathway to $PGH_2$ production (e.g., exogenous lipid hydroperoxides) appear to be more effective as stimulators of PGH synthase.[21] In the case of PGI synthase the reverse is true[20,21] and apparently exogenous lipid hydroperoxides are more effective inhibitors of PGI synthase than is $PGG_2$.[21] The inactivation of PGI synthase is thought to be due to $[O_x]$ released by the peroxidase activity of PGH synthase,[22] which catalyzes the reduction of both $PGG_2$ and lipid hydroperoxides not involved in the catalytic sequence leading to $PGH_2$. The evidence suggests that lipid hydroperoxides, exogenous ones in particular, can stimulate PGH synthase and TX production while inhibiting PGI synthase and $PGI_2$ production, with major implications for cardiovascular health. Evidence from dietary vitamin E and Se studies support this interpretation.[5-18]

Since free radical initiation and activation of the cyclooxygenase step of PGH synthase requires lipid hydroperoxide-derived radicals, it is important to consider the formation of this lipid hydroperoxides and the role of copper in their production. Several recent reviews on the mechanism of lipid peroxidation, cellular defense mechanisms, and implications in disease are available.[23-26] In this chapter, emphasis is directed toward the specific role of copper, both exogenously added free copper, $Cu^{2+}$, and endogenously derived (i.e., dietary copper) in lipid peroxidation and cellular oxygen-derived radical metabolism. In the sequence of events leading to lipid peroxidation, the initiation of peroxidation is considered to be an activated species of $O_2$, the superoxide anion $O_2^-$ which by virtue of an unpaired electron is itself a radical. $O_2^-$ production is mainly due to "leaky" electron transport systems, particularly those associated with the endoplasmic reticulum[26] and the respiratory burst of leukocyte and macrophage NADPH oxidase.[26] Single electron addition to ground state molecular oxygen produces $O_2^-$. Apparently this $O_2^-$ is not of itself damaging but is a major and effective reductant of the small cellular pools of free $Fe^{3+}$ or low molecular weight ligand-bound $Fe^{3+}$, allowing for $Fe^{2+}$ production. By Fenton chemistry $Fe^{2+}$ produces the highly reactive hydroxyl radical, OH·;

$$Fe^{2+} + H_2O_2 \rightarrow Fe^{3+} + OH\cdot + OH\cdot$$

Once produced OH· can cause lipid peroxidation by classical radical propagation;

$$OH\cdot + R\text{–}H \rightarrow R\cdot + H_2O$$

$$R\cdot + O_2 \rightarrow ROO\cdot$$

$$ROO\cdot + RH \rightarrow ROOH + R\cdot, \quad \text{etc.}$$

Lipid hydroperoxides can also undergo Fenton-like reactions with further contributions to lipid peroxidation due to alkoxy, RO·, or peroxy, ROO·, radicals;

$$ROOH + Fe^{2+} \rightarrow RO\cdot + OH^- + Fe^{3+}$$
$$2ROOH + Fe \rightarrow RO\cdot + ROO\cdot + H_2O$$

The sum of $O_2^-$ reduction and the Fenton reaction is referred to as the Fe catalyzed Haber Weiss reaction,

$$O_2^- + H_2O_2 \xrightarrow[\text{catalysis}]{\text{Fe}} OH\cdot + OH^- + O_2$$

Free copper, $Cu^{2+}$, is also effective in radical production since it also reduced to $Cu^{1+}$ which can catalyze the Fenton reaction. However, evidence suggests that iron is the physiologically significant species in the Fenton reaction, and that free $Cu^{2+}$ (or low molecular weight ligand-bound $Cu^+$) is not as important in this chemistry except where $Cu^{2+}$ is added *in vitro*[23,25] or in copper toxicosis where copper intakes exceed the cellular copper binding protein capacity and free $Cu^{2+}$ appears. Recent evidence suggests that contrary to long accepted belief, the 5% free or histidine bound $Cu^{2+}$ pool in plasma may be an artifact of ceruloplasmin proteolytic degradation.[27] Thus the role of $Cu^{2+}$ in lipid peroxidation *in vivo* is not as well established as that of $Fe^{3+}$, although there does appear to be some free $Cu^{2+}$ detectable in tissues. However, *in vitro* experiments clearly show the $Cu^{2+}$ or $Fe^{3+}$ additions are equally effective,[23] although the physiological relevance can be questioned. A further feature of *in vitro* $Fe^{3+}$ or $Cu^{2+}$ additions is that they may not initiate lipid peroxidation in the sense that $O_2^-$ and OH· arc the peroxidative initiators, but may rather stimulate peroxidation by decomposition of already formed ROOH to produce RO·.[25] Another potentially important feature of *in vitro* experiments is that tissue disruption produces xanthine oxidase which is an excellent source of $O_2^-$.[28] However, in intact tissues there appears to be little xanthine oxidase, but rather xanthine dehydrogenase which used $NAD^+$ as electron acceptor rather than molecular oxygen used by xanthine oxidase.[28] Consequently, *in vitro* experiments involving tissue disruption and proteolytic formation of xanthine oxidase from xanthine dehydrogenase can cause production of $O_2^-$ and also release cellular pools of Fe and perhaps Cu as $Fe^{3+}$ and $Cu^{2+}$. In such an *in vitro* situation, lipid peroxidation would be likely.

Superoxide dismutase (SOD) catalyzes the dismutation of $O_2^-$:

$$2O_2^- + 2H^+ \xrightarrow{\text{SOD}} O_2 + H_2O_2$$

The cytosolic enzyme (CuSOD) contains both Cu and Zn but copper is the catalytically important metal.[23,29] SOD functions to limit $O_2^-$ concentration such that reduction of the small cellular $Fe^{3+}$ pools by $O_2^-$ to form $Fe^{2+}$ is decreased.[23] Although $O_2^-$ dismutates spontaneously at an appreciable rate, SOD increases this rate by several orders of magnitude.[23] Hence, Fenton reactions will in turn be limited and lipid peroxidation decreased. The paradoxical increase in $H_2O_2$ due to SOD activity does not contribute to increased OH· formation by Fenton chemistry since catalytic $Fe^{2+}$ concentrations are decreased. Furthermore, $H_2O_2$ is removed by both catalase and Se-dependent glutathione peroxidase (SeGSP). Dietary copper deficiency decreases SOD activity[30] and in addition decreases SeGSP in some tissues[31,32] by an unknown mechanism. Despite the role of dietary copper in pathways of oxygen derived radical metabolism, evidence for changes in oxygen-derived radical production in copper deficiency has been limited to increased amounts of *in vitro* lipid peroxidation in liver microsomes and mitochondria,[29,33,34] alterations in neutrophil function,[35] and changes in erythrocyte and lymphocyte membrane proteins.[36,37] However, recent studies suggest a role for dietary copper in PG and TX production.[38,39]

Ceruloplasmin, the plasma copper transport protein, also appears to have antioxidant properties.[27] This is most probably due to its ferroxidase activity which would serve to decrease $Fe^{2+}$ pools and hence the Fenton reaction.[27]

# II. EXPERIMENTAL APPROACH

## A. Modulation of Copper Status

Studies involving copper and PG and TX production fall into two categories. Several studies have investigated the effect of exogenously added $Cu^{2+}$, or $Cu^{2+}$ bound to low molecular weight ligands such as histidine, and CuSOD additions to incubations of tissue slices and homogenates, and to cells such as macrophages and platelets. In studies such as these it is assumed, although not stated, that the animals from which these tissues and cells were obtained were copper adequate in terms of diet. More recent studies have directly examined the role of dietary copper in tissue homogenate and in *in vitro* incubations of tissues such as whole blood and aortic rings.

### *1. Additions of Ionic Copper and CuSOD In Vitro*

The majority of these studies have involved $Cu^{2+}$ additions to tissue homogenates or incubated tissue slices followed by analysis of the cyclooxygenase products after timed incubation periods. The tissues used have been ovine and bovine vesicular gland homogenates, rat and human uterine (myometrium and endometrium) homogenates, and rabbit kidney medullary slices.[40-45] In some studies an attempt has been made to partially purify the PGH synthase obtained from the tissue homogenate, but in the majority of studies no such purification was attempted. Consequently, the enzymes involved in PG catabolism, in addition to specific prostaglandin synthases, may be present and thus interpretations of the results cannot confidently be made only on the basis of PG synthesis (PGH synthase) or subsequent conversion of $PGH_2$ to specific PGs.

Other studies have involved $Cu^{2+}$ addition to incubations of intact cells such as macrophages and platelets, incubation of tissue explants, and arterial preparations.[46-49] In such cases the accelerated radical production and lipid peroxidation associated with tissue disruption *(vide infra)* would be limited. In certain studies intrauterine $Cu^{2+}$ infusions have been employed in an attempt to mimic $Cu^{2+}$ release from copper-containing intrauterine contraceptive devices (IUD).[50] In these studies a pharmacological basis of action of IUD is suspected. CuSOD additions have also been employed in PG and TX production by tissue homogenates[38] and in *in vitro* incubation of cells such as platelets and macrophages.[51,52]

### *2. Dietary Copper*

The model used in our studies is the acute copper deficiency model. Weanling rats were depleted of copper stores by feeding a copper-deficient diet containing 0.1% of the copper chelator *N,N′*-bis (2-aminoethyl)-1,3-propanediamine tetrahydrochloride. Following partial copper depletion, animals are then randomly assigned to dietary copper treatments. The acute copper deficiency model serves to shorten experimental duration and studies are typically performed in 5 to 6 weeks. *N,N'*-bis (2-aminoethyl)-1,3-propanediamine is a specific copper chelator which does not produce changes in tissue zinc or iron.[53,54] Following completion of the dietary stage tissues and tissue homogenates are obtained for incubations followed by PG and TX analysis.[38,39,55] Liver copper analyses and tissue Cu-dependent SOD assays have been used to verify the copper status of experimental animals at the completion of the studies.

**B. Prostaglandin and Thromboxane Synthesis Models**

Incubations of tissue homogenates, tissue slices, and partially purified PGH synthase were performed using routine incubations in buffers. Incubation times range from 2 to 30 min prior to PG and TX analysis. In many studies such as these the linearity of PG production with time has not been established. In our liver homogenate study, a 10-min incubation was used and both $PGE_2$ and $PGF_{2\alpha}$ production was linear up to 20 min.[38]

*In vitro* incubation studies have employed washed platelets. Fresh blood is collected in sodium citrate (4% by weight) and platelet-rich plasma (PRP) obtained by centrifugation. Pelleted platelets are obtained by centrifugation and platelets are washed using buffer followed by repelleting. Additions to buffered suspensions of washed platelets at 37° are then made and incubation times are typically less than 10 min, followed by PG and TX analysis.[47,51]

Macrophages were isolated by Geys Balanced Salt Solution (GBSS) washings of rat peritonea, 24 h after injection of 1% starch suspension, followed by centrifugation in a Ficoll gradient. *In vivo* incubations of macrophages are performed in GBSS at 37° for up to 1 h.[56]

Our recalcified whole blood model has been developed to examine TX production in response to dietary copper.[55] This model was developed since it closely represents the *in vivo* situation, whereas methods involving PRP or washed platelets may not. Size and density heterogeneity of platelets is well documented. Hence, the centrifugation steps may unintentionally sediment subpopulations of heavier, possibly younger, platelets. It has been demonstrated that TX release increases in proportion to platelet volume. A more physiologically relevant approach is to use a model where centrifugation for platelet separation is avoided. All methods involving platelets have the additional problem of controlling extracellular $Ca^{2+}$ concentrations in order to prevent coagulation. This is typically achieved by collecting blood in citrate or EDTA. However, $Ca^{2+}$ is important in platelet phospholipid arachidonate release *(vide supra)*. This problem has either been ignored or circumvented by washing the platelets to remove the calcium chelators. Our approach was to recalcify the citrated whole blood.

Blood from the abdominal aorta of anesthetized rats was collected in 0.38% (final concentration) trisodium citrate and diluted twofold with phosphate buffered saline (PBS). Platelet numbers were determined using a hemocytometer. Triplicate samples of diluted citrated blood (final citrate concentration, 5.6 m*M*) were pipetted into siliconized cylindrical glass cuvettes containing siliconized magnetic stirrers, then preincubated at 37° for 1 min in a multicavity heated stirrer. Prior to collagen or thrombin challenge the blood was recalcified (5.0 m*M* $Ca^{2+}$). Incubation was continued for an additional 4 min and 0.1 vol of 4.2 m*M* aspirin is added to stop TX synthesis. Samples are then allowed to clot for 1 h at room temperature and serum obtained and assayed for TX. The recalcified whole blood protocol has been validated.[55]

We have developed an aortic ring incubation protocol for $PGI_2$ synthesis by vascular endothelium.[39] Rat thoracic aorta are excised, and sections cut from the mid section. Aortic rings were incubated at 37° with gentle shaking in 1.0 ml of 10 m*M* HEPES (*N*-2-hydroxyethyl piperazine-*N'*-2-ethane-sulfonic acid) buffer, pH 7.4, containing Krebs-Henseleit salts. Aortic ring incubations were performed for 15 min with aliquots of incubation buffer removed at 5, 10, and 15 min for PGI analysis. Results were expressed on a protein basis. Details and validation of this procedure are published.[39]

Typically early studies have employed thin layer chromatography (TLC) identification of PGs by reference to known standards. Quantitative estimates of PG production have been determined by gas-liquid chromatography (GLC) or by radioactive counting of identified TLC spots following exogenous $^{14}C$- or $^{3}H$-arachidonate additions. Validation of these assays has been by use of known PG standards for TLC and GLC and by mass spectrometry. High pressure liquid chromatography (HPLC) has also been used to quantitate PG production. Recent studies typically involve radioimmunoassays (RIA) for analysis of PG and TX produced from both endogenous or exogenous arachidonate. The techniques and validation of RIA for assay of PG and TX have recently been reviewed[57] and in the cases of $TXA_2$ and $PGI_2$, their spontaneous (nonenzymatic) degradation products, $TXB_2$ and 6-keto $PGF_{1\alpha}$, respectively, are the analytes.

## III. STUDIES INVOLVING IONIC COPPER ADDITIONS

The initial report of an effect of $Cu^{2+}$ additions on PG synthesis was that of Nugteren et al.[58] using a crude microsomal preparation of sheep vesicular gland cyclooxygenase activity. Preincubation with $Cu^{2+}$ at 10 to 25 $\mu M$ caused pronounced inhibition of PGE synthesis.[58] Subsequent reports by Lee and Lands[41] and Maddox[40] confirmed this. Additions of 0.5 $\mu M$ $Cu^{2+}$ to sheep vesicular gland particulate fractions stimulated $PGF_{2\alpha}$ production while diminishing that of $PGE_2$.[41] This effect was further enhanced by equimolar additions of dithiols.[41] In this study the increase in $PGF_{2\alpha}/PGE_2$ ratio was interpreted as due to a shift in precursor intermediate (presumably $PGG_2$ or $PGH_2$) metabolism in favor of $PGF_{2\alpha}$ production.[41] The oxidation of added arachidonate substrate remained constant. The results of Maddox confirmed this finding.[40] In the $Cu^{2+}$ addition range of 10 to 100 $\mu M$ $PGE_2$ production decreased and $PGF_{2\alpha}$ increased. Above 1.0 m*M* $Cu^{2+}$ inhibition of both $PGE_2$ and $PGF_{2\alpha}$ production occurred.[40] The interpretations of these $Cu^{2+}$ addition studies have suggested that $Cu^{2+}$ dithiol complexes provide reducing equivalents, either enzymatic or nonenzymatic which favors $PGF_{2\alpha}$ formation. Alternatively, it has been suggested that copper participates in a process which leads to inactivation of $PGE_2$ synthesis from the common intermediary (again, presumably $PGH_2$) with a concomitant activation of $PGF_{2\alpha}$ synthesis.

Vargaftig et al. demonstrated a free $Cu^{2+}$ and aminothiol influence on platelet aggregation, which potentiated arachidonic acid-induced platelet aggregation and the release of platelet constituents.[47] A further study by Vargaftig et al.[59] demonstrated that 1 to 5 m*M* $Cu^{2+}$ potentiated the arachidonic acid-mediated platelet aggregation. Interpretation of these results suggested that $H_2O_2$ or a related chemical species is produced as a result of $Cu^{2+}$ additions which serves to stimulate PG synthesis. Such an interpretation implies lipid peroxidation mediated by free $Cu^{2+}$, a known feature of copper chemistry.

Using a partially purified microsomal bovine vesicular gland preparation of PGH synthase, Chan et al.[45] concluded that the enhancement of $PGF_{2\alpha}$ synthesis, and the accompanying depression of $PGE_2$ synthesis by $Cu^{2+}$-dithiol complexes was due to chemical, nonenzymatic, reduction of PGG or PGH and not an enzymatic mechanism requiring $Cu^{2+}$, or $Cu^{2+}$ dithiols, as a cofactor. $PGF_{2\alpha}$ production from $PGH_2$ occurs by the enzyme endoperoxide F reductase, or by chemical means as outlined above. In addition, $PGF_{2\alpha}$ can arise from $PGE_2$ via the action of 9-keto reductase, but not from $PGE_2$ by chemical (nonenzymatic) reduction. $PGE_2$ production from $PGH_2$ occurs only via the enzyme endoperoxide E isomerase. Hence, $Cu^{2+}$-dithiol complex additions and enhancement of $PGF_{2\alpha}$ production at the expense of $PGE_2$ are best explained as being due to chemical reduction of $PGH_2$, the common precursor.[45]

It is important to bear in mind that $Cu^{2+}$, or $Cu^{2+}$-dithiol, additions to tissue homogenates, or partially purified PGH synthase preparations may not be physiologically relevant. Although not mentioned, all of their preparations were obtained from presumably dietary copper-adequate animals and hence cannot be interpreted as involving dietary copper in PG production or metabolism. Furthermore, even in intact animal tissue it is questionable as to whether $Cu^{2+}$, free or low molecular weight ligand bound, occurs. However, these $Cu^{2+}$ addition studies do show an influence of ionic copper on PG metabolism, but not synthesis as evidenced by changes in $PGF_{2\alpha}/PGE_2$ ratios with essentially constant total PG production (sum of $PGE_2$ and $PGF_{2\alpha}$) or arachidonate oxidation. Hence the evidence from their studies cannot be interpreted as involving $Cu^{2+}$ in PGH synthase. However, above a 1.0 m*M* concentration of $Cu^{2+}$ inhibited both $PGE_2$ and $PGF_{2\alpha}$ production in the study of Maddox,[40] but not in that of Chan et al.[45] A possible explanation for this effect is that lipid peroxidation induced by high exogenous $Cu^{2+}$ may produce radicals sufficient to inactivate PGH synthase.

Several studies have examined $Cu^{2+}$ additions to intact cells. $Cu^{2+}$ addition studies have recently been reported by Fujita et al. using incubations of rabbit kidney medulla slices.[43] Once again, $Cu^{2+}$ addition of from 0.2 to 1.6 m*M* increased $PGF_{2\alpha}$ production and decreased $PGE_2$ production. Most of the change in $PGF_{2\alpha}/PGE_2$ ratios in this study were attributable to decreases in $PGE_2$ formation with smaller increases in $PGF_{2\alpha}$ production than observed by others,[40,41,45] but despite these differences the results are quite consistent.

Macrophage PG and TX synthesis has also been examined in response to added exogenous $Cu^{2+}$.[46] The action of ionic $Cu^{2+}$ additions to starch-elicited peritoneal macrophages appears to be biphasic. $Cu^{2+}$ at concentrations between 1 and 10 $\mu M$ inhibited macrophage $PGE_2$ and TX formation in a dose-dependent manner, but was without effect on PGI production. Above 10 $\mu M$ $Cu^{2+}$ the synthesis of all three was stimulated. Exogenous arachidonate abolished the effect of $Cu^{2+}$ on macrophage PG and TX synthesis. At 100 $\mu M$, $Cu^{2+}$-caused release of arachidonate from macrophages prelabeled with $^{14}C$-arachidonate was increased, and this increase occurred in conjunction with the release of macrophage lysosomal enzymes. The decrease in $PGE_2$ production at low $Cu^{2+}$ additions is consistent with the studies mentioned previously.[40,41] The increased PG and TX production observed at higher $Cu^{2+}$ additions was interpreted as due to increased arachidonate release from macrophage phospholipid pools perhaps due to membrane lipid peroxidation. The influence of macrophage $PGE_2$ on inflammation is thought to occur by $PGE_2$ suppression of macrophage lysosomal enzyme release and inhibition of lymphocyte T cell function. Hence at low (1 to 10 $\mu M$) $Cu^{2+}$ additions these results could be considered as proinflammatory, but the increased $PGE_2$ observed at higher $Cu^{2+}$ additions as not associated with a decrease in lysosomal enzyme release. Hence, it is difficult to consider the overall effect of $Cu^{2+}$ addition as anti-inflammatory. The anti-inflammatory potential of copper, or copper complexes, has received considerable attention. A related study[60] has reported quite similar results and has suggested that the increased $PGE_2$, TX, and PGI production, and elevated macrophage lysosomal enzyme released observed with 100 $\mu M$ $Cu^{2+}$ addition is associated with decreased macrophage cAMP production.

The involvement of $Cu^{2+}$ on the human uterus and PG production by uterine tissue has also been investigated. The rationale studies has been based on changes in tissue $PGE_2$ and $PGF_{2\alpha}$ production caused by added $Cu^{2+}$. These studies have attempted to established the mechanism by which copper-containing IUD function, and the possible role of PGs in this action. It has been known for some time that $Cu^{2+}$ influence rat uterine contractions.[61,62] $Cu^{2+}$ in the range 1 to 10 $\mu M$ caused regular uterine contractions, but at greater than 100 $\mu M$ $Cu^{2+}$ decreased contractions were observed. Endometrial stimulation of PG production may be a mechanism by which IUD function, both for the inert and copper-containing devices, or alternately $Cu^{2+}$ ions may directly stimulate or inhibit myometrial contractile response.[61,62] The

$Cu^{2+}$ concentration-dependent stimulation and inhibition of rat uterus observed by Salgo and Oster[62] could be interpreted on the same basis as the studies of Maddox[40] and Lee and Lands.[41] Such an interpretation would involve preferential formation of $PGF_{2\alpha}$ at low $Cu^{2+}$ additions. Copper-containing IUD dissolve at a rate of 50 μg/d, liberating $Cu^{2+}$.[62] Copper concentrations in uterine fluid and cervical mucus of women fitted with copper-containing IUD are approximately 10 to 50 μ*M*,[44] and this is in the concentration range where $PGF_{2\alpha}$ production is preferentially stimulated by $Cu^{2+}$ in vesicular gland homogenates. $PGF_{2\alpha}$ produces pronounced uterine contractions which may inhibit implantation by suppression of the blood supply necessary for implantation.[61] The inhibitory effects of higher $Cu^{2+}$ concentrations (100 μ*M* $Cu^{2+}$) on rat uterus contraction could be due to inhibition of $PGF_{2\alpha}$ synthesis as observed by Maddox with vesicular gland homogenates.[40]

Kelly and Abel have suggested that the $Cu^{2+}$ release as a result of dissolution of IUD affects the catabolism of $PGE_2$ and $PGF_{2\alpha}$ selectively. Metabolism of both $PGE_2$ and $PGF_{2\alpha}$ involves $NAD^+$ requiring 15-hydroxy PG dehydrogenase and subsequent 13-PG reductase to produce the 15-hydroxy-13,14-dihydro derivatives of $PGE_2$ and $PGF_{2\alpha}$. Using homogenates of human endometrium and myometrium both $PGE_2$ and $PGF_{2\alpha}$ metabolism were inhibited by 10 μ*M* $Cu^{2+}$,[44] a concentration similar to that produced by dissolution of copper-containing IUD.[44,62] The inhibition of $PGF_{2\alpha}$ metabolism was greater than that of $PGE_2$ metabolism. Hence, this inhibition of PG metabolism would accentuate preferential production of $PGF_{2\alpha}$ suggested by other studies.[40,41] Apparently the mechanism of inhibition appears to be a specific effect of $Cu^{2+}$ on 15-hydroxy PG dehydrogenase with no effect on 13-PG reductase.[44] Consequently, the effect of copper appears to be selective and is not merely generalized inactivation of proteins.

Peplow and Hurst[42] have also examined PG production in rat uterine homogenates in response to $Cu^{2+}$. Their results are, in general, similar to these $Cu^{2+}$ additions to vesicular gland homogenates. Addition of 10 to 100 μ*M* $Cu^{2+}$ stimulated $PGF_{2\alpha}$ production. At 1 μ*M* $Cu^{2+}$ no effect was observed and at 1 m*M* $Cu^{2+}$ the stimulating $Cu^{2+}$ effect was decreased. In this study, the influence of $Cu^{2+}$ additions was limited to changes in $PGF_{2\alpha}$ production with no complementary inhibition of $PGE_2$ and $PGI_2$ production.[42]

Using human myometrial strips and portions of uterine artery, Laudanski et al.[50] have shown that while 100 μ*M* $Cu^{2+}$ caused a significant inhibition of myometrial contractile activity, no effect was seen on the uterine artery. In addition, 5 to 50 μ*M* $Cu^{2+}$ inhibited the myometrial contractile response due to both $PGE_2$ and $PGF_{2\alpha}$. Hence, it is unlikely that the inhibition of contractile response seen with 100 μ*M* $Cu^{2+}$ is due to suppression of PG synthesis. *In vivo* infusion of $Cu^{2+}$ into the uterine cavity in amounts similar to those released by dissolution of copper-containing IUD stimulated human myometrial contraction, and this effect was abolished by naproxen, a PG synthesis inhibitor.[50] Hence, the *in vivo* stimulating effect of intrauterine $Cu^{2+}$ infusions was interpreted as due to endogenous PG synthesis and release. The $Cu^{2+}$ infusions in this experiment are quite relevant to the mode of action of copper-containing IUD whereas uterine homogenate preparations are perhaps less physiologically or pharmacologically relevant due to the changes produced by tissue disruption *(vide supra)*. Copper-containing IUD may have a pharmacological basis of action due to $Cu^{2+}$ release.

Studies have also suggested an interaction between $Cu^{2+}$ and receptors involved in PG action. $Cu^{2+}$-histidine complex (200 $\mu M$) amplified the $PGE_2$-mediated stimulation of luteinizing hormone-releasing hormone release in incubation of rat median eminence explants.[48] It was suggested that the $Cu^{2+}$-histidine complex amplified the functional state of a $PGE_2$ postreceptor component, and may involve $Cu^{2+}$-caused peroxidation of membrane lipids with changes in membrane fluidity and responsiveness to the action of the $PGE_2$ receptor complex. In mesenteric vascular preparations, $Cu^{2+}$ has been shown to inhibit vascular smooth muscle pressor response to noradrenaline and angiotensin in a dose-dependent manner in the range 0.4 to 3.2 $\mu M$ $Cu^{2+}$.[49] The effect of $Cu^{2+}$ in this process was similar to that of $PGI_2$ and the results were interpreted as possibly involving the interaction of $Cu^{2+}$ with a $PGI_2$ receptor.[49] If this interpretation is correct then the action of $Cu^{2+}$ may involve membrane lipid peroxidation with changes in membrane fluidity or structure and hence agonist response. In both of these studies, the mechanism of action of $Cu^{2+}$ does not appear to be attributable to changes in PG production or metabolism in contrast to the studies involving $Cu^{2+}$ additions to tissue homogenates and macrophages, and $Cu^{2+}$ uterine infusions.

## IV. DIETARY COPPER, PG, AND TX PRODUCTION

Dietary copper studies in rats have attempted to determine the role of endogenously supplied copper on both liver homogenate PG production and *ex vivo* PGI and TX production in aortic ring and whole blood incubations.[38,39,55]

### A. Liver Homogenate PG Production in Dietary Copper Deficiency[38]

In view of the reports showing an influence of $Cu^{2+}$ additions on PG metabolism in tissue homogenates, $PGE_2$ and $PGF_{2\alpha}$ production was assessed in liver homogenates from copper-deficient (0.6 $\mu g$ Cu/g diet) and control rats (10.7 $\mu g$ Cu/g diet). Copper deficiency significantly decreased both $PGE_2$ and $PGF_{2\alpha}$ production by 56 to 74%. Inclusion of 10 or 100 $\mu M$ $Cu^{2+}$, added as $CuSO_4$, in liver homogenates was without effect on these results. Addition of purified CuSOD (bovine liver) to rat liver homogenates had a major influence on both $PGE_2$ and $PGF_{2\alpha}$ production. Inclusion of added CuSOD, at a concentration approximately equal to that present in control liver, in both copper-deficient and control liver homogenates significantly stimulated both

$PGE_2$ and $PGF_{2\alpha}$ production. The effect of this physiological amount of CuSOD inclusion was larger in copper-deficient homogenates than in control homogenates and caused a three- to fourfold stimulation of $PGE_2$ and $PGF_{2\alpha}$ production while a twofold stimulation was observed in control liver homogenates. When CuSOD was included in copper-deficient homogenates at a concentration approximately three times that found in control liver, $PGE_2$ and $PGF_{2\alpha}$ production was unchanged from liver homogenates without any additions. All of these incubation experiments were also performed with inclusion of 100 $\mu M$ arachidonic acid. In all cases, added arachidonic acid increased $PGE_2$ and $PGF_{2\alpha}$ production twofold, but did not affect the responses to dietary copper, added $Cu^{2+}$, or added CuSOD. Thus, the observed changes could not be interpreted as due to alterations in arachidonic acid precusor pools due to dietary copper. When the ratio of $PGE_2/PGF_{2\alpha}$ was examined, there was a significant difference between copper-deficient and control homogenates indicating more $PGE_2$ produced relative to $PGF_{2\alpha}$, in control animals. The $PGE_2/PGF_{2\alpha}$ ratio was unaffected by $Cu^{2+}$, CuSOD, or arachidonic acid inclusions. The reason for the apparent difference favoring $PGF_{2\alpha}$ production in copper deficiency is not clear, but is obviously opposite to the effect observed by Lee and Lands[41] and Maddox[40] with ionic $Cu^{2+}$ additions. Furthermore, this dietary copper liver homogenate study indicates that total PG production (sum of $PGE_2$ and $PGF_{2\alpha}$) varies in response to dietary copper and CuSOD additions, but not $Cu^{2+}$ additions. In contrast ionic $Cu^{2+}$ additions in the previously mentioned studies indicate that $Cu^{2+}$ affects PG metabolism, as illustrated by $Cu^{2+}$-induced changes in $PGE_2/PGF_{2\alpha}$ ratios, but changes in total $PGE_2$ and $PGF_{2\alpha}$ production did not appear to be significant. Thus, dietary copper does appear to influence the relative amounts of $PGE_2$ and $PGF_{2\alpha}$ produced in liver homogenates but in a manner opposite to that observed in previous $Cu^{2+}$ addition studies. More importantly, the major influence of dietary copper would appear to be total PG production by liver homogenates which was attributed to changes in CuSOD activity due to dietary treatment. When total SOD activity, the sum of endogenous SOD plus added CuSOD, was examined, there was a significant positive linear correlation with both $PGE_2$ ($r = 0.71, p < 0.001$) and $PGF_{2\alpha}$ ($r = 0.81, p < 0.001$) production. These significant correlations occurred in both samples with and without added arachidonic acid, and occurred below the inhibitory amount (approximately three times physiological) of SOD added to the liver homogenates. The significance of these correlations implies a close association between PG production by liver homogenates and the SOD activity over the range 0 to 500 U/mg of liver.

Tissue homogenate PG production is clearly a nonphysiological model. However, the liver homogenate investigation did demonstrate that dietary copper has a quite different mode of action than $Cu^{2+}$ additions to tissue homogenates. The results suggest that dietary copper functions by changing SOD activity which in turn influences PG production. Interpretation of these results was based on changes in lipid peroxidation and radical production as a result of tissue homogenization, but analysis of lipid hydroperoxides or oxyradical formation was not made. Tissue disruption would be expected to produce considerable release of iron, and to a lesser extent copper, from liver storage pools. In copper-deficient animals this metal release would be particularly enhanced due to the significantly larger iron storage pools caused by dietary copper deficiency. Furthermore, proteolytic alteration of $NAD^+$-dependent xanthine dehydrogenase to form xanthine oxidase would, perhaps, cause significant $O_2^-$ production. Hence, $O_2^-$ production and iron release from liver would be expected to cause increased oxyradical production and lipid peroxidation. In copper-deficient homogenates this effect may be larger due to the increased iron pools, and may be sufficiently large to inactivate PGH synthase which would account for the significant reduction in $PGE_2$ and $PGF_{2\alpha}$ production observed in copper deficiency. If this interpretation is correct, it is not surprising that ionic $Cu^{2+}$ additions were without effect since this amount of $Cu^{2+}$ would be small in comparison to the $Fe^{2+}$ released by tissue homogenization. The effect of added CuSOD on $PGE_2$ and $PGF_{2\alpha}$ synthesis may be due to alterations in homogenate oxyradical production and lipid peroxidation due to limiting $O_2^-$ concentrations. Although both copper-deficient and control homogenate PG production was increased by inclusion of physiological amounts of CuSOD, the effect was much larger in the copper-deficient homogenates. Excess CuSOD addition (approximately three times physiological amounts) decreased liver homogenate PG production possibly by reductions in lipid hydroperoxides sufficient to inhibit the accelerative phase (activation) of PG synthase. Since dietary copper deficiency also decreases liver homogenate SeGSP activity by an unknown mechanism,[31,32] it is possible to interpret some of these results based on reduced SeGSP activity. Recent work has emphasized the importance of SeGSP in controlling tissue PG production by limiting the concentration of lipid hydroperoxide activators.[19] However, this interpretation cannot explain the results obtained with CuSOD additions.

Although arachidonic acid additions to liver homogenates stimulated both $PGE_2$ and $PGF_{2\alpha}$ production in all cases, such arachidonic additions were without effect on $PGE_2/PGF_{2\alpha}$ ratio, or the responses to $Cu^{2+}$ and CuSOD additions. Hence, interpretations based on changes in arachidonate precursor pools as a result of copper deficiency cannot account for the results.

### B. Aortic PGI[39] and Platelet TX[55] Synthesis in Dietary Copper Deficiency

Since tissues typically synthesize PG and TX in response to cell surface stimuli, more physiologically relevant information is obtained with *ex vivo* tissue incubations. Using the recalcified whole blood model, and rats fed either copper-deficient (0.6 μg Cu/g) or control (10.7 μg Cu/g) diets, copper deficiency significantly increased TX synthesis by blood challenged with either collagen or thrombin. In two separate experiments, $TXB_2$ production (the spontaneous nonenzymatic degradation product of $TXA_2$) increased by 15 to 47% in response to the challenges. Interpretation of these results was based on lipid peroxidation and compromised oxyradical metabolism — however, neither of these was measured — by platelets due to dietary copper deficiency. As outlined previously, lipid hydroperoxides influence both the accelerative (autoactivation) phase and irreversible inactivation of PGH synthase. However, they have no influence on thc activity of TX synthase. The increase in challenged blood TX production in copper deficiency is very similar to the same effect observed in both vitamin E and Se deficiency, and mechanistically consistent with the roles of these nutrients in lipid hydroperoxide production and oxyradical metabolism. Consequently, it is likely that platelet CuSOD, and perhaps SeGSP, were reduced by copper deficiency and hence lipid hydroperoxide production increased to the concentration range where activation of PGH synthase occurs, 0.01 to 1.0 $\mu M$, but below the concentration range where irreversible inactivation of PGH synthase occurs, $>1.0$ $\mu M$. Other interpretations based on changes in membrane phospholipid unsaturated fatty acids (see Section V) may be applicable.

The influence of added SOD on arachidonate metabolism in both intact and broken isolated washed cat platelets has been examined.[51] SOD addition to either intact or broken platelets had no influence on arachidonate metabolism; $TXB_2$ and 12-hydroxyperoxyeicosatetraenoic acid (12-HPETE) production (a product of platelet 12-lipoxygenase) were unchanged. In this study, no indication was made as to whether the added SOD was the Cu- or Mn-dependent enzyme. Furthermore, platelets were obtained from normal cats, presumably adequate in dietary copper. Hence, this study does not address the role of endogenous SOD activity as changed by dietary copper. A similar study[52] using phagocyting rat peritoneal leukocytes (presumably mainly macrophages/monocytes) suggests that CuSOD additions inhibited PG formation in incubations of these leukocytes by inhibiting phospholipase activity, but not by inhibition of cyclooxygenase activity.

Rat aortic ring incubations have indicated that dietary copper can influence vascular endothelium $PGI_2$ production. Dietary copper deficiency (0.5 $\mu g$ Cu/g) resulted in a significant 47% reduction in $PGI_2$ synthesis rates in comparison to control animal (6.0 $\mu g$ Cu/g) incubations. Marginal copper (1.6 $\mu g$ Cu/g) was without effect on $PGI_2$ synthesis. Aortic SOD specific activity was measured using a sample of the same aorta. Both copper-deficient and marginal copper diets significantly decreased aortic SOD activity by 46% and 32%, respectively, in comparison to control aortae. However, there was no significant difference between copper-marginal and -deficient aortic SOD. In this study, $PGI_2$ synthesis was examined in a basal (unchallenged) incubation system. No agonists, such as angiotensin, nor aortic stimuli were employed. The decreased $PGI_2$ production observed in copper deficiency is quite similar to that observed with deficiencies of Se and vitamin E. The results suggest that the decreased $PGI_2$ synthesis is linked to decreases in aortic oxyradical metabolism due to decreased SOD, just as compromised oxyradical metabolism due to deficiencies of Se or vitamin E produced the same result. Such an interpretation would suggest that, due to decreased SOD activity in copper deficiency, lipid hydroperoxide concentration perhaps would be $>1.0\ \mu M$, in the range where inhibition of PGI synthase and PGH synthase occurs. Although SOD was depressed in marginal copper animals, $PGI_2$ synthesis was unchanged. In this case, lipid hydroperoxides may be in the concentration range of approximately $1.0\ \mu M$ where PGI synthase is inhibited despite stimulation of PGH synthase.

These dietary copper studies indicate a significant two- to threefold increase in the TX to PGI ratio due to dietary copper deficiency. Such a change in the TX to PGI ratio would favor aggregation and vasoconstriction and may be of significance in cardiovascular disease.

## V. FATTY ACIDS IN COPPER DEFICIENCY

Dietary copper deficiency has been shown to influence fatty acids in tissues.[63-68] Despite conflicting data showing both increases and decreases in tissue phospholipid arachidonate more recent studies suggest a 25 to 35% decrease in arachidonate in these tissue phospholipids.[64,65] Decreased phospholipid arachidonate may influence tissue PG and TX production. However, the results from the dietary copper deficiency liver homogenate study cannot be attributed to changes in precursor arachidonate since arachidonate additions ($100\ \mu M$) did not influence the results. Also, it is difficult to reconcile increased TX production and decreased $PGI_2$ production based on decreased phospholipid arachidonate. Changes in membrane fluidity due to changes in a variety of unsaturated fatty acids in membrane phospholipids have been reported recently in copper deficiency.[64] This may be more important since cell surface interactions trigger phospholipid arachidonate release. Changes in membrane fluidity may alter agonist receptor interactions and hence phospholipid arachidonate release.

## VI. SUGGESTIONS FOR FURTHER RESEARCH

In order to clearly establish the role of dietary copper in PG and TX production, lipid hydroperoxide concentrations in tissue incubations should be assessed. If the role of dietary copper in PG and TX production is due to CuSOD activity changes, then changes in PG and TX synthesis by tissue homogenates should parallel changes in tissue lipid hydroperoxide concentrations. Several methods are available for estimating lipid peroxidation,[23] but measuring endogenous (tissue) lipid hydroperoxide concentrations is far more difficult and new methodology is needed.

Since phosphoinositide arachidonate concentrations may be particularly relevant to precursor arachidonate release, attention should be directed toward the influence of dietary copper on unsaturated fatty acids in this phospholipid pool.

Recent work has suggested the SeGSP is a major determinant of lipid hydroperoxide activators of PGH synthase.[19] Since dietary copper deficiency has been shown to decrease SeGSP activity in liver and lung,[31,32] it may also cause this effect in blood platelets and the vascular endothelium. Numerous analyses of human diets have shown that copper intakes are below the 2 to 3 mg/d considered necessary to replace daily losses.[69-73] It is doubtful that frank copper deficiency is a problem in humans, but marginal copper nutriture may be more likely. Consequently, studies involving marginal copper diets and PG and TX production by tissue incubations may be of more relevance to human cardiovascular health. In such studies, physiologically relevant agonist stimuli would be appropriate.

## REFERENCES


1. **Majerus, P. W., Connolly, T. M., Deckmyn, H., Ross, T. S., Bross, T. E., Ishii, H., Bansal, V. S., and Wilson, D. B.,** The metabolism of phosphoinositide-derived messenger molecules, *Science,* 234, 1519, 1986.
2. **Gale, P. H. and Egan, R. W.,** Prostaglandin endoperoxide synthetase-catalyzed oxidation reactions, in *Free Radicals in Biology,* Vol. 6, Pryor, W. A., Ed., Academic Press, New York, 1984, 1.
3. **Lands, W. E. M., Kulmacz, R. J., and Marshall, P. J.,** Lipid peroxide actions in the regulation of prostaglandin biosynthesis, in *Free Radicals in Biology,* Vol. 6, Pryor, W. A., Ed., Academic Press, New York, 1984, 39.
4. **Hemler, M. E. and Lands, W. E. M.,** Evidence for a peroxide-initiated free radical mechanism of prostaglandin biosynthesis, *J. Biol. Chem.,* 255, 6253, 1980.
5. **Carpenter, M. P.,** Antioxidant effects on the prostaglandin endoperioxide synthetase product profile, *Fed. Proc.,* 40, 189, 1981.
6. **Falanga, A., Doni, M. G., Delanini, F., deBellis Vitti, G., Imberti, L., Donati, M. B., and deGaetano, G.,** Unbalanced plasma control of TXA2 and PGI2 synthesis of vitamin E-deficient rats, *Am. J. Physiol.,* 245, H867, 1983.
7. **Hamelin, S. StJ. and Chan, A. C.,** Modulation of platelet thromboxane and malonaldehyde by dietary vitamin E and lineolate, *Lipids,* 18, 267, 1983.
8. **Hope, W. C., Dalton, C., Machlin, L. J., Filipski, R. J., and Vane, R. M.,** Influence of dietary vitamin E on prostaglandin biosynthesis in rat blood, *Prostaglandins,* 10, 557, 1975.
9. **Hwang, D. H. and Donovan, J.,** In vitro and in vivo effects of vitamin E on arachidonic acid metabolism in rat platelets, *J. Nutr.,* 112, 1233, 1982.
10. **Karpen, C. W., Merola, A. J., Trewyn, R. W., Cornwell, D. G., and Panganamala, R. V.,** Modulation of platelet thromboxane A2 and arterial prostacyclin by dietary vitamin E, *Prostaglandins,* 22, 651, 1981.
11. **Okuma, M., Takayama, H., and Uchino, H.,** Generation of prostacyclin like substance and lipid peroxidation in vitamin E-deficient rats, *Prostaglandins,* 19, 527, 1980.

12. **Spitz, B., Deckmyn, H., Van Bree, R., Pijnenborg, R., Vermylen, J., and Van Assche, F. A.,** Influence of a vitamin E-deficient diet on protacyclin production by mesometrial triangles and aortic rings from nondiabetic and diabetic pregnant rats, *Am. J. Obstet. Gynecol.,* 151, 116, 1985.
13. **Likoff, R. O., Guptill, D. R., Lawrence, L. M., McKay, C. C., Mathias, M. M., Nockels, C. F., and Tengerdy, R. P.,** Vitamin E and aspirin depress prostaglandins in protection of chickens against *Escherichia coli* infection, *Am. J. Clin. Nutr.,* 34, 245, 1981.
14. **Lawrence, L. M., Mathias, M. M., Nockels, C. F., and Tengerdy, R. P.,** The effect of vitamin E on prostaglandin levels in the immune organs of chicks during the course of an *E. coli* infection, *Nutr. Res.,* 5, 497, 1985.
15. **Meydani, S. N., Meydani, M., Verdon, C. P., Shapiro, A. A., Blumberg, J. B., and Hayes, K. C.,** Vitamin E supplementation suppresses prostaglandin E2 synthesis and enhances the immune response of aged mice, *Mech. Ageing Dev.,* 34, 191, 1986.
16. **Schoene, N. W., Morris, V. C., and Levander, O. A.,** Altered arachidonic acid metabolism in platelets and aortas from selenium-deficient rats, *Nutr. Res.,* 6, 75, 1986.
17. **Doni, M. G., Bonaccorso, G., and Piva, E.,** High glutathione peroxidase and prostacyclin-like activity generation in rat aorta, *Haemostasis,* 13, 248, 1983.
18. **Wu-Wang, C. Y., Craig-Schmidt, M. C., and Faircloth, S. A.,** Conversion of arachidonate to prostanoids by lung microsomes from rats fed varying amounts of vitamin E, *Prostaglandins Leukotrienes Med.,* 26, 291, 1987.
19. **Marshall, P. J., Kulmacz, R. J., and Lands, W. E.,** Constraints on prostaglandin biosynthesis in tissues, *J. Biol. Chem.,* 262, 3510, 1987.
20. **Moncada, S., Gryglewski, R. J., Bunting, S., and Vane, J. R.,** Lipid peroxide inhibits enzyme in blood vessel microsomes that generates from prostaglandin endoperoxides substance (prostaglandin-X) which prevents platelet aggregation, *Prostaglandins,* 12, 715, 1976.
21. **Kent, R. S., Diedrich, S. L., and Whorton, A. R.,** Regulation of vascular prostaglandin synthesis by metabolites of arachidonic acid in perfused rabbit aorta, *J. Clin. Invest.,* 72, 455, 1983.
22. **Ham, E. A., Egan, R. W., Soderman, D. D., Gale, P. H., and Kuehl, F. A., Jr.,** Peroxidase-dependent deactivation of prostacyclin synthase, *J. Biol. Chem.,* 254, 2191, 1979.
23. **Halliwell, B. and Gutteridge, J. M. C.,** The importance of free radicals and catalytic metal ions in human disease, *Mol. Aspects Med.,* 8, 89, 1985.
24. **Tate, R. M. and Repine, J. E.,** Phagocytes, oxygen radicals, and lung injury, in *Free Radicals in Biology,* Vol. 6, Pryor, W. A., Ed., Academic Press, New York, 1984, 199.
25. **Gutteridge, J. M. C.,** Lipid peroxidation: some problems and concepts, in *Oxygen Radicals and Tissue Injury,* Proc. Upjohn Symp., Halliwell, B., Ed., FASEB, Bethesda, MD, 1988, 9.
26. **Fisher, A. B.,** Intracellular production of oxygen-derived free radicals, in *Oxygen Radicals and Tissue Injury,* Proc. Upjohn Symp., Halliwell, B., Ed., FASEB, Bethesda, MD, 1988, 34.
27. **Gutteridge, J. M. C., Winyard, P. G., Blake, D. R., Lunec, J., Brailsford, S., and Halliwell, B.,** The behavior of caeruloplasmin in stored human extracellular fluids in relation to ferroxidase II activity, lipid peroxidation and phenanthroline-detectable copper, *Biochem. J.,* 230, 517, 1985.
28. **Simpson, P. J., Fantone, J. C., and Lucchesi, B. R.,** Myocardial ischemia and reperfusion injury: oxygen radicals and the role of the neutrophil, in *Oxygen Radicals and Tissue Injury,* Proc. Upjohn Symp., Halliwell, B., Ed., FASEB, Bethesda, MD, 1988, 63.
29. **Paytner, D. I.,** The role of dietary copper, manganese, selenium, and vitamin E in lipid peroxidation in tissues of the rat, *Biol. Trace Elem. Res.,* 2, 121, 1980.

30. **Chow, C. K.,** Nutritional influence on cellular antioxidant defense systems, *Am. J. Clin. Nutr.*, 32, 1066, 1979.
31. **Jenkinson, S. G., Lawrence, R. A., Burk, R. F., and Williams, D. M.,** Effects of copper deficiency on the activity of the selenoenzyme glutathione peroxidase and on excretion and tissue retention of 75-SeO3, *J. Nutr.*, 112, 197, 1982.
32. **Allen, K. G. D., Arthur, J. R., Morrice, P. C., Nicol, F., and Mills, C. F.,** Copper deficiency and tissue glutathione concentration in the rat, *Proc. Soc. Exp. Biol. Med.*, 187, 38, 1988.
33. **Davies, N. T. and Sarkozy, P.,** The effects of copper deficiency on lipid peroxidation in rat liver microsomes, in *Trace Elements in Man and Animals,* Vol. 5, Mills, C. F., Bremner, I., and Chesters, J. K., Eds., Commonweath Agricultural Bureaux, Farnham Royal, U.K., 1985, 39.
34. **Fields, M., Ferretti, R. J., Smith, J. C., and Reiser, S.,** Interaction between dietary carbohydrate and copper nutriture on lipid peroxidation in rat tissues, *Biol. Trace Elem. Res.*, 6, 379, 1984.
35. **Boyne, R. and Arthur, J. R.,** Effects of selenium and copper deficiency on neutrophil function in cattle, *J. Comp. Pathol.*, 91, 271, 1981.
36. **Korte, J. J. and Prohaska, J. R.,** Dietary copper deficiency alters protein and lipid composition of murine lymphocyte plasma membranes, *J. Nutr.*, 117, 1076, 1987.
37. **Johnson, W. T. and Kramer, T. R.,** Effect of copper deficiency on erythrocyte membrane proteins of rats, *J. Nutr.*, 117, 1085, 1987.
38. **Lampi, K. J., Mathias, M. M., Rengers, B. D., and Allen, K. G. D.,** Dietary copper and copper dependent superoxide dismutase in hepatic prostaglandin synthesis by rat liver homogenates, *Nutr. Res.*, 8, 1191, 1988.
39. **Mitchell, L. L., Allen, K. G. D., and Mathias, M. M.,** Copper deficiency depresses rat aortae superoxide dismutase activity and prostacyclin synthesis, *Prostaglandins,* 35, 977, 1988.
40. **Maddox, I. S.,** The role of copper in prostaglandin synthesis, *Biochim. Biophys. Acta,* 306, 74, 1973.
41. **Lee, R. E. and Lands, W. E. M.,** Cofactors in the biosynthesis of prostaglandins E and F2a, *Biochim. Biophys. Acta,* 260, 203, 1972.
42. **Peplow, P. V. and Hurst, P. R.,** Prostaglandin synthesis by rat uterine homogenates in the presence of copper ions, and the effects of indomethacin and salicylic acid, *Prostaglandins Leukotrienes Med.*, 21, 15, 1986.
43. **Fujita, T., Ohtani, N., Aihara, M., Nishioka, K., and Fujimoto, Y.,** Comparison of the effects of Fe2 + and Cu2 + on prostaglandin synthesis in rabbit kidney medulla slices, *J. Pharm. Pharmacol.*, 39, 230, 1987.
44. **Kelly, R. W. and Abel, M. H.,** Copper and zinc inhibit the metabolism of prostaglandin by the human uterus, *Biol. Reprod.*, 28, 883, 1983.
45. **Chan, J. A., Nagawawa, M., Takeguchi, C., and Sih, C. J.,** On agents favoring prostaglandin F formation during biosynthesis, *Biochemistry,* 14, 2987, 1975.
46. **Elliott, G. R., van Batenburg, M. J., and Bonta, I. L.,** Copper modulation of macrophage cyclooxygenase metabolite synthesis, *Prostaglandins,* 34, 657, 1987.
47. **Vargaftig, B. B., Tranier, Y., and Chignard, M.,** Inhibition by sulfhydryl agents of arachidonic acid-induced platelet aggregation, and release of potential inflammatory substances, *Prostaglandins,* 8, 133, 1974.
48. **Ayalla, B. and Gloria, C.,** Copper amplification of prostaglandin E2 stimulation of the release of luteinizing hormone-releasing hormone is a postreceptor event, *Proc. Natl. Acad. Sci. U.S.A.*, 84, 580, 1987.
49. **Cunnane, S. C., Zinner, H., Horrobin, D. F., Manku, M. S., Morgan, R. O., Karmali, R. A., Ally, A. I., Karmazyn, M., Barnete, W. E., and Nicolaou, K. C.,** Copper inhibits pressor responses to noradrenaline but not potassium. Interactions with prostaglandins E1, E2, and I2 and penicillamine, *Can. J. Physiol. Pharmacol.*, 57, 35, 1979.

50. **Laudanski, T., Kostrzewska, A., and Akerlund, M.,** Involvement of prostaglandins in the effect of cupric ions on the human uterus, *Prostaglandins,* 32, 33, 1986.
51. **Ellis, E. F. and Cockrell, C. S.,** The effect of superoxide dismutase and catalase on metabolism of 3H-arachidonic acid by washed platelets, *Prostaglandins Leukotrienes Med.,* 13, 153, 1984.
52. **Parente, L.,** Study on the effect of superoxide dismutase on arachidonic acid metabolism, *Prostaglandins,* 23, 725, 1982.
53. **Allen, K. G. D., Twedt, D. C., and Hunsaker, H. A.,** Tetramine cupruretic agents: a comparison in dogs, *Am. J. Vet. Res.,* 48, 28, 1987.
54. **Twedt, D. C., Hunsaker, H. A., and Allen, K. G. D.,** Use of 2,3,2-tetramine as a hepatic copper chelating agent for treatment of copper hepatotoxicosis in Bedlington Terriers, *J. Am. Vet. Med. Assoc.,* 192, 52, 1988.
55. **Allen, K. G. D., Lampi, K. J., Bostwick, P. J., and Mathias, M. M.,** Increased thromboxane production in clotting whole blood from copper deficient rats, submitted, 1988.
56. **Riley, S. C. and Poyser, N. L.,** Effects of oestradiol, progesterone, hydrocortisone and oxytocin on prostaglandin output from the guinea-pig endometrium maintained in tissue culture, *Prostaglandins,* 34, 535, 1987.
57. **Mathias, M. M. and Dupont, J.,** Effects of dietary fat on eicosanoid production in normal tissues, in *Carcinogenesis and Dietary Fat,* Abraham, S., Ed., Martinus Nijhoff, Boston, in press, 1988.
58. **Nugteren, D. H., Beerthuis, R. K., and van Dorp, D. A.,** The enzyme conversion of all-cis 8,11,14-eicosatrienoic acid into prostaglandin E1, *Recl. Trav. Chim.,* 85, 405, 1966.
59. **Vargaftig, B. B., Tranier, Y., and Chignard, M.,** Blockade by metal complexing agents and by catalase of the effects of arachidonic acid on platelets: relevance to the study of anti-inflammatory mechanisms, *Eur. J. Pharmacol.,* 33, 19, 1975.
60. **Elliott, G. R., Adolphs, M. J. P., Van Batenburg, M., and Bonta, I. L.,** Copper regulation of macrophage eicosanoid and cyclic-AMP synthesis, in *Biology of Copper Complexes,* Sorenson, J. R. J., Ed., Humana Press, Clifton, NJ, 1987, 229.
61. **Oster, G. and Salgo, M. P.,** The copper intrauterine device and its mode of action, *N. Engl. J. Med.,* 293, 432, 1975.
62. **Salgo, M. P. and Oster, G.,** Copper stimulation and inhibition of the rat uterus, *Fertil. Steril.,* 25, 113, 1974.
63. **Singh, N. P. and Medeiros, D. M.,** Effect of copper deficiency and sodium intake upon liver lipid and mineral composition in the rat, *Biol. Trace Elem. Res.,* 6, 423, 1984.
64. **Lei, K. Y., Rosenstein, F., Shi, F., Hassel, C. A., Carr, T. P., and Zhang, J.,** Alterations in lipid composition and fluidity of liver plasma membranes in copper-deficient rats, *Proc. Soc. Exp. Biol. Med.,* 188, 335, 1988.
65. **Ovecka, G. D., Miller, G., and Medeiros, D. M.,** Fatty acids of liver, cardiac and adipose tissues from copper deficient rats, *J. Nutr.,* 118, 480, 1988.
66. **Cunnane, S. C., Horrobin, D. F., and Manku, M. S.,** Contrasting effects of low or high copper intake on rat tissue lipid essential fatty acid composition, *Ann. Nutr. Metab.,* 29, 103, 1985.
67. **Cunnane, S. C., McAdoo, R. K., and Prohaska, J. R.,** Lipid and fatty acid composition of organs from copper-deficient mice, *J. Nutr.,* 116, 1248, 1986.
68. **Balevska, P. S., Russanov, E. M., and Stanchev, P. I.,** Copper deficiency-induced changes in the fatty acid composition of mitochondrial and microsomal membranes of rat liver, *Biol. Trace Elem. Res.,* 8, 211, 1985.
69. **Klevay, L. M., Reck, S. J., and Barcome, D. F.,** Evidence of dietary copper and zinc deficiencies, *JAMA,* 241, 1916, 1979.
70. **Holden, J. M., Wolf, W. R., and Mertz, W.,** Zinc and copper in self-selected diets, *J. Am. Diet. Assoc.,* 75, 23, 1979.

71. **Klevay, L. M., Reck, S. J., Jacob, R. A., Logan, G. M., Munoz, J. M., and Sandstead, H. H.,** The human requirement for copper. I. Healthy men fed conventional American diets, *Am. J. Clin. Nutr.*, 33, 45, 1980.
72. **Patterson, K. Y., Holbrook, J. T., Bodner, J. E., Kelsay, J. L., Smith, J. C., and Veillon, C.,** Zinc, copper and manganese intakes and balance for adults consuming self-selected diets, *Am. J. Clin. Nutr.*, 40, 1397, 1984.
73. **Food and Nutrition Board, and National Research Council,** *Recommended Dietary Allowances,* 9th ed., National Academy of Sciences, Washington, D.C., 1980.

Chapter 10

# COPPER AND THE ARTERY

**Kenneth G. D. Allen**

## TABLE OF CONTENTS

## I. INTRODUCTION

The earliest reference to a role for copper in the cardiovascular system is that of Bennetts et al. in 1942 which describes "falling disease" in copper-deficient cattle.[1] Dietary copper deficiency has structural consequences in both the myocardium and the vascular system. This review concentrates on changes in the arterial system due to copper deficiency. Alterations in arterial morphology may have consequences for subsequent atherosclerotic disease. The structural changes observed in the arteries of frankly copper-deficient animals show some links with the generally accepted events in the pathogenesis of atherosclerosis.[2,3] In addition, marginal copper deficiency studies have demonstrated structural changes in the artery that may have bearing on the atherosclerotic process.

The structure of the normal artery comprises three morphologically distinct layers.[2,3] The intima, or innermost area, consists of a narrow region with a single layer of endothelial cells separating this region from the lumen of the artery. The internal elastic lamina separates the intima from the media, or middle layer, of the artery. This internal elastic lamina consists of a fenestrated sheet of elastic fibers. The media comprises smooth muscle cells, and associated extracellular connective tissue components synthesized by these smooth muscle cells such as collagen, elastin, and glycosoaminoglycans (proteoglycans) — the extracellular matrix. Elastic arteries, such as the aorta, contain larger amounts of elastic tissue than that found in the smaller muscular arteries. In such elastic arteries, the smooth muscle cells arrange in small groups and spiral between the elastic fibers supporting these arteries. In small muscular arteries, such as the coronary artery, there are less elastic fibers and proportionally more smooth muscle cells. The adventitia, or outermost area of the artery, serves to anchor the artery to the surrounding tissue and contains fibroblasts, some smooth muscle cells and extracellular connective tissue matrix components. The adventitia is separated from the media by the external elastic lamina which is a discontinuous sheet of elastic tissue, more poorly defined than the internal elastic lamina.

Elastic fibers, such as found in the elastic laminae of arteries, are composed of the connective tissue protein elastin. This protein imparts an elastic, or compliant, property to blood vessels that allows the vessels to expand and contract to accommodate the pumping of blood.[4] Collagen, a major constituent of the extracellular matrix of the media and also of the intima, is a connective tissue component which imparts plastic and tensile properties to tissues.[4] Proteoglycans of the connective tissue matrix serve to maintain water balance within the connective tissue matrix and possibly as plasticizers for collagen and elastin.[5,6] The role of nutrients in synthesis of connective tissue proteins has recently been reviewed.[6]

Structural defects in collagen and elastin caused by copper deficiency are attributable to the copper-dependent enzyme lysyl oxidase (protein lysine 6-

oxidase; EC 1.4.3.13). This enzyme catalyzes the conversion of the amine side chains of specific peptidyl lysine and hydroxylysine residues to their corresponding aldehydes, allysine, and hydroxyallysine, respectively.[7] Lysyl oxidase is a copper metalloenzyme and copper is required for both enzymatic activity and the synthesis of the apoenzyme at the transcriptional or translational level.[8-10] Oxidation of specific lysyl and hydroxylysyl residues in collagen and elastin is the major post-translational modification of these connective tissue proteins allowing for the formation of lysine derived cross-links. Steps in the biosynthesis of both collagen and elastin have been recently reviewed.[6] Proline and lysine hydroxylation occurs intracellularly during procollagen transcription whereas peptidyl lysine oxidation occur subsequent to triple helix formation and occurs as an extracellular event.[6] In elastin biosynthesis tropoelastin is secreted from cells and cross-linking again occurs as an extracellular event.[6] Elastin turnover rates appear to be much longer than those of collagen. Turnover times for elastin appear to be in the range of months or years.[6] This slow turnover may be a reflection of the unique amino acid sequences present in elastin which imparts an apolar character and resistance to hydrolysis.[6]

Considerable effort has been directed to identifying the cross-links present in both collagen and elastin.[7,11] In both proteins all cross-links are derived from peptide lysine and hydroxylysine aldehydes (allysine and hydroxyallysine) resulting from lysyl oxidase activity. Several reviews have summarized the cross-links present in collagen and elastin.[6,7,11] In general collagen cross-links fall into three categories: aldol condensation products arising from lysine and allysine residues, condensation products formed from hydroxylysine residue, and the more recently identified pyridinoline (3-hydroxypyridinium) cross-links. In elastin, which lacks hydroxylated lysine, cross-links consist mainly of lysyl derived desmosine (pyridinium) isomers. Other chemically stable derivatives found in elastin are lysinonorleucine (also involved in collagen cross-linking) and merodesmosine. The pyridinium isomers (desmosine and isodesmosine) of elastin arise from condensation of four peptidyl lysine residues whereas the 3-hydroxypyridinium (pyridinoline) cross-links in collagen arise from condensation of three peptidyl lysine residues.[6] Mechanisms of formation of collagen and elastin cross-links and the analysis of cross-linking components have been reviewed by Robins.[11]

Studies on the arterial changes caused by copper deficiency have in part been motivated by a possible role for dietary copper in the development of atherosclerosis and coronary heart disease.[12,13] The pathogenesis of atherosclerosis, a disease of the intima, has recently been reviewed and an updated response-to-injury hypothesis has been formulated.[2,3] Basic events in the pathogenesis of atherosclerosis postulated by this response-to-injury hypothesis include endothelial damage (either denuding or nondenuding) as the initial and most important feature of atherosclerosis induction. However, such endothelial damage does not have to be limited to denuding injury as was originally thought to be the case.[2] The sequence of events in the development of atherosclerosis postulates that endothelial changes, such as produced by diet-induced hypercholesterolemia, involves monocyte, and possibly platelet, adherence to the endothelium as the first event. Subsequent macrophage development and chemoattractant and mitogen synthesis by the macrophage leads to smooth muscle cell accumulation in the intima—a characteristic of atherosclerosis. The fatty streak of arteries is considered to be an early stage of atherosclerosis.[3] The synthesis of chemoattractants by endothelial cells, platelets, and macrophages is thought to account for the movement and proliferation of smooth muscle cells within the intima.[3] Subsequent accumulation of cholesterol by macrophages, and perhaps smooth muscle cells, leads to the progression of the mature lesion which consists of intimal smooth muscle cells surrounded by connective tissue matrix and variable amounts of extracellular lipid. Events in the pathogenesis of atherosclerosis have been described in month-by-month progression in the aorta and iliac arteries of hyperlipidemic monkeys.[14,15]

## II. EXPERIMENTAL APPROACH

### A. Dietary Copper Deficiency

Dietary copper deficiency has been achieved in a variety of animal models using frankly copper-deficient diets, usually $< 1.0$ μg Cu/g. Diets employed for producing copper deficiency have also included milk-based diets,[16] which are low in copper, and also by feeding zinc to pregnant sows to reduce copper absorption and produce copper-deficient offspring.[17] Supplemental vitamin C (ascorbic acid), which reduces copper absorption, has also been used to achieve copper deficiency.[18] In some early studies, removal of dietary copper as the insoluble sulfide has also been employed.[18] More recent studies on dietary copper deficiency use AIN (American Institute of Nutrition) based diets providing $<1.0$ μg Cu/g diet.

Marginal copper deficiency has also been examined in the rat.[19] The dietary requirement for copper for the rat is 5.0 μg Cu/g diet for growth, reproduction, and lactation.[20] The AIN recommendation for dietary copper is 6.0 μg Cu/g diet.[21] However, recent work has suggested that 9.0 μg Cu/g diet is necessary for optimal support of gestation and lactation in the rat.[22] Marginal copper deficiency has been achieved in the rat using a 2.0 μg Cu/g diet fed to female rats, breeding these rats, and feeding the male offspring the same diet for up to 117 d.[19] Marginal copper deficiency is characterized by normal serum copper concentrations, but with reduced hepatic copper concentrations. Consequently, marginal copper-deficient rats appear normal in terms of growth, reproductive success, and serum copper and cholesterol concentrations, are without the cardiac rupture characteristic of severe (frank) copper deficiency in rats, but have significantly decreased liver copper stores although the hepatic copper concentrations are much higher than those of frank deficiency.[19]

### B. Tissue Fixation and Preparation

In most of the early studies involving dietary copper and arterial structure, artery samples obtained at autopsy were removed and subsequently fixed by immersion in formalin or glutaraldehyde, 2 to 10%. Fixed tissues were embedded in paraffin, sectioned and stained for histological examination. Typical stains have been hematoxylin and eosin and various stains such as Richardson's trichrome, Verhoeff's elastic-Van Greson, Masson's trichrome, alcian blue in order to discriminate among collagen, elastic tissue, and glycosoaminoglycans. Perfusion fixation has also been employed.[19] Typically the vasculate is perfused with an electrolyte solution prior to perfusion and fixation with glutaraldehyde.[19] Such an approach is thought to limit glutaraldehyde-caused vascular changes that occur when blood is not first removed which can complicate morphological interpretation.[23] Lock's physiological saline and Earles balanced salt solution have been used for rinsing the systemic vasculature.[19,23] Perfusion is accomplished at physiological pressure, 100 to 130 mm Hg. Perfusate is introduced into the aorta via the left ventricle and the femoral artery is severed. Approximately 50 ml of physiological electrolyte solution is sufficient to remove blood from the vasculature. The fixture is perfused until 100 ml has flowed, and the formaldehyde/glutaraldehyde used (0.5%/1.5%) is typically prepared in cacodylate buffer.[19] Following fixation the aortae are excised, dissected, and submerged in aldehyde fixative for a further period.[19] Postfixation in buffered 1% osmium tetroxide is followed by stepwise dehydration using increased ethanol concentration (50 to 100%) and embedding. Thin sections are stained with uraryl acetate and lead citrate prior to transmission electron microscopy.[19]

In the majority of studies, the consequences of dietary copper deficiency or marginality have been descriptive rather than quantitative. However, pathological grading of arterial lesions has been reported[25] and quantitation of cellularity and elastin and collagen synthesis has also been reported.[25] Morphometric analysis has not been reported.

## III. STRUCTURAL CHANGES IN COPPER-DEFICIENT ARTERIES

### A. Severe (Frank) Copper Deficiency

The first observations of aortic lesions were reported in copper-deficient chickens [26] and swine.[27] In chickens [26] dissecting aneurysms of the aorta were reported with death due to hemorrhaging. Thickening of the wall of the aorta was observed and this was reported as occurring in the medial region of the artery. The luminal cross-sections were greatly reduced by copper deficiency. The elastic fibers in copper-deficient chick aortae showed characteristic focal breaks, fragmentation, and increased separation between fibers. It was also suggested that increased amounts of collagen were present in the media of copper-deficient chicks. Aortic lesions of a similar nature were also reported in copper-deficient turkey poults.[28]

The characteristics of arteries from copper-deficient swine included ruptures of the coronary and pulmonary arteries and the ascending aorta.[27] Some thickening of the thoracic aortic media was also observed. Histological examination of the aortae showed numerous defects at sites not involved by gross lesions. The defects reported were accumulations of amorphous material between elastic laminae of the media, rupture or abnormal fenestration of the elastic laminae. Tensile strength of the descending thoracic aorta was reduced nearly threefold by copper deficiency. In muscular arteries such as the coronary arteries, defects in elastic components appear more pronounced in the internal elastic lamina.[27] In addition, the subendothelial space of copper-deficient muscular arteries appears to be wider than in control animals with more electron-dense microfibrils and collagen fibrils.[27] Evidence also suggests that medial-derived smooth muscle cells protrude through the internal elastic lamina of copper-deficient muscular arteries, including the coronary arteries, which results in endothelial cell elevations. Although protrusion of these medial smooth muscle cells is documented, evidence for migration and proliferation of these smooth muscle cells in the intima is not compelling in this study. Stainable lipid was not reported in the intimal lesions of the copper-deficient pig.[27]

Increased proteoglycans have been reported in copper-deficient aortas in pigs[29] and rats.[30] Linker et al. have reported a threefold increase in copper-deficient swine aorta proteoglycan content.[29] In particular, dermatan sulfate and chondroitin 4- or 6-sulfate, or both 4- and 6-sulfate increased while heparin sulfate levels remained unaltered. Apparently this increase in aortic proteoglycan content is not reflected by changes in articular cartilage proteoglycans. The change in proteoglycans in copper-deficient swine aorta does not explain the diminished tensile strength of these aortae.[31] Increased proteoglycan production in copper-deficient rat aortae has also been reported.[30] Copper deficiency altered the concentration, size, and quantity of certain proteoglycans in rat aortae. Total uronate in the aortae of copper-deficient rats increased 25%, and proteoglycans were of larger molecular weight.[30] Concentrations of isomeric chondroitin sulfates, particularly dermatan sulfate, were greater in copper-deficient aortae. A similarity to proteoglycans in atherosclerotic aortae was noted.[30,32] Proteoglycans may play a role in extracellular lipid accumulation, particularly from low density lipoproteins.[2,33] This increase in isomeric chondroitin sulfates may reflect a generalized increase in extracellular connective tissue matrix in the arteries of copper-deficient swine and rats since evidence of increased collagen and elastin production has also been indicated.[25]

In a subsequent study with miniature swine, copper restriction was achieved by periodic copper supplementation of severely copper-deficient swine.[34] This periodic supplementation was used to prevent death due to arterial rupture in severe copper deficiency, and to examine the repair of small arterial tears. Copper-restricted animals showed progressive vascular lesions characterized by loss of elastic fibers with intimal and medial fibrosis in both muscular and elastic arteries.[34] Intimal fibrous plaques contained occasional foamy macrophages (histiocytes). This report is the first indicating the presence of what is presumably a lipid laden macrophage in copper-deficient arterial tissue. The importance of both medial fibrosis and the lipid laden macrophage in the development of atherosclerosis has been reviewed.[2,3] The copper-restricted pigs in this study survived with scarred arteries. Many foamy macrophages were present in the copper-restricted animals at sites of healed vascular lesions. Intimal fibrosis has also been reported in young copper-deficient swine following copper supplementation.[35] The results in the copper-restricted swine model, involving periodic copper supplementation of frankly copper-deficient animals, may more accurately reflect a copper marginal treatment. Serum and liver copper values for copper-restricted and copper-deficient animals support this interpretation since there is a clear indication that copper concentrations in copper-restricted swine are higher than in copper-deficient animals, but lower than those of control animals.[34]

In another study of copper-deficient swine, the arterial lesions of both frankly deficient and copper-supplemented frankly deficient swine were examined.[35] Characteristic arterial features of frank copper deficiency included endothelium elevation due to medial cells protrusion, presumably smooth muscle cell protrusion, through breaks in the internal elastic lamina. Proliferation of these subendothelial cells was observed with plaque-like thickening of the intimal region. In copper-deficient swine treated with copper for 2 weeks after deficiency was established, plaque-like morphology was examined in detail. The plaques consisted of intimal smooth muscle cells well separated by amorphous material containing collagen and elastin. Endothelial cells remained normal and intact and intimal smooth muscle cells showed elastin formation on the cell surface. No lipid was observed in these intimal lesions. In copper-deficient animals supplemented with copper for longer periods, 3 to 6 weeks, intimal plaques were much thicker with larger amounts of collagen and elastin. Apparently the largest plaques were associated with repair of the widest breaches in the internal elastic laminae and occasionally protruded to partially block the artery.[35] Two studies have detailed the elastin defect and the repair process in copper-deficient pigs.[36,37]

In both of these copper-supplemented copper-deficient swine studies,[34,35] intimal plaque formation was thought to be a repair-associated phenomena. If such is the case, interpretation is difficult since in both studies the animals were initially frankly copper deficient before copper supplements were provided. As mentioned previously, this copper-supplemented treatment of copper-deficient animals may, in some respects, be a marginal copper deficiency study. However, any copper marginality is complicated by a prior frankly deficient state. The repair process in arteries normally involves smooth muscle cells proliferation in the intima.[2] Such a situation is apparently a normal response to injury[2] and once repair is achieved lesion progressions halts.[2] It is only in situations with chronic injury that this normal reparative process can lead to classical atherosclerosis lesions.[2] Hence, it is not clear that these fibromuscular lesions[34,35] are progenitors of a classical atherosclerotic lesion. In one study, no lipid was evident,[35] while in the other foamy histiocytes (macrophages) were reported in the intimal lesion but no indication of lipid accumulation was presented.[34] The role of the macrophage in lipid accumulation has been emphasized in the atherosclerotic process.[3]

Arterial lesions in copper-deficient rabbits have also been characterized.[18] These arterial lesions were of three main types. Separation and fragmentation of the elastic lamellae of the media in elastic arteries, calcification of the internal elastic lamina in both muscular and elastic arteries with frequent breaks and disruptions in this internal elastic lamina, and calcification and fragmentation of the internal elastic lamina in the media of muscular arteries. Irregular intimal thickening was also reported in the muscular arteries.[18] Nystrom has also reported defects in the coronary arteries of copper-deficient rabbits.[38] Fibroblast and lipid accumulations occurred close to breaks and ruptures of the coronary vessels. This report concluded that changes in lipid metabolism were the primary cause of arterial lesions.[38]

In copper-deficient cattle a reduction in aorta wall thickness was observed.[39] This was accompanied by a decrease in collagen present between the elastic laminae. These changes in copper-deficient cattle are quite different from those observed in swine, chicks, and rabbits. In copper-deficient cattle, decreases were also noted in smooth muscle cells in both the aorta and the vena cava. Atrophy of smooth muscle cells was also observed, but there was a surprising lack of defective elastogenesis in copper-deficient cattle aortae.[39]

The first investigation of cardiovascular defects in copper-deficient rats was reported by Kelly et al.[40] Although extensive myocardial lesions were found they reported no evidence of diet-related phathological changes in arteries such as the aorta and coronary arteries, or any internal visceral or peripheral vessels.[40] In a subsequent copper-deficient rat study, by Allen and Klevay, changes in aortal morphology were reported.[41] Marked depletion and distortion of elastic fibers was evident throughout the aortae of copper-deficient rats but no changes were seen in intramyocardial arteries. This study [41] was the first indication of arterial damage in rats due to frank copper deficiency. No aortic ruptures occurred in copper-deficient rats despite significant changes in elastic tissue. Apparently, rats are resistant to aortic rupture in contrast to copper deficiency in swine and chicks. Lesions in the arteries of copper-deficient rats have been confirmed by Petering et al.[24,42] Arterial lesions were present mostly in the branches of the distal aorta and also the major brain arteries. Arterial lesions involved proliferation of smooth muscle cells with damage and separation of the internal elastic laminae from the media. Arterial lesions were present in 85 to 100% of copper-deficient rats.[24]

Kitano has also investigated the structural defects in aortae from copper-deficient rats.[43] In copper deficiency, a characteristic widening of the subendothelial space was reported but endothelial cells appeared normal. Internal elastic laminae were disrupted by copper deficiency with only irregular clumps of elastin remaining. Smooth muscle cells were not observed in the intima of copper-deficient rats in this study. The widening of the subendothelial space was associated with the appearance of amorphous reticular material. In the media of copper-deficient animals, collagen fibers were reduced and fragments of elastic fibers surrounded by microfibrils were also observed. No changes in the number or morphology of medial smooth muscle cells was noted. The reduction in tensile strength of copper-deficient aortae was attributed to decreases in collagen and elastin. Copper deficiency also produced reductions in aorta norepinephrine content. Changes in resting membrane potential of copper-deficient pulmonary arteries were also observed.[43]

In a recent study with copper-deficient pig aorta, there is evidence of induction of increased collagen and elastin biosynthesis.[25] In this study, there was a twofold increase in thickness and diameter of the thoracic aortae in copper deficiency. In addition, there were large increases in the number of smooth muscle cells per unit area throughout the length of copper-deficient aortae. The increases in smooth muscle cells were quantitated by a fluorometric DNA assay and were in the range of 1.3- to 1.6-fold throughout the aortae. It is not clear in this study if smooth muscle cell hyperplasia occurred in all areas of the artery or if it was more pronounced in the intimal or medial regions. Explants of aortic segments indicated increases in both relative and absolute synthesis of collagen and elastin. Collagen synthesis appeared to be increased along the length of the copper-deficient aortae with maximal increases in the thoracic segment. Elastin synthesis increased in the thoracic aortae, but not in the abdominal aortae, of copper-deficient animals. Increases in relative collagen and elastin synthesis were in the range 1.3- to 4-fold, with larger increases being observed in relative elastin synthesis than relative collagen synthesis. Absolute production of both collagen and elastin also increased 4- to 15-fold and the increase was again greater for elastin production than for collagen production. When changes in cellularity were taken into account, the net collagen production and elastin (microgram synthesis per gram wet tissue per hour) increased by over 15-fold for collagen and to 30-fold for elastin in copper-deficient thoracic aortae.[25] Thus, copper deficiency produced smooth muscle cell hyperplasia and a striking increase (induction) of connective tissue proteins.

### B. Marginal Copper

Frank copper deficiency, even in those studies where supplemental copper is provided intermittently, typically produces severe derangements in arterial structure. Furthermore, frankly copper-deficient animals have serum, or plasma, copper valves which are extremely low (0.1 μg Cu/ml or less), anemia, and markedly depressed growth. Extreme copper deficiency is an unlikely outcome in humans. A more likely situation is a marginal deficit of copper which does not result in very low plasma (serum) copper concentrations or anemia. The marginally copper-deficient animal is a model where clinical copper indices, such as circulatory copper, are normal but there is a deficit of hepatic copper stores.[19] Such a model may be more relevant to human copper status since there is considerable evidence of insufficient dietary copper intake in humans,[44-48] but little evidence for frank (extreme) copper deficiency.[19]

Copper marginality has been produced in rats by feeding weanling females a 2.0 μg/Cu diet and then breeding the females.[19] Male offspring are fed the same dietary copper content. Control females, and their pups are fed 10.0 μg Cu/g diets. This procedure results in male offspring with depressed liver copper stores, although hepatic copper is much higher than in frankly deficient rats, normal serum copper concentrations, normal weights, and normal breeding success and pup mortality. Consequently, such rats can be confidently described as copper marginal rather than frankly copper deficient.[19] Aortal morphology in marginal copper rats is not as severe as in copper-deficient animals. Ultrastructural changes observed in the aorta of copper marginal rats were focally abnormal features of endothelial cells. These abnormalities comprised thinned and narrowed portions of endothelial cells suggesting endothelial damage. These indications of endothelial damage, or changes, are the first indication of endothelial cell changes attributable to dietary copper insufficiency. Although copper deficiency studies have suggested an elevation of endothelial cells due to medial cell protrusion through the internal elastic laminae, the endothelial cells remained morphologically normal.[27] Endothelial thinning and narrowing was associated with cytoplasmic extensions of smooth muscle cells to the endothelial layer and occurred at regions where the internal elastic laminae lacked normal morphological integrity. Defects were also observed in the internal elastic laminae. These deficits were not as severe as those in frank copper deficiency and included discontinuous and narrowed regions of this lamina with elastin changes and associated microfibrils. Smooth muscle cells were often found in the subendothelial space in copper marginal animals which may indicate either a repair process or a response-to-injury mediated intimal proliferation of these cells. Large aggregations of collagen fibers and elastic masses were associated with subendothelial smooth muscle cells perhaps indicating increased smooth muscle cell synthesis. Medial damage was much less pronounced than that of severe copper deficiency. Elastic laminae in the media were uneven, irregular, and gapped. The extracellular space between elastic laminae was more extensive, more dispersed. Smooth muscle cells exhibited pockets of elastic masses. Many of these elastic masses appeared to be condensing onto the surface of the elastic laminae and may indicate enhanced elastogenesis.[19] There was no evidence of lipid accumulation in any of the lesions. Thus, marginally copper-deficient rats show aortal changes similar to, but not as severe as, those of frank copper deficiency. Novel findings of marginal copper aortae were evidence of endothelial cell damage and intimal smooth cell appearance with these cells elaborating increased amounts of collagen and elastin. These findings are indicative of some of the events leading to the fibromuscular lesion of atherosclerosis.[2,3] However, there was no evidence of monocyte/macrophage involvement or progression of these changes to recognizable atherosclerotic lesion. It should be emphasized that these changes occurred in normolipemic rats at 117 days of age, a relatively short period despite the *in utero* induction of copper marginality.

A recent report has extended the marginal copper rat model[19] to examine the influence of diet-induced (cholesterol/cholic acid feeding) hypercholesterolemia on arterial (aorta) endothelial cell morphology examined by scanning electron microscopy.[49] Hypercholesterolemia in copper marginal rats caused leukocyte and platelet adherence to endothelial cells, and endothelial cell damage. It appears that hypercholesterolemic marginal copper rats have endothelial cell damage. However, marginal copper rats with normal cholesterol and hypercholestolemic copper normal rats do not show evidence of endothelial cell damage.[49]

### C. Menkes' Disease

Menkes' disease is an inborne error of copper metabolism in humans where intestinal copper absorption is impaired and copper deficiency results.[50] Defective arteries of Menkes' disease patients are quite similar to those of dietary copper deficiency in animals.[50,51] Typically absent or diminished elastic laminae are observed in the media, split and fragmented internal elastic laminae are also characteristic. The increased amounts of collagen observed may indicate a generalized repair response. Intimal smooth muscle cell proliferation has been reported but endothelial cell changes have not been noted.[51] Rupture of the arteries is reported.[50] The intima is often thickened with intimal proliferation and subintimal plaques present.[50] This human copper deficit produces arterial changes quite similar to those in copper-deficient animals.

## IV. RELEVANCE OF DIETARY COPPER DEFICIENCY-INDUCED ARTERIAL CHANGES TO ATHEROSCLEROSIS

The pathogenesis of atherosclerosis has been reviewed by Ross.[2,3] The fibrous plaque is representative of various forms of advanced atherosclerosis and consists of smooth muscle cells with elaborated extracellular connective tissue matrix (collagen, elastin, and proteoglycans). The lipid content is variable and consists of both intracellular and extracellular lipids. The luminal surface of this lesion is often calcified with a dense fibrous cap of smooth muscle cells and connective tissue components. The fatty streak, commonly found in children, is now considered the earliest form of atherosclerosis and it is recognized that it is a potential progenitor of more advanced lesions.[14,15] Macrophages predominate in fatty streaks while smooth muscle cells are characteristic of advanced lesions. Loss of endothelial cells and exposure of the subendothelial intima is not a prerequisite for the development of atherosclerosis[3] as was thought a decade ago.[2] Thus, damage of the intima with plaque formation can occur in fully endothelial covered intima by platelet and monocyte interaction followed by smooth muscle cell migration and intimal proliferation. Subsequent endothelial retractions, and platelet interaction with exposed intima can accelerate or exacerbate the process. A key feature for atherosclerosis development is the presence or release of chemoattractants and cellular growth factors which cause intimal smooth muscle cell proliferation. The major cells involved in atherosclerosis such as platelets, monocytes, macrophages, endothelial cells, and smooth muscle cells contain or release chemoattractants, for example leukotrienes, or growth factor such as platelet-derived growth factor.[3]

The presence of fibromuscular plaques in the arteries of copper-deficient animals[34,35] and evidence for smooth muscle cell proliferation[25,34,35] suggests a link with the development of atherosclerosis. Furthermore evidence for increases in the extracellular connective tissue matrix components has been shown for collagen, elastin, and proteoglycans.[25,29,30,34,35] Intimal thickening and smooth muscle cell hyperplasia are also characteristic in copper-deficient animals. Taken together, these changes do parallel those observed in atherosclerosis.[2,3] However, gross visible lesions of atherosclerosis have not been reported in copper deficiency studies. Several lines of evidence suggest that lipid accumulation by arterial tissue occurs in copper deficiency. A study in vervet monkeys suggested that the degree of aortic sudanophilia (gross visible lesion) was greatest in the monkeys with lowest hepatic copper stores.[52] It should be noted these animals were not copper deficient nor was the study a copper deficiency study but rather a dietary fiber study.[52] Harvey and Allen[53] observed a 15% increase in total aortic cholesterol in copper-deficient rats. These rats were hypercholesterolemic due to copper deficiency and hence cholesterol accumulation by the arterial tissue could be due to the hypercholesterolemia or the aortic lesions (not examined in this study), or both. These aortic cholesterol increases were based on a dry weight basis. Nystrom has interpreted arterial changes in copper-deficient rabbits due to alternations in lipid metabolism.[38] Studies have reported foamy histiocytes (lipid laden macrophages) in fibrous intimal lesions.[34]

In copper deficiency studies, there is no evidence for endothelial cell abnormalities, despite elevations in endothelial cells due to intimal smooth muscle cell intrusion. The characteristic feature of copper-deficient arteries is major changes in elastin morphology. This has been reported in virtually every study. There is no evidence to suggest that the initial or important lesion of atherosclerosis involves, as its first stage, damage to elastic fibers. Elastic fiber changes are characteristic of advanced atherosclerosis and increased elastin amounts may be involved in extracellular lipid accumulation just as some of the proteoglycans are involved in this process.[19,33] Defects in arterial structure due to copper deficiency may represent a normal repair process.

This is particularly likely in copper-deficient animals provided with occasional copper supplements to prevent death. If such is the case and if this repair process is chronic or repeated due to chronic copper deprivation, it is possible that the normal repair process can lead to pathological lesions. An analogous situation would be the chronic injury and atherosclerosis development hypothesized by Ross and Glomset.[2]

Marginal copper deficiency is much more likely in humans than frank (extreme) copper deficiency. Considerable dietary evidence suggests human copper intakes in the U.S. are below those necessary to replace daily losses.[44-48] However, evidence for extreme copper deficiency in humans is lacking. The marginal copper rat model shows depressed hepatic copper stores but normal circulatory copper concentrations.[19] Aortal changes are similar, but less severe, than those of frank copper deficiency. In particular, elastic fiber changes are subtle. Despite this, there is direct ultrastructural evidence for endothelial damage, intimal smooth muscle appearance and intimal elaboration of collagen and elastin. However, there is no indication of lipid accumulation, either intracellular or extracellular lipid.[19]

## V. SUGGESTIONS FOR FUTURE RESEARCH

Copper deficiency and copper marginality studies have been conducted for quite short periods—several weeks to 3 months. Future research should be directed to arterial changes in marginally copper-deficient animals over longer periods of time. Furthermore, the interplay of hyperlipidemia and marginal copper status should be investigated. Evidence suggests that dietary copper through its role in copper-dependent superoxide dismutase (SOD) influences eicosanoid formation (See Chapter 9) which is known to be dependent on oxy-radical production. Several of the effects of copper deficiency have recently been shown to be decreased by chronic treatment with a hydroxyl radical scavenger.[54] The role of lipid peroxidation in atherosclerosis development has received attention recently.[55,56] Copper deficiency causes both hypercholesterolemia[41,53] and arterial lesions[41,42] and may provide a good model to examine these effects. However, the marginal copper rat model does not exhibit hypercholesterolemia[19] and hence requires dietary lipids to cause cholesterolemia.[49] Elaboration of arterial proteoglycans[29,30] in copper deficiency and elastin in both copper deficiency and marginality[19,25] may enhance lipid accumulation with longer studies since both of these components bind cholesterol.[19,33,57]

# REFERENCES


1. **Bennetts, H. W., Harley, R., and Evans, S. T.,** Studies on copper deficiency of cattle: the fatal termination ("falling disease"), *Aust. Vet. J.*, 18, 50, 1942.
2. **Ross, R. and Glomset, J. A.,** The pathogenesis of atherosclerosis, *N. Engl. J. Med.*, 295, 369 and 420, 1976.
3. **Ross, R.,** The pathogenesis of atherosclerosis — an update, *N. Engl. J. Med.*, 314, 488, 1986.
4. **Gallop, P. M. and Paz, M. A.,** Posttranslational protein modifications, with special attention to collagen and elastin, *Physiol. Rev.*, 55, 418, 1975.
5. **Muir, H.,** Proteoglycans as organizer of the intercellular matrix, *Biochem. Soc. Trans.*, 11, 613, 1983.
6. **Tinker, D. and Rucker, R. B.,** Role of selected nutrients in synthesis, accumulation and chemical modification of connective tissue proteins, *Physiol. Rev.*, 65, 607, 1985.
7. **Eyre, D. R., Paz, M. A., and Gallop, P. M.,** Cross-linking in collagen and elastin, *Annu. Rev. Biochem.*, 53, 717, 1984.
8. **Siegel, R. C., Pinnell, S. R., and Martin, G. R.,** Cross-linking of collagen and elastin. Properties of lysyl oxidase, *Biochemistry,* 9, 4486, 1970.
9. **Harris, E. D.,** Copper-induced activation of aortic lysyl oxidase in vivo, *Proc. Natl. Acad. Sci. U.S.A.*, 73, 371, 1974.
10. **Rayton, J. K. and Harris, E. D.,** Induction of lysyl oxidase with copper, *J. Biol. Chem.*, 254, 621, 1979.
11. **Robins, S. P.,** Anaylsis of the crosslinking components in collagen and elastin, *Methods Biochem. Anal.*, 28, 329, 1982.
12. **Klevay, L. M.,** Coronary heart disease: the zinc/copper hypothesis, *Am. J. Clin. Nutr.*, 28, 764, 1975.
13. **Klevay, L. M.,** Ischemic heart disease. A major obstacle to becoming old, *Clin. Geriatr. Med.*, 3, 361, 1987.
14. **Faggiotto, A., Ross, R., and Harker, L.,** Studies of hypercholesterolemia in the nonhuman primate. I. Changes that lead to fatty streak formation, *Arteriosclerosis,* 4, 323, 1984.
15. **Faggiotto, A. and Ross, R.,** Studies of hypercholesterolemia in the nonhuman primate. II. Fatty streak conversion to fibrous plaque, *Arteriosclerosis,* 4, 341, 1984.
16. **Sandberg, L. B. and Wolt, T. R.,** Production and isolation of soluble elastin from copper-deficient swine, *Methods Enzymol.*, 82, 657, 1982.
17. **Hill, G. M., Ku, P. K., Miller, E. R., Ullrey, D. E., Losty, T. A., and O'Dell, B. L.,** A copper deficiency in neonatal pigs induced by a high zinc maternal diet, *J. Nutr.*, 113, 867, 1983.
18. **Hunt, C. E. and Carlton, W. W.,** Cardiovascular lesions associated with experimental copper deficiency in the rabbit, *J. Nutr.*, 87, 385, 1965.
19. **Hunsaker, H. A., Morita, M., and Allen, K. G. D.,** Marginal copper deficiency in rats. Aortal morphology of elastin and cholesterol values in first generation adult males, *Atherosclerosis,* 51, 1, 1984.
20. **National Research Council Subcommittee on Laboratory Animal Nutrition,** Nutrient Requirement of Laboratory Animals (Nutrient Requirement of Domestic Animals, Number 10), 3rd ed., National Academy of Sciences, Washington, D.C., 1978, 17 and 23.
21. **Bieri, J. G., Stoewsand, G. S., Briggs, G. M., Phillips, R. W., Woodard, J. C., and Knapka, J. J.,** Report of the American Institute of Nutrition Ad Hoc Committee on Standards for Nutritional Studies, *J. Nutr.*, 107, 1340, 1977.
22. **Cerklewski, F. L.,** Determination of a copper requirement to support gestation and lactation for the female albino rat, *J. Nutr.*, 109, 1529, 1979.
23. **Lee, R. M. K. W.,** Preservation of in vivo morphology of blood vessels for morphometric studies, *Scanning Microscopy,* 1, 1287, 1987.

24. **Petering, H. G., Murthy, L., Stemmer, K. L., Finelli, V. N., and Menden, E. E.,** Effects of copper deficiency on the cardiovascular system of the rat, *Biol. Trace Element Res.*, 9, 251, 1986.
25. **Hill, K. E. and Davidson, J. M.,** Induction of increased collagen and elastin biosynthesis in copper-deficient pig aorta, *Arteriosclerosis*, 6, 98, 1986.
26. **O'Dell, B. L., Hardwick, B. C., Reynolds, G., and Savage, J. E.,** Connective tissue defect in the chick resulting from copper deficiency, *Proc. Soc. Exp. Biol. Med.*, 108, 402, 1961.
27. **Shields, G. S., Coulson, W. F., Kimball, D. A., Carnes, W. H., Cartwright, G. E., and Wintrobe, M. M.,** Studies on copper metabolism. XXXII. Cardiovascular lesions in copper-deficient swine, *Am. J. Pathol.*, 41, 603, 1962.
28. **Savage, J. E., Ross, D. A., Reynolds, G., and O'Dell, B. L.,** Copper deficiency and beta-aminopropionitrile toxicity in turkeys, *Fed. Proc.*, 21, 311 (abstr.), 1962.
29. **Linker, A., Coulson, W. F., and Carnes, W. H.,** Cardiovascular studies on copper-deficient swine. VI. The mucopolysaccharide composition of aorta and cartilage, *J. Biol. Chem.*, 239, 1690, 1964.
30. **Radhakrishnamurthy, B., Ruiz, H., Dalferes, E. R., Klevay, L. M., and Berenson, G. S.,** Composition of proteoglycans in the aortas of copper-deficient rats, *Proc. Soc. Exp. Biol. Med.*, 190, 98, 1989.
31. **Coulson, W. F. and Linker, A.,** Cardiovascular studies on copper deficient swine. XI. Mucopolysaccharides and the mechanical properties of the aorta, *Biochim. Biophys. Acta*, 158, 117, 1968.
32. **Wight, T.,** Proteoglycans in pathological conditions: atherosclerosis, *Fed. Proc.*, 44, 381, 1985.
33. **Iverius, P. H.,** Possible role of glycosoaminoglycans in the genesis of atherosclerosis, *Ciba Found. Symp.*, 12, 185, 1973.
34. **Waisman, J., Cancilla, P. A., and Coulson, W. F.,** Cardiovascular studies of copper-deficient swine. XIII. The effect of chronic copper deficiency on the cardiovascular system of miniature pigs, *Lab. Invest.*, 21, 548, 1969.
35. **Carnes, W. H., Coulson, W. F., and Albino, A. M.,** Intimal lesions in muscular arteries of young copper-deficient swine, *Ann. N.Y. Acad. Sci.*, 127, 800, 1965.
36. **Coulson, W. F. and Carnes, W. H.,** Cardiovascular studies on copper-deficient swine. IX. Repair of vascular defects in deficient swine treated with copper, *Am. J. Pathol.*, 50, 861, 1967.
37. **Waisman, J. and Carnes, W. H.,** Cardiovascular studies on copper-deficient swine. X. The fine structure of the defective elastic membranes, *Am. J. Pathol.*, 51, 117, 1967.
38. **Nystrom, S. H. M.,** Studies of the effects of experimental hypocupremia on the rabbit with special reference to the vessels of the neck and the base of the brain, *Acta Neuropathol.*, 10, 209, 1968.
39. **Mills, C. F., Dalgrano, A. C., and Wehnam, G.,** Biochemical and pathological changes in tissues of Friesian cattle during the experimental induction of copper deficiency, *Br. J. Nutr.*, 35, 309, 1976.
40. **Kelley, W. A., Kesterson, J. W., and Carlton, W. W.,** Myocardial lesions in the offspring of female rats fed a copper deficient diet, *Exp. Mol Pathol.*, 20, 40, 1974.
41. **Allen, K. G. D. and Klevay, L. M.,** Cholesterolemia and cardiovascular abnormalities in rats caused by copper deficiency, *Atherosclerosis*, 29, 81, 1978.
42. **Stemmer, K. L., Petering, H. G., Murthey, L., Finelli, V. N., and Menden, E. E.,** Copper deficiency effects on cardiovascular system and lipid metabolism in the rat; the role of dietary proteins and excessive zinc, *Ann. Nutr. Metab.*, 29, 332, 1985.
43. **Kitano, S.,** Membrane and contractile properties of rat vascular tissue in copper-deficient conditions, *Circ. Res.*, 46, 681, 1980.
44. **Klevay, L. M., Reck, S. J., and Barcome, D. F.,** Evidence of dietary copper and zinc deficiencies, *JAMA*, 241, 1916, 1979.

45. **Holden, J. M., Wolf, W. R., and Mertz, W.,** Zinc and copper in self-selected diets, *J. Am. Diet. Assoc.*, 75, 23, 1979.
46. **Klevay, L. M., Reck, S. J., Jacob, R. A., Logan, G. M., Munoz, J. M., and Sandstead, H. H.,** The human requirement for copper. I. Healthy men fed conventional American diets, *Am. J. Clin. Nutr.*, 33, 45, 1980.
47. **Patterson, K. Y., Holbrook, J. T., Bodner, J. E., Kelsay, J. L., Smith, J. C., and Veillon, C.,** Zinc, copper and manganese intakes and balance for adults consuming self-selected diets, *Am. J. Clin. Nutr.*, 40, 1397, 1984.
48. Food and Nutrition Board and National Research Council, *Recommended Dietary Allowances,* 9th ed., National Academy of Sciences, Washington, D.C., 1980.
49. **Allen, K. G. D. and Allen, C. B.,** Marginal copper intakes and rat aorta endothelium morphology by scanning electron microscopy, *FASEB J.*, 3, A1061 (abstr.), 1989.
50. **Danks, D. M.,** Hereditary disorders of copper metabolism in Wilson's disease and Menkes' disease, in *The Metabolic Basis of Inherited Disease,* 5th ed., Stanbury, J. B., Wyngaarden, J. B., Fredrickson, D. S., Goldstein, J. L., and Brown, M. S., Eds., McGraw-Hill, New York, 1983, 1251.
51. **Oakes, B. W., Danks, D. M., and Campbell, P. E.,** Human copper deficiency: ultrastructural studies of the aorta and skin in a child with Menkes' syndrome, *Exp. Mol. Pathol.*, 25, 82, 1976.
52. **Klevay, L. M., Mendelsohn, D., Vanderwalt, J. J., Davidson, L. M., and Kritchevsky, D.,** Aortic sudanophilia and zinc-copper ratios in the liver of vervet monkeys fed different types of dietary fiber, *S. Afr. Med. J.*, 59, 605, 1981.
53. **Harvey, P. W. and Allen, K. G. D.,** Lipoproteins and liver lipids in copper-deficient rats, *Nutr. Res.*, 5, 511, 1985.
54. **Saari, J. T.,** Chronic treatment with dimethyl sulfoxide protects against cardiovascular defects of copper deficiency, *Proc. Soc. Exp. Biol. Med.*, 190, 121, 1989.
55. **Haberland, M., Fong, D., and Cheng, L.,** Malondialdehyde-altered protein occurs in atheroma of Watanabe heritable hyperlipidemic rabbits, *Science,* 241, 215, 1988.
56. **Brown, M. S. and Goldstein, J. L.,** Lipoprotein metabolism in the macrophage: implications for cholesterol deposition in atherosclerosis, *Annu. Rev. Biochem.*, 52, 223, 1983.
57. **Urry, D. M.,** Molecular perspectives of vascular wall structure and disease—the elastic component, *Perspect. Biol. Med.*, 21, 265, 1978.

Chapter 11

# CARDIAC LIPID METABOLISM AS AFFECTED BY DIETARY COPPER

**Denis M. Medeiros**

## TABLE OF CONTENTS

# I. INTRODUCTION

Dietary copper is required for the normal function of the cardiovascular system. Rats raised on purified diets without added copper exhibit several cardiovascular defects, including morphological alterations of both the heart and blood vessels, as well as changes in lipid metabolism which may have deleterious effects. Cardiac hypertrophy, aneurysms, abnormal EKG pattern, blood pressure changes, impaired glucose tolerance, and hypercholesterolemia are some of the reported copper deficiency symptoms.[1]

The effects that dietary copper has upon lipoprotein and cholesterol metabolism have been investigated by several groups.[2-8] Recently more attention has been devoted to the role dietary copper has upon fatty acid metabolism as reviewed briefly in this article. However, less research has been conducted to examine the effect dietary copper may have upon cardiac lipid metabolism. This lack of information is surprising when one considers the changes known to occur in lipid metabolism in other tissues, coupled with morphological and functional changes in the hearts of copper-deficient rats.

Changes in lipid metabolism among cardiac tissue may give us a better understanding as to the origin of some of the cardiac pathological and functional changes resulting from copper deficiency. Since most of the studies on cardiac lipid metabolism as affected by copper are limited to fatty acid metabolism, an overview of the present state of knowledge pertaining to copper and fatty acid metabolism is presented, followed by the limited studies published on cardiac tissue fatty acids. This article presents some recent findings from our laboratory on this latter subject. Suggestions and techniques used in typical copper deficiency feeding trials are given in detail to allow the interested investigator to pursue such studies on their own. Finally, areas needed for further investigation are identified.

# II. FATTY ACID METABOLISM AS AFFECTED BY COPPER DEFICIENCY

A possible role in which copper could effect fatty acid metabolism may be the effect of Δ9 desaturase activity. Figure 1 outlines the various pathways in the desaturation and elongation of fatty acids. The n-6 and n-3 fatty acids tend to be controlled by the Δ5 and Δ6 desaturases and the n-3 series will also utilize Δ4 desaturase in the conversion of docosapentaenoic acid (C22:5n-3) to docosahexaenoic acid (C22:6n-3). In contrast, the n-9 series will also utilize Δ9 desaturase for the conversion of both palmitic (C16:0) and stearic (C18:0) acids to palmitoleic (C16:1n-7) and oleic (C18:1n-9) acids, respectively. Delta 5 and 6 desaturases are also utilized for the the n-9 series. Wahle and Davies[9] reported that for C16 and C18 fatty acids of subcutaneous tissues from copper-deficient rats, decreased monounsaturated:saturated fatty acids were observed. Decreased microsomal Δ9 desaturase activity was observed

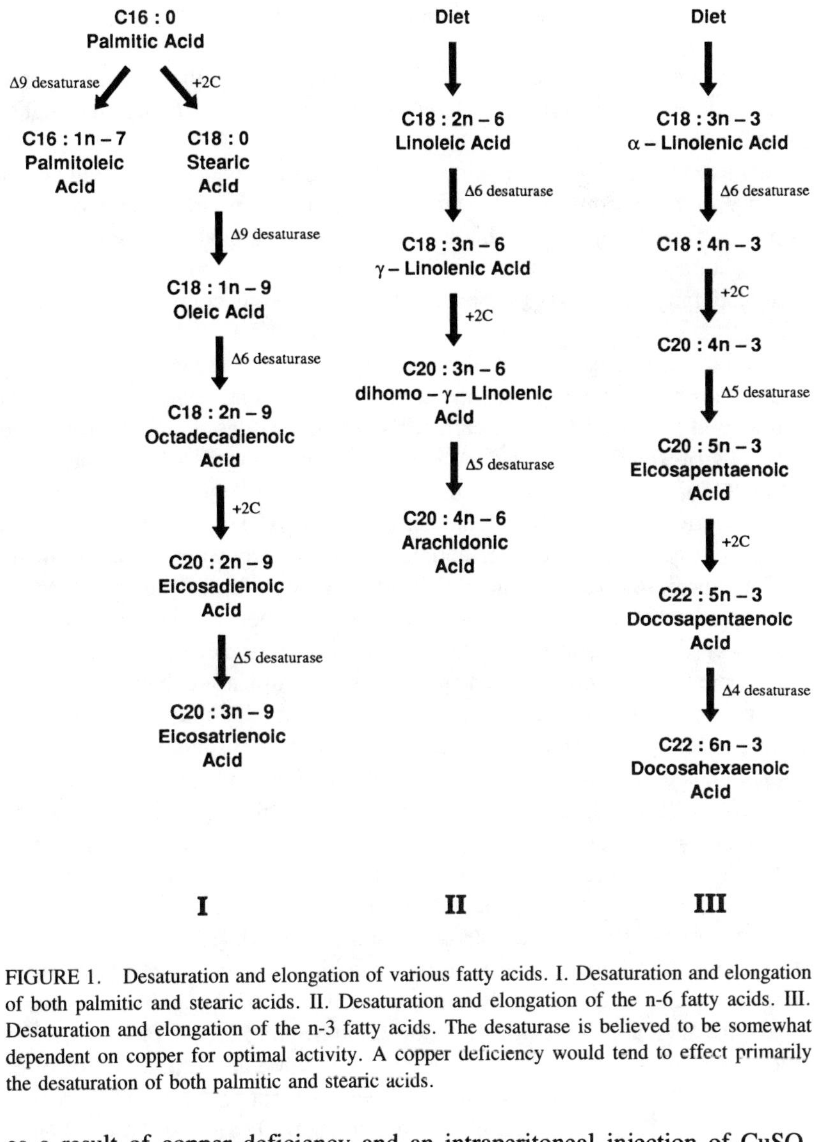


FIGURE 1. Desaturation and elongation of various fatty acids. I. Desaturation and elongation of both palmitic and stearic acids. II. Desaturation and elongation of the n-6 fatty acids. III. Desaturation and elongation of the n-3 fatty acids. The desaturase is believed to be somewhat dependent on copper for optimal activity. A copper deficiency would tend to effect primarily the desaturation of both palmitic and stearic acids.

as a result of copper deficiency and an intraperitoneal injection of $CuSO_4$ returned the desaturase activity to control values.

The study of Wahle and Davies[9] tended to support other work on the effects dietary copper had upon fatty acid composition. For instance, Moore et al.[10] had earlier reported that pigs supplemented with 250 mg Cu/kg diet had higher oleic:stearic acids (18:1n-9/18:0) in whole back fat of pigs. Thompson et al.[11] observed similar results with pigs supplemented with 250 mg Cu/kg diet and noted the liver stearic acid desaturase activity was higher with copper supplementation. Elliot and Bowland[12] also observed decreased percent of 16:0

and 18:0 fatty acids and a concomitant increase in the percent weight of 16:1 and 18:1 fatty acids in the back fat of pigs supplemented with 250 mg Cu/kg diet as compared to nonsupplemented pigs. Ho and Elliot[13] demonstrated enhanced desaturase activity in liver and adipose tissues of pigs fed higher amounts of copper.

In rats, Ovecka et al.[14] demonstrated a greater level of 18:0 but lower 18:1n-9 and consequently elevated 18:0/18:1n-9 in perirenal adipose from copper-deficient rats. Thus, this is consistent with a possible impairment of Δ9 desaturase activity. However, increased levels of longer chain polyunsaturated fatty acids were observed in adipose from copper-deficient rats: 20:3n-6, 20:4n-6, and 22:5n-6.

The fatty acid composition of other tissues besides adipose as affected by copper has been studied. Cunnane et al.[15] reported that copper-deficient rats had altered plasma, liver, and cardiac fatty acid profiles. Palmitic (16:0) and oleic (18:1n-9) acids were decreased whereas docosahexaenoic (22:6n-3) and stearic (18:0) acids were increased in the phospholipid fraction of tissues from copper-deficient rats. Arachidonic acid (20:4n-6) was increased in liver and plasma phospholipids of copper-deficient rats. The liver from copper-adequate rats had increased palmitic and oleic acids in phospholipids and triglycerides, but the essential fatty acids were decreased. Copper supplementation (250 mg Cu/kg diet) tended to have the reverse. For example, palmitic and oleic acids were higher in both liver and plasma phospholipids and liver triglycerides of rat supplemented with 250 mg Cu/kg diet. Arachidonic and docosahexaenoic acids (22:6n-3) were decreased in plasma phospholipids of copper-supplemented rats. Other studies have not shown consistent results. For copper-deficient mice, Cunnane et al.[16] reported variable results among selected organs, except dihomo-gamma-linoleic acid (20:3n-6) was lower in copper-deficient mice as compared to copper-adequate mice. Singh and Medeiros[17] reported elevated liver total palmitic (16:0) and oleic (18:1n-9) acid levels among copper-deficient rats, but decreased total linoleic (18:2n-6) acid concentrations. In addition, liver total oleic:stearic acids were higher in copper-deficient rats. Ovecka et al.[14] observed increased 20:4n-6 of liver phospholipids and increased 18:2n-6, 22:5n-6, 22:5n-3, and 22:6n-3 of liver triglycerides from copper-deficient Long-Evans rats. Korte and Prohaska[18] reported that splenic lymphocyte plasma membranes of copper-deficient mice had lower 16:0, 18:0, and 20:3n-6 and higher 18:1n-9, 18:2n-6, and 20:4n-6. Isolated erythrocyte membranes from copper-deficient mice had lower 20:3n-6 and 20:4n-6, but higher 18:1n-9 and 18:2n-6. The 18:1/18:0 was higher in membranes of copper-deficient mice.

## III. FATTY ACID METABOLISM IN CARDIAC TISSUE AS AFFECTED BY COPPER DEFICIENCY

Cunnane et al.[15] reported higher levels of 16:0 but lower levels of 18:1n-9 in cardiac phospholipids of copper-deficient rats. Rats supplemented with

250 mg Cu/kg diet had higher 16:0, 18:1n-9, and 20:3n-9 levels as compared to copper-adequate and copper-deficient rats. The n-6 fatty acids, except for 18:2n-6, were lower in copper-deficient rats as compared to copper-adequate rats. On the other hand, n-3 fatty acids (22:5n-3 and 22:6n-3) were higher in copper-deficient rats. Supplementation with 250 mg Cu/kg diet yielded higher n-6 and lower n-3 fatty acids when compared to copper-adequate rats. These results are in contrast to another study by Cunnane et al.[16] in which phospholipid fatty acids from copper-deficient mice had lower levels of 20:3n-6, 18:3n-3, and 22:6n-3. Also, among the copper-deficient group, 16:0 was lower but 18:0 higher among cardiac phospholipid fatty acids. Total cardiac phospholipid levels did not differ by copper status.

In order to clarify what effect copper deficiency had upon the fatty acid composition of cardiac tissue, our laboratory undertook a study which incorporated a group of pair-fed rats to control for decreases in food intake arising from the copper-deficient state. The previous studies reviewed did not incorporate such a group, making interpretation as to whether the results were due to copper or simply less calories more difficult.

## IV. MATERIALS AND METHODS

### A. Experimental Procedure

The male Long-Evans hooded rat has been used as our animal model of choice in assessing various effects due to copper deficiency. This strain of rat has an added advantage with respect to the presence of black fur. Midway during the course of a typical 8 week feeding trial, animals fed a copper-deficient diet will develop hypopigmentation of the fur. The "greying" effect allows monitoring of the severity of the copper-deficient state in conjunction with weight gain data in a noninvasive manner.

Male weanling Long-Evans rats purchased from Blue-Spruce Farms (Altamount, NY) are individually housed in stainless steel cages with a 12-h light-dark cycle. Diets fed to the rats are formulated after recommendations from the American Institute of Nutrition.[19] The diet consists of 50% sucrose, 15% cornstarch, 20% casein, and 5% corn oil by weight as the energy sources. The diet also contains a vitamin and mineral mix. In the case of the copper-deficient diet, cupric carbonate is omitted. In the copper-adequate diet, cubric carbonate is added to give a final copper content of 8.0 mg Cu/kg diet. Due to the basal amounts of copper in the diet ingredients analysis of the diets by atomic absorption spectrophotometry (AAS) normally reveals approximately 0.4 to 0.7 mg Cu/kg diet and approximately 8.5 to 9.0 mg Cu/kg diet in the copper-deficient and copper-adequate diets, respectively. Rats are given deionized-distilled water *ad libitum*. Normally rats given *ad libitum* access to these diets begin to show symptoms for deficiency as indicated by change in weight gain 3 weeks later. Hypopigmentation begins 4 to 5 weeks after instituting the feeding trial. A typical feeding trial may last from 7 to 10 weeks, at which time the rats fed the copper-deficient diet show symptoms of copper deficiency with the mortality rate increasing due to the copper-deficient state. Instituting the copper restriction after weaning may not produce copper deficiency symptoms. In one study with Wistar rats[20] we instituted copper restriction 2 weeks post-weaning after rats had consumed an adequate copper commercial ration. Liver copper levels were half the level of control values 20 weeks later, but not what would be considered deficient (less than 1 μg Cu/g wet weight tissue). Other symptoms of copper deficiency were not apparent. Apparently short-term exposure to a copper-adequate diet may allow rats sufficient copper reserves to prevent the appearance of deficiency symptoms in a typical feeding trial.

In a study to examine the influence of copper deficiency upon the fatty acid composition of heart muscle a typical feeding trial as described above was conducted. The rats were fed the purified diets for 8 weeks. The three treatments were designated as adequate copper (8.5 mg Cu/kg diet, group AC), no copper added (0.4 mg Cu/kg diet, group NC), and a pair-fed group (group PF) fed daily (1500 h) an adequate diet of copper in an amount based on the average intake of group NC. Groups AC and NC were fed *ad libitum*. Groups AC and PF each contained 10 rats, while group NC contained 18. It is always advisable to assign extra animals to the copper-deficient group because of the potential of death resulting from the copper-deficient state. In this experiment only nine rats in group NC were alive at the completion of the experiment.

On completion of week 8, the rats were anesthetized by the use of petroleum ether. A 3- to 5-ml 24-h fasting blood sample was obtained by cardiac puncture with a syringe. The heart and liver were removed and rinsed in deionized-distilled water. Livers were wet-ashed with trace element grade nitric and perchloric acids (Fisher Scientific Co., Fairlawn, NJ), diluted with deionized-distilled water, and analyzed for copper by AAS. This is routinely done in copper deficiency trials to verify the copper-deficient state. A bovine liver standard purchased from the National Bureau of Standards (Washington, DC) is normally treated in a similar manner to verify that the analyzed values are similar to the certified values. Heart weights are also recorded at this time to determine the presence of cardiac enlargement due to copper deficiency. The blood is typically used to determine hematocrit values, again as another index of copper deficiency. Decreased hematocrit values are expected in copper-deficient animals relative to animals fed a copper-adequate diet. In this experiment heart samples removed were frozen at −25°C in 50-ml Nalgene containers until lipid analysis.

### B. Tissue Lipid Extraction

Lipid was extracted from cardiac tissue as two major fractions: (1) polar or phospholipid containing, and (2) neutral or triglyceride containing. The technique used was a dry column procedure developed by Marmer and Maxwell[21] and modified by Zubillaga and Maerker[22] The method allows for the separation of these fractions by sequential elution. The method was originally developed for the analysis of food lipid. As Marmer and Maxwell[21] state, the method was developed as a substitute for the methods developed by Folch et al.[23] and Bligh and Dyer[24] which utilized various mixture ratios of chloroform to methanol. These methods were pointed out as having several drawbacks, among them the possibility that chloroform is a suspected carcinogen. These procedures also required large amounts of solvent and were time consuming, and emulsion problems were often encountered. Furthermore, and perhaps more important from a time point of view, the separation of lipid into the various classes required chromatographic procedures.

In the current study, columns of 16 mm × 25 cm were used for the extraction of 1 g cardiac muscle samples. A trap was made by placing 0.2 g calcium phosphate, 1.6 g Celite 545 (Accurate Chemical and Scientific Corp., Westbury, NY), and a small amount of magnesium oxide (less than 1 g) into the column. The trap serves to prevent nonlipid impurities from being eluted from the column. Magnesium oxide traps pigmented materials in particular. Cardiac muscle (1g) was ground with 4 g of anhydrous sodium sulfate in a mortar and pestle. After removal of excess water from the muscle, 3 g of Celite 545 was added to the mortar and gently ground. This mixture was transferred to the columns and gently packed.

Dichloromethane (30 ml) was added to the columns to elute the triglycerides. After the last of the dichloromethane reached the top of the packing material the collecting flask was changed and the phospholipids were eluted by addition of 30 ml of a dichloromethane/methanol (9:1) mixture. The extracted solvents were evaporated in a Buchs rotary evaporator (Brinkman Instruments, Westbury, NY) and the lipid redissolved with 10 ml hexane. Samples were transferred to 100-mm × 15-mm screw top test tubes, capped, and stored in a freezer at −25°C until further analysis.

### C. Fatty Acid Analysis

After warming the lipid-hexane solution to room temperature, an internal standard was added to each tube. Heneicosanoic acid (21:0) was used as the internal standard and 300 μl at a concentration of 1.78 mg/ml was added to each tube. The tubes were vortexed and the solvent evaporated under reduced pressure with a bell jar and water vacuum pump.

Added to each tube was 4 ml of 95% ethanol and 1 ml of 33% potassium hydroxide. The tubes were capped, vortexed, and placed in a water bath at 90°C. The tubes were vortexed frequently until the solution became clear. After 1 h the tubes were cooled at room temperature and 4 ml of deionized-distilled water and 3 ml of hexane were added. The capped tubes were placed in a shaker for 3 min and then centrifuged for 10 min at 1000 $\times g$. The top separated layer contained hexane and was discarded since it contained nonsaponifiable substances. Concentrated hydrochloric acid (1 ml) was added to the bottom aqueous layer and vortexed. After the addition of 3 ml of hexane the tubes were again shaken and centrifuged. The top hexane layer containing the free fatty acids was saved.

The hexane was evaporated under reduced pressure and 1 ml of boron trichloride methanol (10%) added. The tubes were capped and placed in a warm bath for 30 min and vortexed. After cooling, 7 ml of deionized-distilled water and 2 ml of hexane were added. The tubes were shaken for 3 min and centrifuged for 10 min. The top hexane layer contained the methyl-esters of the fatty acids and was stored in a 12-mm × 75-mm capped tube.

Methyl-esters of fatty acids were analyzed by a Varian 2100 gas chromatograph (Palo Alto, CA) using a 1.2-μl sample. The chromatographic conditions included: a 1.83-m packed glass column (2 mm I.D.) containing 10% SP2330 on 100/120 mesh Chromosorb W/AW; injector temperature of 170°C, detector temperature of 225°C; column temperature 170 to 210°C at an increase of 4°C/min, and the nitrogen carrier had a flow rate of 20 ml/min.

Identification of fatty acids was made by co-chromatrography and comparison of retention times of the standards with those of the samples. Data were expressed as milligrams of fatty acids per 100 g of wet tissue. A Hewlett Packard 3390 A integrator (Avondale, PA) was used. Data were analyzed by analysis of variance using a BMDP7D computer software program.[25,26]

### D. Verification of Copper Deficiency

During the 8-week feeding trial, nine rats from group NC died. Changes in body weight among the three groups were not detected until the third week of the experiment. Group AC weighed significantly more ($p \leq 0.05$) than group PF or group NC at the completion of the 8 weeks (Table 1). Both the AC and PF rats had significantly higher ($p \leq 0.05$) hematocrit values than did the NC group. NC rats had greater heart weights than AC rats, and the heart weight of AC rats were higher than those of the PF rats ($p \leq 0.05$). Heart:body weight was significantly higher ($p \leq 0.05$) in the NC rats than in the AC or PF groups. The total liver copper content whether expressed on a wet or dry weight basis was significantly greater ($p \leq 0.05$) in AC rats followed by PF and NC rats, respectively. Hypopigmentation was observed in the fur of NC rats. These changes among the NC rats as compared to the other two groups suggested that the animals were copper-deficient by these criteria.

## V. CARDIAC FATTY ACIDS AS AFFECTED BY COPPER DEFICIENCY

Cardiac FAME composition was altered by the copper-deficient state as 1 of 12 of the phospholipid fatty acids and 5 of 12 of the triglyceride fatty acids

**TABLE 1**
**Final Body Weight, Blood, and Tissue Values of Adequate Copper (AC), Copper Deficient (NC), or Pair-Fed (PF) Rats[1,2]**

| | AC (n = 10) | NC (n = 9) | PF (n = 10) |
|---|---|---|---|
| Final body weight (g) | 333 ± 9.51[a] | 195 ± 9.52[b] | 209 ± 1.19[b] |
| Hematocrit (%) | 46.7 ± 1.51[a] | 25.2 ± 2.28[b] | 47.4 ± 1.42[a,3] |
| Heart weight (g) | 1.11 ± 0.044[b] | 1.81 ± 0.240[a] | 0.694 ± 0.011[c] |
| Heart:body weight | 0.003 ± 0.001[b] | 0.009 ± 0.001[a] | 0.003 ± 0.01[b] |
| Liver weight (g) | 9.81 ± 0.390[a] | 7.98 ± 0.304[b] | 6.51 ± 0.445[c] |
| Liver:body weight | 0.029 ± 0.001[c] | 0.041 ± 0.002[a] | 0.031 ± 0.002[b] |
| Copper (μ/g wet weight) | 5.06 ± 0.153[a] | 0.587 ± 0.081[c] | 4.25 ± 0.185[b] |
| Copper (μ/g dry weight) | 16.3 ± 0.485[b] | 1.98 ± 0.277[a] | 14.0 ± 0.622[c] |

[1] Mean ± S.D.
[2] Where significant differences exist, means followed by different superscript letters are significantly different at $p \leq 0.05$.
[3] Values based on nine animals.

From Ovecka, G. D. Miller, G., and Medeiros, D. M., *J. Nutr.*, 118, 480, 1988. With permission.

were significantly different between groups AC and NC (Tables 2 and 3). PF rats had the highest level of palmitic (16:0) acid in the cardiac phospholipids followed by group AC which had a larger amount than did group NC ($p \leq 0.50$). The following fatty acids were in greater concentration in the triglycerides of NC rats than in either AC or PF rats ($p \leq 0.05$): 18:0, 20:3n-6, 20:4n-6, 22:5n-3, and 22:6n-3 fatty acids. Other differences as related to pair-feeding were apparent as discussed below.

Cardiac total fatty acids from triglycerides were lower in PF rats than in NC rats (Table 4). Cardiac total fatty acids from phospholipids were higher in PF rats than in AC or NC rats ($p \leq 0.05$).

## VI. INTERPRETATION

The interpretation of the fatty acid data was facilitated by reporting changes in fatty acid levels due to a copper-deficient state as opposed to those that were altered due to decreased food intake. For example, fatty acids that exhibited significant differences when AC and PF rats were compared to NC rats were probably a result of copper deficiency. All three groups could differ from each other as well. On the other hand, alterations in fatty acid levels of NC rats due to decreased food intake could be observed when AC rats were different from NC and Pf rats, but only when the latter did not differ from one another. Normally, decreased food intake has little effect upon body composition of rats. Thonney et al.[27] evaluated the effect of food restriction

**TABLE 2**
**Fatty Acid Composition of Cardiac Phospholipids of Adequate Copper (AC), Copper Deficient (NC), and Pair-Fed (PF) Rats[1,2]**

| | mg fatty acid per 100-g tissue | | |
|---|---|---|---|
| **Fatty acid** | **AC (n = 10)** | **NC (n = 9)** | **PF (n = 10)** |
| 14:0 | 0.86 ± 0.332 | 1.81 ± 0.678 | 1.15 ± 0.633 |
| 16:0 | 180 ± 4.44[b] | 138 ± 7.39[c] | 231 ± 8.42[a] |
| 16:1n-7 | 1.70 ± 0.285 | 1.73 ± 0.348 | 2.66 ± 0.428 |
| 18:0 | 327 ± 8.07[b] | 319 ± 5.88[b] | 412 ± 11.3[a] |
| 18:1n-9 | 54.1 ± 8.51 | 45.0 ± 6.83 | 65.9 ± 12.0 |
| 18:2n-6 | 87.0 ± 15.7 | 72.2 ± 7.9 | 125.0 ± 21.0 |
| 18:3n-3 | 0.70 ± 0.136[a,b] | 0.52 ± 0.091[b] | 1.50 ± 0.403[a] |
| 20:3n-6 | 1.08 ± 0.106[b] | 1.27 ± 0.103[b] | 1.96 ± 0.137[a] |
| 20:4n-6 | 119 ± 21.7 | 77 ± 10.1 | 131 ± 17.9 |
| 22:5n-6 | 29.4 ± 5.23 | 18.6 ± 3.44 | 30.7 ± 4.38 |
| 22:5n-3 | 0.636 ± 0.118[b] | 0.622 ± 0.097[b] | 1.25 ± 0.163[a] |
| 22:6n-3 | 6.93 ± 1.39[b] | 4.72 ± 0.661[b] | 11.0 ± 1.46[a] |

1 Mean ± S.D.
2 Where significant differences exist, means followed by different superscript letters are significant at $p \leq 0.05$.

From Ovecka, G. D., Miller G., and Medeiros, D. M., *J. Nutr.*, 118, 480, 1988. With permission

in rats of five different ages and observed that for each additional unit of weight gain per unit time, similar values of percent body water, protein, lipid, and ash were observed. Although pair feeding may result in a meal eating pattern, Ozelchi et al.[28,29] observed little influence of meal pattern upon body fat deposition. However, the actual fatty acid composition of the lipid caused by food restriction via pair feeding was not addressed in any of these studies.

Pair feeding of rats may result in alterations of fatty acids in liver. Kramer et al.[30] and Cunnane et al.[31] reported that pair-fed rats had reduced 20:4n-6 and increased 18:2n-6 levels among liver phospholipids. These studies, however, did not evaluate cardiac fatty acids. Data from our laboratory would indicate that food restriction via pair feeding did have an effect upon cardiac fatty acids. Among phospholipids, pair-fed rats had higher levels of 16:0, 18:0, 18:3n-3, 20:3n-6, 22:5n-3, and 22:6n-3 as compared to *ad libitum*-fed copper-adequate and copper-deficient rats. On the other hand, the effect of pair feeding upon the triglyceride fatty acids resulted in lower levels of 18:0, 18:2n-6, 20:4n-6, 22:5n-6, and 22:6n-3.

Table 3 shows that a higher level of stearic acid (18:0) and therefore a higher ratio of 18:0 to 18:1n-9 was evident for cardiac triglycerides in NC rats. Thus for C18 fatty acids the changes were consistent with the changes

## TABLE 3
## Fatty Acid Composition of Cardiac Triglycerides of Adequate Copper (AC), Copper Deficient (NC), and Pair-Fed (PF) Rats[1,2]

| | mg fatty acid per 100-g tissue | | |
|---|---|---|---|
| **Fatty acid** | **AC (n = 10)** | **NC (n = 9)** | **PF (n = 10)** |
| 14:0 | 6.42 ± 1.01 | 10.7 ± 6.34 | 4.39 ± 2.12 |
| 16:0 | 201 ± 20.9 | 210 ± 54.8 | 166 ± 57.6 |
| 16:1n-7 | 21.4 ± 2.83 | 19.3 ± 10.1 | 17.0 ± 5.64 |
| 18:0 | 269 ± 9.83[b] | 429 ± 45.9[a] | 138 ± 18.4[c] |
| 18:1n-9 | 184 ± 22.3 | 246 ± 98.4 | 132 ± 34.7 |
| 18:2n-6 | 206 ± 26.3[a,b] | 333 ± 77.6[a] | 131 ± 17.5[b] |
| 18:3n-3 | 2.41 ± 0.342 | 4.57 ± 1.70 | 1.64 ± 0.309 |
| 20:3n-6 | 1.98 ± 0.168[b] | 3.54 ± 0.729[a] | 2.00 ± 0.172[b] |
| 20:4n-6 | 185 ± 19.4[b] | 263 ± 39.7[a] | 86 ± 9.22[c] |
| 22:5n-6 | 80.5 ± 9.09[a] | 99.2 ± 13.2[a] | 35.5 ± 3.22[b] |
| 22:5n-3 | 1.46 ± 0.146[b] | 3.30 ± 0.452[a] | 1.34 ± 0.119[b] |
| 22:6n-3 | 21.5 ± 2.34[b] | 28.4 ± 2.81[a] | 14.5 ± 1.52[c] |

1 Mean ± S.D.
2 Where significant differences exist, means followed by different superscript letter are significant at $p \leq 0.05$.

From Ovecka, G. D., Miller G., and Medeiros, D. M., *J. Nutr.*, 118, 480, 1988. With permission.

## TABLE 4
## Total Cardiac Fatty Acids From Rats Fed Adequate Copper (AC), No Copper (NC), and Pair-Fed (PF) Diets[1,2]

| | g/100 g tissue | | |
|---|---|---|---|
| **Lipid class** | **AC (n = 10)** | **NC (n = 9)** | **PF (n = 10)** |
| Triglycerides | 1.20 ± 0.186[a,b] | 1.68 ± 1.007[a] | 0.74 ± 0.452[b] |
| Phospholipids | 0.81 ± 0.178[b] | 0.69 ± 0.106[b] | 1.02 ± 0.186[a] |

1 Mean ± S.D.
2 Where significant differences exist, means followed by different superscript letters are significant at $p \leq 0.05$.

From Ovecka, G. D., Miller, G., and Medeiros, D. M., *J. Nutr.*, 118, 480, 1988. With permission.

reported by others for different tissues.[9-12] A clear trend was observed in that NC rats had greater levels of essentially longer chain fatty acids, including n-3 fatty acids, in the triglycerides as compared to the other two groups. Specifically, cardiac triglycerides of NC rats had higher levels of 20:3n-6, 20:4n-6, 22:5n-3, and 22:5n-6. This pattern was not evident among the cardiac phospholipids which is in agreement with a study conducted by Cunnane et al.[16] for mice hearts. In fact, Cunnane et al.[16] observed that copper-deficient mice had increased 16:0 and 18:0 levels, but decreased levels of 20:3n-6, 18:3n-3, and 22:6n-3 among cardiac phospholipids.

Viestenz and Klevay[32] observed that copper-deficient rats developed abnormal electrocardiograms (EKG). In particular abnormal ST segment, bundle branch block, supraventricular beats, ventricular beats, and wandering pacemaker were observed in copper-deficient rats. However, the degree of EKG abnormality did not correlate closely with the degree of pathology. Copper treatment of deficient rats did not reverse pathological changes but did tend to improve EKG pattern in four out of six copper-treated ''deficient'' rats. The authors suggested several possible mechanisms by which lack of copper may result in abnormal EKG patterns. They suggested that copper-deficient rats may have an abnormal response to norepinephrine or angiotensin. Another suggested mechanism for arrhythmias was the effect of prostaglandins (PG) in the absence of copper. Cunnane et al.[33] suggested that copper had action similar to those of $PGI_2$ (prostacyclin) and suggested that it may act in a similar way to $PGI_2$ at the receptor level. Swift et al.[34] reported that in isolated rat hearts made arrythmic by perfusion of $PGE_2$ or $PGF_{2\alpha}$, addition of copper to the perfusing buffer stabilized the heart. Also perfusion of copper prior to PG perfusion blocked the arrhythmias. Apparently these observations were suggested by Cunnane[35] as a blocking of PGE synthesis by copper and a concomitant stimulation of $PGI_2$ action.

Dietary copper deficiency could result in some of the cardiac disturbances by changes in prostaglandin metabolism via an effect upon fatty acid metabolism. This suggestion is reviewed in detail by Cunnane.[35] An increase in a dietary zinc to copper ratio, as would be the case in the rats fed a copper-deficient diet, could favor Δ6 desaturase of linoleic acid, which could lead to increased production of 18:3n-6, 20:3n-6, and 20:4n-6 fatty acids, respectively. The latter two fatty acids are also precursors to series 1 and 2 prostaglandins, respectively. On the other hand, a decrease in the dietary zinc to copper ratio, as would be the case in our group AC rats, would favor the conversion of stearic acid to oleic acid via Δ9 desaturase. However, it is not clear if a copper-deficient state would result in a significant uptake of zinc, as would be the case of increased copper uptake in a zinc-deficient state.[36] If a low copper diet favors the desaturation of linoleic acid by Δ6 desaturase, an increase in the level of 20:3n-6 and 20:4n-6 fatty acids might be expected in group NC rats. This is the case among the cardiac triglycerides but not the phospholipid fatty acids. Production of prostaglandins is normally a result of the recruitment of fatty acids from membrane phospholipids. Our laboratory has observed decreased 20:4n-6 levels in liver phospholipids of copper-deficient rats.[14] Whether such a decrease could result in a decrease of $PGI_2$, which would appear possible, awaits further investigation.

The potential consequences of alterations in fatty acid composition of various tissues as they may affect PG synthesis in a copper-deficient state deserve further attention. Lampi et al.[37] reported that copper-deficient rats developed increased thromboxane 2 ($TXB_2$) levels in collagen or thrombin challenged whole blood while at the same time a decrease in liver homogenate $PGE_2$ and $PGF_2$ synthesis was reported. The effect of copper deficiency upon $TXB_2$ levels is of particular interest since this compound from blood platelets is known to be a potent vasoconstrictor. Both our laboratory[20] and Klevay[38] have reported that hypertension occurs in rats given a copper-deficient diet post-weaning. The role of fatty acid metabolism and PG synthesis resulting from copper deficiency may help explain such observations.

Changes in enzymatic activity in a copper-deficient state may also be responsible for changes in fatty acid composition. Copper-deficient rats are known to have decreased post-heparin lipoprotein lipase acitivity.[6] Valsala and Kurup[39] have reported that cardiac and adipose lipoprotein lipase activity among copper-deficient rats is reduced compared to copper-adequate rats. Such a reduction could lead to decreased uptake of chylomicrons and very low density lipoproteins by cardiac tissue, thereby altering the fatty acid composition. Valsala and Kurup[39] also observed increased serum triglycerides as did Wu et al.[8] in copper-deficient rats, which would be consistent with a decline in lipoprotein-lipase acitivity.

## VII. RECOMMENDATIONS FOR FURTHER RESEARCH

Most of the research reviewed dealt with fatty acid profiles of cardiac tissue from copper-deficient rats. It would appear that a broader spectrum of research pertaining to other cardiac lipids would be appropriate. The relationship of the various change in blood cholesterol and lipoprotein metabolism upon cardiac tissue deserves special attention. The consequence of lipid alterations upon cardiac tissue needs to address the effects upon: (1) coronary blood vessels, (2) cardiac connective tissue, and (3) cardiac myocytes. Studies examining histological, biochemical, and, perhaps most important, the functional consequences of changes in lipid metabolism would be fertile ground for further research. It would also appear that research should focus on the mechanism(s) operating in fatty acid metabolism that relate to the presence of copper. Are the changes in fatty acids among tissues due to a copper-deficient state a result of a shift in lipid pools, a decrease or increase in lipid accumulation in tissues, or do the changes represent alterations in the synthesis or degradation of fatty acids?

## REFERENCES

1. **Medeiros, D. M.,** The copper:zinc hypothesis and cardiovascular disease, *Biochem. Arch.* 1, 67, 1985.
2. **Klevay, L. M.,** Hypercholesterolemia in rats produced by an increase in the ratios of zinc to copper ingested, *Am. J. Clin. Nutr.* 12, 1060, 1973.
3. **Lei, K. Y.,** Cholesterol metabolism in copper-deficient rats, *Nutr. Rep. Int.,* 15, 597, 1977.
4. **Lei, K. Y.,** Oxidation, excretion, and tissue distribution of 26-$^{14}$C cholesterol in copper deficient rats, *J. Nutr.,* 108, 232, 1978.
5. **Lei, K. Y.,** Alterations in plasma lipid, lipoprotein and apolipoprotein concentrations in copper deficient rats, *J. Nutr.,* 113, 2178, 1983.
6. **Lau, B. W. C. and Klevay, L. M.,** Postheparin plasma lipoprotein lipase in copper deficient rats, *J. Nutr.,* 112, 928, 1982.
7. **Allen, K. G. D. and Klevay, L. M.,** Hyperlipoproteinemia in rats due to copper deficiency, *Nutr. Rep. Int.,* 22, 295, 1980.
8. **Wu, B. N., Medeiros, D. M., Lin, K. N., and Thorne, B. M.,** Long term effects of dietary copper and sodium upon blood pressure in the Long-Evans rat, *Nutr. Res.,* 4, 305, 1984.
9. **Wahle, K. W. J. and Davies, N. T.,** Effects of copper deficiency in the rat on fatty acid composition of adipose tissue and desaturase activity of liver microsomes, *Br. J. Nutr.,* 34, 105, 1975.
10. **Moore, J. H., Christie, W. W., Braude, R., and Mitchell, K. G.,** The effect of dietary copper on the fatty acid composition and physical properties of pig adipose tissues, *Br. J. Nutr.,* 23, 281, 1969.
11. **Thompson, E. H., Allen, C. E., and Meade, R. J.,** Influence of copper on stearic acid desaturation and fatty acid composition in the pig, *J. Anim. Sci.,* 36, 868, 1973.
12. **Elliot, J. I. and Bowland, J. P.,** Effects of dietary copper sulfate and protein on the fatty acid composition of porcine fat, *J. Anim. Sci.,* 30, 923, 1970.
13. **Ho, S. L. and Elliot, J. I.,** Fatty acid composition of porcine depot fat as related to the effect of supplemental dietary copper on the specific activities of fatty acyl desaturase systems, *Can. J. Anim. Sci.,* 54, 23, 1974.
14. **Ovecka, G. D., Miller, G., and Medeiros, D. M.,** Fatty acids of liver, cardiac and adipose tissues from copper deficient rats, *J. Nutr.,* 118, 480, 1988.
15. **Cunnane, S. C., Horrobin, D. F., and Manke, M. S.,** Contrasting effects of low or high copper intake on rat tissue lipid essential fatty acid composition, *Ann. Nutr. Metab.,* 29, 103, 1985.
16. **Cunnane, S. C., McAdoo, K. R., and Prohaska, J. R.,** Lipid and fatty acid composition of organs from copper-deficient mice, *J. Nutr.,* 116, 1248, 1986.
17. **Singh, N. P. and Medeiros, D. M.,** Effect of copper deficiency and sodium intake upon liver and mineral composition in the rat, *Biol. Trace Elem. Res.,* 6, 423, 1984.
18. **Korte, J. J. and Prohaska, J. R.,** Dietary copper deficiency alters protein and lipid composition of murine lymphocyte plasma membranes, *J. Nutr.,* 117, 1076, 1987.
19. **American Institute of Nutrition,** Report of the AIN ad hoc committee on standards for nutritional studies, *J. Nutr.,* 107, 1340, 1977.
20. **Medeiros, D. M.,** Hypertension in the Wistar rat as a result of post-weaning copper restriction, *Nutr. Res.* 7, 231, 1987.
21. **Marmer, W. N. and Maxwell, R. J.,** Dry column method for the quantitative extraction and simultaneous class separation of lipids from muscle tissue, *Lipids,* 16, 365, 1981.
22. **Zubillaga, M. P. and Maerker, G.,** Modified dry column procedure for extraction of lipids from cured meats, *J. Food Sci.,* 49, 107, 1981.
23. **Folch, J., Lees, M., and Sloane-Stanley, G. A.,** Simple method for the isolation and purification of total lipids from animal tissues, *J. Biol. Chem.,* 226, 497, 1957.

24. **Bligh, E. G. and Dyer, W. J.,** A rapid method of total lipid extraction and purification, *Can. J. Biochem. Physiol.*, 37, 911, 1957.
25. **Steel, R. D. G. and Torrier, J. H., Eds.,** *Principles and Procedures of Statistics: A Biometrical Approach,* 2nd ed., McGraw-Hill, New York, 1980.
26. **Dixon, W. J., Ed.,** *BMDP Statistical Software,* University of California Press, Berkeley, CA, 1983.
27. **Thonney, M. L., Oberbauren, A. M., Duhaime, D. J., Jenkins, T. C., and Firth, N. L.,** Empty body component gain of rats grown at different rates to a range of final weights, *J. Nutr.*, 114, 1777, 1984.
28. **Ozelchi, A., Romsos, D. R., and Leveille, G. A.,** Influence of initial food restriction on subsequent body weight gain and body fat accumulation in rats, *J. Nutr.*, 108, 1724, 1978.
29. **Ozelchi, A., Romsos, D. R., and Leveille, G. A.,** Influence of diet composition on nitrogen balance and body composition in meal-eating and nibbling rats, *J. Nutr.*, 107, 1768, 1977.
30. **Kramer, T. R., Briske-Anderson, M., Johnson, S. B., and Holman, R. T.,** Influence of reduced food intake on polyunsaturated fatty acid metabolism in zinc-deficient rats, *J. Nutr.*, 114, 1224, 1984.
31. **Cunnane, S. C., Horrobin, D. F., and Manku, M. S.,** Essential fatty acids in tissue phospholipids and triglycerides of the zinc-deficient rat, *Proc. Soc. Exp. Biol. Med.*, 177, 441, 1984.
32. **Viestenz, K. E. and Klevay, L. M.,** A randomized trial of copper therapy in rats with electrocardiographic abnormalities due to copper deficiency, *Am. J. Clin. Nutr.*, 35, 258, 1982.
33. **Cunnane, S. C., Zinner, H., Horrobin, D. F., Manku, M. S., Morgon, R. O., Karmali, R. A., Ally, A. I., Karmazyn, M., Barnette, W. E., and Nicolaou, K. C.,** Copper inhibits pressor responses to noradrenaline but not potassium. Interactions with prostaglandin $E_1$, $E_2$, and $I_2$ and penicillamine, *Can. J. Physiol. Pharmacol.*, 57, 35, 1979.
34. **Swift, A., Karmazyn, M., Horrobin, D. F., Manku, M. S., Karmali, R. A., Morgan, R. O., and Ally, A. I.,** Low prostaglandin concentration cause cardiac rhythm disturbances. Effect reversed by low levels of copper or chloroquine, *Prostaglandins,* 15, 651, 1978.
35. **Cunnane, S. C.,** Differential regulation of essential fatty acid metabolism to the prostaglandins: possible basis for the interaction of zinc and copper in biological systems, *Prog. Lipid Res.*, 21, 73, 1982.
36. **Schwarz, F. J. and Kirchgessner, M.,** Intestinal absorption of copper, zinc, and iron after dietary depletion, in *Trace Element Metabolism in Animals,* Vol. 2, Hoekstra, W. G., Suttie, J. W., Ganther, H. E., and Mertz, W., Eds., University Park Press, Baltimore, 1974, 519.
37. **Lampi, K., Mathias, M. M., and Allen, K. G. D.,** The role of dietary copper in prostaglandin synthesis, *Fed. Proc. Fed. Am. Soc. Exp. Biol.*, 45, 237 (abstr.), 1986.
38. **Klevay, L. M.,** Hypertension in rats due to copper deficiency, *Nutr. Rep. Int.*, 35, 999, 1987.
39. **Valsala, P. and Kurup, P. A.,** Investigations on the mechanisms of hypercholesterolemia observed in copper deficiency in rats, *J. Biosci.*, 12, 137, 1987.

Chapter 12

# ISCHEMIC HEART DISEASE: TOWARD A UNIFIED THEORY

**Leslie M. Klevay**

## TABLE OF CONTENTS

## I. INTRODUCTION

The heart is a remarkable organ to which most people pay little attention. Weighing less than half a pound, "each ventricle puts out 150,000 tons of blood" in a lifetime. "Three thousand million beats beat in a man who gets old."[1]

The human heart begins its development in the third or fourth week of gestation. By the sixth week it "exhibits the general external shape and markings that characterize it permanently."[2] The blood vessels develop more or less contemporaneously.

Early in life the human cardiovascular system develops rather well with low morbidity and mortality. There were 92 deaths from congenital anomalies of the heart and circulatory system (categories 745 to 747) per 100,000 live births in the U.S. in 1984;[3] this death rate was based on 3379 deaths.

In contrast, at about age 30, ischemic heart disease begins its ascent to prominence in the U.S. until it exceeds all other causes of death in the fifth decade of life.[4] Ischemic heart disease may not be the leading cause of death in advanced age, however.[5] In 1984 there were 541,288 deaths from ischemic heart disease (categories 410 to 414) in the U.S. with a rate of 228.9 per 100,000; ischemic heart disease accounted for 26.5% of all deaths.[3]

Atherosclerosis is the most frequent cause of myocardial infarction which, in turn, is one of the more common modes of dying of ischemic heart disease.[6] The atherosclerotic process begins early in life, perhaps in infancy or *in utero*.[7-11] This process, which often culminates in ischemic heart disease, is a problem of connective tissue metabolism, endocrinology, lipid metabolism, physiology, thrombus formation, and uric acid metabolism.[4] This process also includes arterial endothelial damage, endothelial and vascular permeability, platelets, smooth muscle cells, and vasoactive amines.[4] Other contributors to the encrustation-insudation of atherosclerosis may include: arterial spasm, foam cells, growth factors, lysosomes, macrophages, matrix carbohydrates, monocytes, and pinocytosis.[12]

## II. COPPER AND CHOLESTEROL METABOLISM

### A. The Earliest Data

The first data that joined the metabolism of copper with that of cholesterol were obtained in 1970. When the experiments were published in 1973[13] an increase in the concentration of the cholesterol in plasma due to an increase in the ratio of zinc to copper ingested had been found in five experiments, in two environments during 3 years. The results were unprecedented; no reference common to cholesterol, copper, and zinc had been found in 31 years of nutrition reviews.[14,15]

#### *1. The Original Diet*

The diet, locally known as the "1968 diet," used in these experiments is shown in Table 1. It is based on 62% sucrose, 20% egg white, and 10% corn oil, it contains all nutrients known to be essential for rats, and it is deficient in copper and zinc.[13,15] Rats fed this diet usually receive most of their copper and zinc as solutions of reagent grade cupric sulfate and zinc acetate in demineralized water;[13,15] sometimes these salts are mixed into the diet after being finely ground with a mortar and pestle. Rats are housed in cages of stainless steel; water bottles have silicone stoppers.[16]

We usually do not measure the dietary copper chemically; we follow the lead of Gallagher et al.[17] and do a biological assay of the diet in each experiment. After hypercholesterolemia and/or other signs of deficiency are found, we proceed with other measurements. Sometimes the addition of test materials to the diet is based on the copper concentration.[18,19] This diet usually contains less than 0.8 μg Cu/g. A concentration of 0.44 μg/g is one of the lower values we have found.[18] Plasma cholesterol usually is measured by fluorescence.[20]

The drinking solutions originally were made with water distilled in glass vessels.[13] Over the years the method of water purification has varied. Generally the resistance fell between 3 and 18 million Ohm-centimeters. Recently demineralized water has been used.[21]

**TABLE 1**
**Composition of the Diet**

| | g |
|---|---|
| Sucrose | 623.5 |
| Fibrous cellulose powder | 30.0 |
| Zn- and Cu-free Jones Foster salt mix | 40.0 |
| Choline chloride | 1.5 |
| Egg white | 200.0 |
| Mix of vitamins soluble in water[a] | 5.0 |
| Mix of vitamins soluble in fat[b] | 10.0 |
| Corn oil | 90.0 |

[a] This mixture consists of the following components in milligrams: thiamin hydrochloride, 400; riboflavin, 800; pyridoxine hydrochloride, 400; calcium pantothenate, 2500; niacinamide, 4000; folic acid, 100; menadione, 10; biotin, 20; cyancobalamin, 15. These were mixed with 500 g sucrose.

[b] This mixture had the following composition: calciferol, 5.70 mg; α-tocopherol, 10 g; vitamin A palmitate, 2 g; mixed with 1000 g corn oil. Diets and vitamin mixtures are prepared in sufficient quantities so that weighing error is less than 1%. Usually mixing for 30 min is sufficient to provide homogeneity.[199]

### *2. Some Modifications*

The standard diet has been changed slightly over the years. An amount of biotin adequate in Cincinnati (0.2 mg/kg of diet) proved inadequate in Grand Forks. The biotin requirement of rats fed egg white was determined;[22] the amount of biotin was increased to 2.00 mg/kg.[23] Beet sugar is more available than cane sugar in Grand Forks and has been substituted.[19] Difficulty in obtaining pure vitamin A palmitate caused a change to vitamin A palmitate in vegetable oil and a decrease to 20 IU/g of diet.[24]

## B. Recent Work

The relationship between the metabolism of copper and that of cholesterol has grown closer in the last 15 years. The simplest explanation for the original experiment[13] with the high ratio of zinc to copper is that the zinc induced mild copper deficiency.[25] Hypercholesterolemia in copper deficiency has been found in at least 11 independent laboratories.[26-36] Lipid abnormalities have been associated with copper in several species,[4] including people.[37,38]

### *1. Negative Experiments*

Occasionally diets low in copper have not produced the expected hypercholesterolemia. With four doses of dietary copper and four doses of dietary zinc supplements in a factorial design, Frimpong and Magee[39] found no relationship of serum cholesterol to dietary copper when rats were fed a diet containing egg white. The amount of biotin (0.2 mg/kg) in the diet sometimes is insufficient in a diet based on egg white.[22] In an earlier experiment [40] with a similar diet, one of the authors noted that hair was lost ''over the entire body'' or around the forehead of the majority of the rats. Hair loss is a sign of biotin deficiency,[41] and biotin deficiency alters cholesterol metabolism.[22]

Two reports of apparently well controlled experiments describe a failure to find hypercholesterolemia in copper deficiency.[42,43] Perhaps the reasons for these failures, if they can be found, will be instructive in understanding the ways in which dietary copper affects lipid metabolism.

Scientists who find significant differences usually publish them and usually do not mention higher probability values if lower ones are found. Considering the statistical probabilities associated with the differences between mean cholesterol concentrations,[13,26-36] at least 15,445 negative experiments either remain undone or are filling some files if the hypercholesterolemia with diets low in copper is due to chance instead of cause.

### 2. *Dietary Diversity*

Considering the experiments above[26-36] and other experiments by these authors, copper deficiency has produced hypercholesterolemia in animals fed at least seven different salt mixes (AIN, Jones-Foster, Lefevre, Lei, Nielsen, O'Dell, Wesson), five different carbohydrates (fructose, glucose, ground corn, starch, sucrose), three different proteins (casein, egg white, or milk), and three different fats (coconut, corn, or cottonseed-coconut). One can conclude from this mélange that the effect on cholesterol metabolism was produced by an alteration in copper metabolism and not by some other dietary component closely associated with dietary copper. An earlier discussion of dietary diversity is available.[15]

## C. The Mechanism

The mechanism by which copper deficiency can produce hypercholesterolemia is understood incompletely. At least three enzymes seem to be involved. The activities of lecithin:cholesterol acyltransferase and lipoprotein lipase are decreased in copper-deficient rats.[28,44,45] The activity of hydroxymethylglutaryl-coenzyme A reductase is increased in copper deficiency.[35] These changes in activity should increase cholesterol in plasma. Other mechanistic possibilities involving kinetics, pool sizes, and receptors are mentioned by Lei in this volume.[46]

# III. NUTRITION

The early results on the hypercholesterolemia of copper deficiency,[13,26,27,29] the hypothesis that these results are relevant to the epidemiology and etiology

**TABLE 2**
**Insufficient Amounts of Dietary Copper for Adult Humans**

| Amount (mg/d) | Criterion | Ref. |
|---|---|---|
| 0.65 | Hypertensive response to hand-grip test | 52 |
| 0.78 | Decreased glucose clearance | 53 |
| 0.83 (1.5[a]) | Hypercholesterolemia | 37 |
| 1.02 (1.4[a]) | Lipoprotein change | 38 |

[a] Perhaps is insufficient, see text.

of ischemic heart disease,[13-15] the high prevalence of this disease and of hypercholesterolemia in the industrialized world, and the realization that diets are lower in copper than previously thought[15,47] prompted a gradual awakening to the importance of the copper requirement of people.

## A. The Adult Copper Requirement

The daily amount of copper required by adults is thought to be 2 mg.[48] This is the amount sufficient to compensate for average fecal and urinary loss as determined by the balance technique, and is largely based on the summary data of Cartwright[49] (for details see Reference 4). For example, our recent balance data[50] are quite similar to those, decades old.[49]

The criteria of nutrient essentiality, the principles of measurement of nutritional requirements, and some imperfections in balance studies have been reviewed.[51] If physiological changes can be induced by a diet low in a nutrient, the amount of the nutrient in the diet is insufficient. The requirement exceeds the dietary amount that produced the change. Depletion experiments are superior to balance experiments in the measurement of requirements. Probably there is little more to be learned about the copper requirement by doing balance studies.

### *1. How Much Copper is Too Little?*

Table 2 summarizes amounts of copper insufficient to maintain normal physiology in depletion experiments of humans. The autonomic response of eight women aged 18 to 36 years was evaluated using a sustained hand grip test.[52] The increase in blood pressure during the test was greater when dietary copper was lower than in either a control or repletion dietary period. The glucose tolerance curves of two men were raised during a period of low copper intake and fell when copper intake was increased.[53] The first person to be depleted of copper experimentally, a 29-year-old man, responded with an increase in total and low density lipoprotein cholesterol and with a short run of ventricular tachycardia.[37] An increase in total cholesterol of 32 mg/dl to 234 mg/dl is sufficient to classify him as hypercholesterolemic by current

**TABLE 3**
**Excessive Amounts of Zinc Supplements for Adult Humans**

| Amount (mg/d) | Criterion | Ref. |
|---|---|---|
| 300 | Increased LDL, decreased HDL[a] | 63 |
| 160 | Decreased HDL | 64 |
| 50 to 75 | Decreased HDL | 65 |
| 17 to 52 | Decreased HDL | 66 |
| 15.5 | Decreased copper retention | 72 |
| 4 | Increased fecal copper[b] | 73 |

[a] Low density and high density lipoprotein cholesterol, respectively.
[b] Adolescent, not adult.

guidelines.[54] On repletion with copper, total cholesterol fell to 198 mg/dl. The copper requirement of this man may have exceeded 1.5 mg/d because he lost more copper than this amount via feces and urine in the control period. Twenty men aged 21 to 57 years responded to a diet low in copper with a decrease in high density lipoprotein (HDL) cholesterol and an increase in low density lipoprotein (LDL) cholesterol.[38] The daily copper requirement of these men may have exceeded 1.4 mg because when copper was restored to the diet, cholesterol concentrations reverted not only to those of the control period, but improved significantly. Some of those men also experienced electrocardiographic abnormalities.[37,55] Copper is far more powerful on a molar basis in controlling cholesterolemia than clofibrate, calcium, or fat.[56] The molar response to copper calculated from more recent data[38] is similar to that found earlier.[56]

### *2. How Much Zinc is Too Much?*

A considerable amount of evidence implicating copper metabolism in the etiology and pathogenesis of ischemic heart disease is based on relative deficiency of copper as influenced by zinc.[13-15,57-61] According to Underwood, high dietary zinc can increase copper requirements and lessen copper toxicity.[62] Table 3 summarizes the effects of excessive amounts of zinc supplements in humans.

High density lipoprotein cholesterol can be decreased and low density lipoprotein cholesterol increased by very large doses of zinc.[63] The first potentially adverse effect of oral zinc on cholesterol metabolism was noticed by Hooper et al.[64] who lowered high density lipoprotein cholesterol of young men with 160 mg zinc daily; similar results with 50 to 75 mg have been noted.[65] Surprised at a ''poor correlation between the level of physical activity...and HDL-cholesterol,'' Goodwin et al.[66] classified their subjects by amount of activity and amount of zinc supplement taken. They found that zinc supplementation (median dose 24 mg/d) could "negate the beneficial effect of exercise on raising HDL-cholesterol." Potentially adverse effects of zinc supplements on lipoprotein metabolism sometimes are transient[67] or undetectable.[68-70] Although the latter study[70] revealed minimal effects on lipoproteins, the zinc supplement decreased the activities of ferroxidase and superoxide dismutase of young women. Zinc supplements increased the total cholesterol in Cynomolgus monkeys.[71] Addition of 15.5 mg of zinc to the diet of elderly people produced decreased copper retention.[72] In adolescent girls, an increase in daily dietary zinc from 11 to 15 mg by the consumption of lemonade containing zinc sulfate produced increased fecal loss of copper.[73]

### B. Usual Copper Intakes

Data on usual intakes of dietary copper in the U. S. and the industrialized world have been reviewed.[4,5,15,60,61] More than 20 cited articles contained evidence that intakes less than 2 mg/d, the minimum estimated as safe and adequate by the Committee on Dietary Allowances,[47] are common. Several more recent articles confirm these low intakes.[74-83] In contrast are mean daily intakes of copper greater than 2 mg in Nepal,[79] Java,[84] and India,[85] three areas where risk of ischemic heart disease is low. Data cited are based on direct, chemical analysis. None of the data were obtained by calculations from food tables; in our experience this method is unreliable for copper.[4,50,56,86] The data from Java[84] were calculated from menus and analyses of the foods contained in them.

Table 4 contains estimates of the frequencies of dietary copper values from ten dietary surveys.[74,75,77,79-81,87-90] All data are based on chemical analyses and are assumed to follow a normal distribution. The daily intakes selected for tabulation approximate the amounts of copper found to be insufficient in depletion experiments with humans (see Section III.A.1). The grand mean for frequency of diets containing less than 1.0 mg of copper daily is 35% (standard deviation, 12.7%). In contrast, the grand mean for frequency of diets containing more than 2.0 mg daily is 14% (standard deviation 14.9%). Consideration of the median values 33 and 10%, respectively, decreases the frequency of diets in each group slightly. These data are more extensive than those used in earlier estimates of dietary frequency.[18,37,53,58,60,61,91,92]

### C. Modifying Factors

Within the past few years more than a dozen chemicals that can modify the utilization of dietary copper have been identified. Definition of the concept of cholesterotropic and cuprotropic chemicals was begun following the observation that clofibrate increased liver copper while lowering the plasma cholesterol of rats fed a diet deficient in copper.[18] This concept has been extended in a series of papers [4,19,51,56,86,93] where the diversity of these chemicals is summarized.

Members of this class of chemicals produce reciprocal changes in plasma

**TABLE 4**
**Frequency of Diets Containing Less Than Specified Amounts of Copper in Various Surveys (%)**

Daily Intakes, mg/d

| | 0.6 | 0.7 | 0.8 | 1.0 | 1.4 | 2.0[a] | Ref. |
|---|---|---|---|---|---|---|---|
| | <2 | 2 | 3 | 7 | 20 | 44 | 88 |
| | 17 | 20 | 23 | 30 | 46 | 30 | 79 |
| | 11 | 15 | 19 | 30 | 58 | 10 | 75 |
| | 7 | 11 | 16 | 31 | 69 | 2 | 80 |
| | 21 | 24 | 27 | 34 | 50 | 27 | 74 |
| | 7 | 11 | 18 | 36 | 79 | <1 | 87 |
| | 16 | 20 | 25 | 37 | 63 | 9 | 77 |
| | 30 | 34 | 39 | 48 | 66 | 13 | 90 |
| | 16 | 23 | 31 | 49 | 82 | <1 | 89 |
| | 18 | 25 | 32 | 50 | 82 | 1 | 81 |
| Mean | 14 | 18 | 23 | 35 | 62 | 14 | |
| S.D. | 8.2 | 9.0 | 10.0 | 12.7 | 19.2 | 14.9 | |

*Note*: Data tabulated in order of frequency at 1.0 mg/d are rounded to the nearest unit.

[a] Frequency of diets containing more than 2.0 mg.

cholesterol and copper nutriture. Some of these chemicals decrease cholesterolemia and enhance copper metabolism. Others increase cholesterolemia and inhibit copper metabolism. In retrospect, zinc was the first of these to be discovered (see Section II.A). It is assumed that these chemicals can modify other actions of copper as well as those related to the metabolism of cholesterol.

Aspirin (see Section IV.C), calcium, and phytate probably are the more commonly ingested of these modifiers with potential benefit; beer, a complex solution, resembles these (see Section IV.E.2). Fructose, sucrose and zinc are those most likely to be harmful. A mixture of cholesterol plus cholic acid has been used for more than half a century to produce hypercholesterolemia and atherosclerosis in experimental animals. Recently cholesterol alone has been found to lower the liver copper of rabbits.[93] Perhaps some of the pathology induced with this method during the last three quarters of a century is from induced copper deficiency.

## IV. ISCHEMIC HEART DISEASE AS COPPER DEFICIENCY

The major consequence of this disparity between intakes of dietary copper and the amount of dietary copper required is likely to be ischemic heart disease. The mechanism(s) by which normal arteries become atherosclerotic and by which atherosclerosis increases until death from ischemic heart disease occurs also is understood incompletely (see Section I). However some characteristics are so closely associated with this process that they can be considered as mechanistic.

**A. Risk Factors**

A few hundred characteristics have been associated with risk of ischemic heart disease.[94,95] Several of these are more important than the others because they "are closely associated with risk,...can be used to predict risk to a reasonable degree and...seem to involve a potentially plausable mechanism of producing disease."[4] Insull summarized them: glucose intolerance, hypercholesterolemia, abnormalities of the electrocardiogram, male predisposition, hypertension, and smoking of cigarettes.[96] Consideration of these characteristics permits the identification of the 20% of the presymptomatic population with 40% of the risk.[97]

**1. Glucose intolerance**—Abnormal glucose metabolism is one of the earlier characteristics of copper deficiency to be discovered.[98] The phenomenon was ignored for nearly a half century.[99,100] Its mechanism is unknown, but it seems to involve the uptake of glucose by skeletal muscle[99] and insulin receptors.[101] It does not seem to involve altered activity of glucose-6-phosphatase.[102] Glycosylated hemoglobin is increased in copper deficiency.[103]

**2. Hypercholesterolemia**—See Sections II.A and II.B; see also Chapter 1.

**3. Abnormal electrocardiograms**—Several electrocardiographic abnormalities have been found in animals deficient in copper.[21,104-107] The most frequent abnormalities are of the ST segment and bundle branch block.[104,105] Altered electrocardiograms may be caused by altered cardiac energy metabolism and mitochondrial electron transport.[106] As was the case when the metabolism of copper and cholesterol first were joined, these results[4,104] were unprecedented. No reference common to copper and electrocardiograms was found in 62 years of Physiological Reviews.

**4. Male predisposition**—Female rats are protected against the fructose-induced mortality of copper deficiency.[108] Weanlings were fed a diet low in copper for 8 weeks. Male rats began dying in the sixth week and 40% were dead by the eighth week. No females were dead then. Responses to few other nutrients are affected by sex; a reviewer did not mention dietary fat as having a greater effect on males.[109]

**5. Hypertension**—The blood pressure of rats can be altered by copper deficiency. Weanling, deficient rats become hypotensive[110-112] from structural[27,105,106,113] (see Section IV.B), physiological[104-107] (see Section IV.D) and endocrine[24] abnormalities. Adult rats deprived of copper become hypertensive.[114,115]

**6. Cigarettes**—Effects of cigarette smoke on the metabolism and physiology of copper have not been studied experimentally. Smokers have higher serum copper than normal.[116]

**TABLE 5**
**Anatomical Similarities Between Animals Deficient in Copper and People with Ischemic Heart Disease**

| | |
|---|---|
| Arterial | Hemopericardium |
| Elastic degeneration | Hemothorax |
| Fibrosis | Pericarditis |
| Foam cells | Ruptured papillary muscle |
| Hemorrhage (intramural) | Small stature |
| Mucopolysaccharide increase | Ventricular |
| Necrosis | Aneurysm |
| Smooth muscle proliferation | Calcification |
| Atrial | Edema |
| Thrombosis | Fibrosis |
| Coronary | Hemorrhage (focal) |
| Elastic fragmentation | Hypertrophy |
| Hyalinization | Infiltration (neutrophil and round cell) |
| Necrosis | Necrosis (myocardial cell) |
| Smooth muscle degeneration | Rupture |
| Sudanophilia | Thrombosis (endocardial) |
| Thrombosis | |

**7. Hyperuricemia**—Hyperuricemia was not included among the risk factors by Insull.[96] Perhaps uric acid should be considered a risk factor because some of the multivariate analyses from Framingham are corrected for uric acid.[117] Risk of ischemic heart disease in this population was related to uric acid value and people with gout had a twofold increased risk.[97] Newland has discussed the association between uric acid and ischemic heart disease.[118] Rats deficient in copper are hyperuricemic.[119] The epidemiologic association between uric acid and other risk factors has been reviewed.[5]

## B. Anatomical Similarities

There are numerous other similarities besides the risk factors between animals deficient in copper and people with ischemic heart disease. Tabulation of these similarities began in 1980. The anatomical similarities are shown in Table 5. Because data have been collected from numerous publications and references have been cited in earlier tabulations,[4,5,60,61] enumeration and citation in this and the next section (Section IV.C) are restricted either to similarities found recently or to recently found references to similarities found earlier.

The earliest association of copper deficiency and abnormal cardiovascular function was that of Bennetts et al.[120] who described "Falling Disease" in cattle. This type of sudden death was "most frequently observed when cows are being brought in to be milked, are standing in the bail, or are being driven out to paddock again. In some cases cows have actually fallen on the milker."[121] The first suggestion that copper could be important in coronary heart disease (now called ischemic heart disease) was in the paper associating copper with

**TABLE 6**
**Chemical and Physiological Similarities Between Animals Deficient in Copper and People with Ischemic Heart Disease**

| | |
|---|---|
| Abnormal electrocardiograms | Glucose intolerance |
| Bundle branch block | Hypercholesterolemia |
| Pathological Q waves | Hypertriglyceridemia |
| ST segment | Hypertension |
| Decreased | Hyperuricemia |
| Myocardial cytochrome oxidase | Low blood pressure under load |
| Dihomo-γ-linolenic acid | Male predisposition |
| Cardiac copper | Mitigated by |
| Longevity | Aspirin |
| Exacerbated by | Beer |
| Salt | Clofibrate |
| | Sudden death |

cholesterol metabolism.[13] In contrast, when discussing disease in muscular arteries of copper-deficient swine,[122] Carnes said, ''I wouldn't suggest that copper deficiency has any bearing on atherosclerotic lesions'' although they found lesions similar to those in coronary arteries of aging swine and in young human subjects thought to be similar to the early phases of atherosclerosis seen in adults. Carnes et al.[122] found ''proliferation of the medial cells'' and ''protrusion and migration of medial cells'' that ''account for the bulk of the initial intimal plaque formed.'' One can conclude that they found smooth muscle cell proliferation because the media of arteries of medium caliber ''consist almost exclusively of smooth muscle cells.''[123] Hunsaker et al. also noted a migration of smooth muscle cells in aortas of marginally copper-deficient rats.[124] Hill and Davidson found increased numbers of smooth muscle cells in aortas of pigs deficient in copper.[125] This smooth muscle proliferation was accompanied by a substantial increase in collagen production. Proliferation and migration of smooth muscle cells are very important in the early stages of the atherosclerotic process;[126] ''smooth-muscle cell proliferation is a key event in atherogenesis.''[127]

Small stature is a recent addition[5] to Table 5. Morris et al.[128] noted the ''incidence of ischaemic heart-disease is greatest among the shortest men'' in their classic study of the London busmen. As copper-deficient infants grow more rapidly after supplementation,[129] one can infer that copper deficiency produces smaller humans as well as smaller rats.[5,13,27]

### C. Chemical and Physiological Similarities

These are summarized in Table 6. Low auricular copper found at open heart surgery[130] and low myocardial copper at autopsy of recent victims of ischemic heart disease[131] confirm earlier reports of low myocardial copper of people who died of myocardial infarction.[60] Copper was measured in presumably normal muscle in all cases.

Hypertension is a risk factor for ischemic heart disease (see Section IV.A.5). "Despite the difficulty in assigning NaCl a role in the development of hypertension, many...recommend reduced NaCl consumption for the American population."[132] Salt restriction remains important in the treatment of the congestive heart failure[133] and the myocardial infarction[134] that often result from ischemic heart disease.[6] Sodium is considered one of four principal factors in the etiology of essential hypertension.[135]

Rats fed extra sodium chloride drank more water, as expected. When they also were deficient in copper, they were unable to excrete the extra salt and water and became edematous.[136] Renal failure was produced with half the salt in half the time with copper deficiency in comparison to the classic experiments of Meneely.[137] Salt fed to sensitive rats of the Dahl strain increased blood pressure and decreased the copper in heart, kidney, and liver.[138]

Menkes' disease is a complex, hereditary disease of copper transport that results in abnormal distribution of copper.[139] Some organs are high in copper, but liver is low. Blackett et al.[140] reported "lipoprotein disorders are variably present in the patients with Menkes' disease." One of three patients also was glucose intolerant. This variability may be similar to that found among three men fed a diet low in copper experimentally. One responded to the diet with hypercholesterolemia and an increase in low density lipoprotein cholesterol (glucose metabolism was not evaluated).[37] Two others responded with an increase in the height of the glucose tolerance curves but showed no change in lipid metabolism under apparently similar conditions[53] (see Section III.A.1).

Wilson's disease is rare, usually is inheritied, and is characterized by excessive retention of dietary copper leading to, among other things, increased copper in liver.[141] Data on lipid metabolism in patients with Wilson's disease have been hard to find.[57,141] Hill et al.[142] reported a mean cholesterol of 148 mg/dl for a group of nine patients, three of whom were untreated. This unusually low value may be related to increased liver copper because in rats there often is an inverse correlation between the concentration of cholesterol in plasma and liver copper[18,27,143,144] or copper in liver microsomes.[19] This correlation also may explain the hypercholesterolemia of Menkes' disease.

A study on the use of aspirin in the prevention of ischemic heart disease[145] provides recent encouraging data in support of several earlier studies (see Reference 19) showing that consumption of aspirin often is associated with lower risk. Aspirin fed to rats as part of a diet deficient in copper[19] decreased the concentration of cholesterol in plasma and raised the copper in liver microsomes. Aspirin was both cholesterotropic and cuprotropic (see Section III.C). The effects of beer are summarized (see Section IV.E.2).

Clofibrate is an antilipemic drug that seems effective in decreasing risk of death due to ischemic heart disease.[146] Clofibrate fed to rats as a component of a diet deficient in copper decreased the concentration of cholesterol in plasma and increased both plasma copper and liver copper.[18] Clofibrate was both cholesterotropic and cuprotropic (see Section III.C).

## D. Physiological Methods and Other Results

Most of the recording of electrocardiograms in rats and mice deficient in copper have been done with rather simple apparatus. Some are more complex, however.[106]

Electrocardiograms are done on rats anesthetized with intraperitoneal barbituate using a Physiograph (Narco Bio-Systems, Houston, TX) and a cardiac preamplifier (E & M Instrument Company, Inc., Houston, TX) with the following settings: sensitivity 10, inner knob at 10 o'clock; balance: vertical, down; chart speed: 10 cm/s. The cardiac preamplifier was set on high. Needle electrodes were inserted subcutaneously: ground, left front leg; negative, right front leg; positive, 2 cm to the left of the sternum at the level of the xiphoid.

Electrode placement is similar for mice, but the sensitivity must be increased approximately 16-fold to compensate for the lower voltage of mouse cardiograms.

Table 6 contains electrocardiographic abnormalities associated with risk of ischemic heart disease and found in animals deficient in copper. Numerous other cardiographic changes have been recorded in animals deficient in copper.[21,104-107] These include atrial and ventricular premature discharges, bradycardia, coupled discharges, first-, second-, and third-degree heart block, inverted or absent T wave, nodel and ventricular escape, ventricular rhythm, and wandering pacemaker.

Blood pressures are measured on unanesthetized rats with an Electro-Sphygmomanometer and pneumatic pulse transducer (Narco Bio-Systems). The blood pressure of each rat is represented by the mean of three measurements obtained within an interval of approximately 2 min. Coefficients of variation for pressure measurements on individual rats average 2.0%.[114]

Ahrens et al. found that "substitution of sucrose for starch at moderate levels" elevates blood pressure of rats.[147] The diet of conventional foods was designed to "simulate the gross composition of the U.S. human diet" and contained 17.6% purified carbohydrate. Approximately 23% of the diet was ground beef, and 58% of the diet was whole milk. Pure sirloin fed to mice induces hypercholesterolemia and high mortality because it is a poor source of copper.[148] Milk has been used in some of the classic inductions of copper deficiency in animals.[149,150] One can conclude that the diet[147] was low in copper. The effects on blood pressure attributed to sucrose may have been mediated by the induction of copper deficiency because sucrose exacerbates deficiency signs when dietary copper is low.[101] Others have found hypertensive effects of sucrose with ill-defined diets.[151-154] Michaelis et al.[155] found a hypertensive response to sucrose with a diet that seems adequate in copper.

Survivors of myocardial infarction had a higher ratio of zinc to copper in hair than healthy subjects.[156] Precautions were taken to exclude from study metal workers and users of shampoos containing zinc.

## E. Links with Other Hypotheses

The theoretical relationship between copper metabolism and ischemic heart disease was initiated in 1973[13] and has been developed in a series of papers.[4,5,14,15,25,56-61] Early emphasis was placed on the adverse effects of a high ratio of zinc to copper and on epidemiology and pathogenesis with the hope of elucidating the etiology of this illness.[13-15,57-61] A disparity between the putative requirement for dietary copper and the usual amounts of dietary copper in the industrialized world has been identified. Recent emphasis has shifted to similarities between animals deficient in copper and people with ischemic heart disease with the hope of clarifying the pathophysiology of the latter.[4,5,56,60,61] Data demonstrating that humans respond to diets low in copper similarly to animals are accumulating (see Section III.A.1). Complete understanding of the copper deficiency theory and its evolution is possible only if all of these papers are read carefully.

The origin(s) of ischemic heart disease is unknown. Its current intricacy rivals that of scurvy in 1772[4] and pellagra in 1915.[15] Hundreds of factors have been suggested as contributing to risk;[94,95] these factors are a bewildering array of paradoxical or apparently dissimilar observations.[14,57,91]

Chamberlin urges "the method of multiple working hypotheses" when studying complex phenomena.[157] The multiple method promotes thoroughness and prevents neglect of important data when emotional partialities prevent serious consideration of all but a single explanation. The proponents of the lipid hypothesis (see Section IV.E.1) have ignored the dictum as they have ignored, and largely are ignoring, many important observations on the biochemistry, chemistry, epidemiology, and pathophysiology (see below) of ischemic heart disease. They have expended little effort to relate these observations to the lipid hypothesis although Kuhn[158] argues that theories incorporating other theories contribute to scientific progress.

Perhaps emotional partialities or nearly instinctive assumptions prompt some to consider ischemic heart disease to be a problem revolving about the toxicology of dietary fat. Medawar's warning[159] "...the intensity of the conviction that a hypothesis is true has no bearing on whether it is true or not" is important in this context. The lipid hypothesis is likely to remain an hypothesis as it has been for decades unless it fully incorporates several diverse observations and concepts.

### *1. The Lipid Hypothesis*

The lipid hypothesis probably is the most widely accepted and most intensively studied hypothesis on the origin of ischemic heart disease. In brief, large amounts of dietary fat are toxic. Some of the nuances of concepts within this hypothesis have been delineated.[5,61] Links between the lipid hypothesis and the copper deficiency theory have been reviewed and summarized.[5] The concepts that follow support other hypotheses. It is impossible to explain these concepts in terms of dietary lipid.

**Diet and risk**—One of the weaknesses of the lipid hypothesis is the general lack of association between dietary fat and either heart disease risk or concentration of cholesterol in plasma when data from *within* single countries are examined.[5,61] These associations with diet occur when data collected *among* countries are considered. In contrast is the direct and significant correlation between the ratio of zinc to copper of milk available in 47 cities of the U.S. with the death rate for coronary heart disease in these cities (r = 0.354, $p < 0.02$).[8] Ethanol and the ratio of zinc to copper of milk are the only chemical characteristics of food or diet that have been associated with risk of heart disease death in an epidemiologic study in the continental U.S.

### 2. *Alcoholic Beverages*

Moderate consumption of alcoholic beverages, especially beer, often is associated with decreased risk of death from ischemic heart disease or coronary occlusion.[160-171] Rats fed a diet deficient in copper that drink beer live five times as long as similar rats that drink water[172] (also Klevay and Moore, unpublished). Beer drinking rats also have lower plasma cholesterol, less cardiac enlargement and higher liver copper. Although beer is a complex mixture, it has the property of being both cholesterotropic and cuprotropic (see Section III.C). Perhaps greater beer consumption in warm weather causes the slightly lower concentration of cholesterol noted in June in comparison to December.[173]

### 3. *The "Water Factor"*

Kobayashi made the first observation on the relationship between the chemistry of river water and the risk of vascular disease.[174] In brief, risk is high where communal water supplies are soft and low where water is hard. The material responsible for this phenomenon has come to be called the water factor.[58] This relationship, which also has been found by others and has been reviewed,[58,175] was an important stimulus to my efforts to find a trace substance that contributed to heart disease risk.[25]

Water hardness is highly correlated (r = 0.96) with calcium in water.[57] According to Romasz et al.,[176] increasing the dietary calcium of rats increased liver copper as cholesterol in serum decreased. The details of the links between the water factor and the copper theory have been reviewed.[58]

### 4. *Magnesium Deficiency and Metabolism*

Magnesium metabolism and magnesium deficiency have been hypothesized to be important in the origin of ischemic heart disease.[177,178] Magnesium in water is correlated with hardness of water (r = 0.90).[57] There is a very high correlation between copper and magnesium in human meals (r = 0.85).[87] Diets low in magnesium probably are low in copper.

### 5. *Sucrose Excess and Disease*

Yudkin[179,180] has developed the hypothesis that diets high in sucrose are important in the etiology of ischemic heart disease. The hypothesis is based on epidemiology and experiments with animals and people. Consideration of the frequency of diets low in copper (see Section III.B, Table 6) and the metabolic interaction between sucrose and low copper diets (see Section III.C)[181] provides a link between the sucrose hypothesis and the copper theory.

#### *6. Disease and Lack of Fiber*

Cleave and Campbell[182] emphasized the potentially adverse effects of diets high in sugar on several other illnesses besides ischemic heart disease. Burkitt and Trowell[183] reconsidered Cleave's hypothesis, revised it, and noted that diets high in sugar (and other highly refined carbohydrates) necessarily were low in dietary fiber. They postulated beneficial effects of fiber including an increase in clot lysis. If diets high in fiber resemble the breakfast cereals we studied, these diets also are high in copper because there are high, positive correlations between various fibrous components and the amount of copper in the cereals.[184]

#### *7. Uric Acid and Disease*

Newland[118] has discussed the association between uric acid and ischemic heart disease. He suggests that uric acid promotes thrombosis and hypertension. Mice deficient in copper have atrial thrombosis,[21] and deficient, adult rats are hypertensive[114,115] (see Section IV.A.7).

#### *8. Aspirin and Disease*

The protective effect of aspirin on risk of ischemic heart disease and its possible relationship to copper metabolism[19] are summarized (see Sections III.C and IV.C).

#### *9. Injury By Free Radicals*

Some of the damage to arteries in the atherosclerotic process, to hearts following myocardial infarction, and to other organs following circulatory shock may be caused by free radicals.[185,186] Copper-zinc superoxide dismutase is found in various tissues, is dependent upon copper for activity,[187] and is protective against free radicals.[185,188] Activity of this enzyme in heart is substantially decreased in copper deficiency.[189] As cardiac copper is low in ischemic heart disease (see Section IV.C), it seems likely that protection against free radicals may be decreased because of decreased activity of superoxide dismutase.

Saari[190] has shown that the cardiac enlargement found in copper deficiency can be decreased by the free-radical scavenger dimethylsulfoxide. Human aortae with aneurysms or occlusive disease have decreased activity of copper-superoxide dismutase.[191]

#### *10. Thrombosis and Ischemic Heart Disease*

The thrombotic process is important in atherogenesis and ischemic heart disease.[6,192-194] The link between thrombosis and the copper theory was forged with the finding of atrial thrombosis in mice deficient in copper.[21] This hypercoagulability may be caused by the high ratio of thromboxane $A_2$ to prostacyclin found in copper deficiency.[195,196] This change in ratio has been suggested to contribute to atherosclerosis via lipid hydroperoxides.[186]

When platelets aggregate on the walls of normal arteries there is a release of endothelium-derived relaxing factor which results in arterial dilation and the washing away of the early thrombus. In animals that are hypertensive or atherosclerotic less of this factor is released. It is avidly degraded by superoxide ion[197] (see Section IV.E.9).

#### *11. Irrelevant Correlations*

Measurements of copper in plasma have been mentioned only rarely in the development of the copper deficiency theory because there is no certainty that plasma copper reflects either copper intake or organ copper.[4,93,138,198-201] Sometimes decreased organ copper is found with either normal[199] or increased copper in plasma.[93,200,201] Sometimes significant correlations between plasma copper and plasma cholesterol are found[202-204] and sometimes not.[205,206]

#### *12. Recapitulation and Remarks*

Animals respond to copper deficiency with chemical and physiological characteristics that resemble the risk factors for ischemic heart disease in humans: glucose intolerance, hypercholesterolemia, hyperuricemia, abnormal electrocardiograms, male predisposition, and hypertension. Similar evidence on glucose, cholesterolemia, cardiograms, and blood pressure is available from experiments with human volunteers.[37,38,52,53] Copper deficiency is the only nutritional insult that reproduces all of these factors. Copper deficiency may be a variable hidden in other explanations of heart disease etiology, e.g., alcoholic beverages, hard water, sucrose, etc.

There are numerous other similarities in addition to the risk factors between animals deficient in copper and people with ischemic heart disease. Some epidemiologic characteristics, e.g., the relationships of risk to consumption of alcoholic beverages or aspirin or the association of risk to the availability of hard water, are subjects of extensive study. Mention of these relationships are egregiously absent from writings by those who study and comment upon dietary fat. Other lesser known characteristics of ischemic heart disease epidemiology have been reviewed,[15,57,61] e.g., the protective effect of cirrhosis and the exacerbating effect of hemodialysis can be explained only in terms of copper metabolism. Similarly the lipid hypothesis cannot explain the effects of oral zinc on lipoprotein metabolism (see Sections III.A.2 and IV.E).

## V. CONCLUSION

Production of hypercholesterolemia in rats more than 15 years ago by a high ratio of zinc to copper ingested has provided a new way of looking at ischemic heart disease. Early emphasis was placed on relative or absolute deficiency of copper as a way of explaining much of the epidemiology. Data from experiments with animals, iatrogenic maneuvers, and experiments of nature were included. Emphasis gradually shifted from etiology to pathophysiology. Nearly 50 similarities between animals deficient in copper and people with ischemic heart disease have been identified. The most important of these—glucose intolerance, hypercholesterolemia, abnormal electrocardiograms, male predisposition, hypertension, and hyperuricemia— provide the beginning of a mechanistic explanation of how copper deficiency can produce heart disease although the mechanistic details of each of these are not known. Cholesterolemia, glucose clearance, electrocardiograms, and blood pressure have been altered in people consuming diets low in copper under controlled conditions. There are links between other hypotheses and the copper deficiency theory. This theory is the most inclusive when all data are considered.

## NOTE ADDED IN PROOF

Work germane to this topic has been found since the manuscript was written. Brief comments are arranged according to the subdivisions of the original manuscript.

### II.B. Recent Work

Hypercholesterolemia has been found in copper deficiency again.[208,209] If these results are due to chance instead of cause, 118 more negative experiments should exist.

### II.C. The Mechanism

Copper deficiency decreases the activities of hepatic lipase[208] and lipoprotein lipase.[45,208] An increase in activity of hydroxymethylglutaryl-coenzyme A reductase in deficiency has been confirmed.[210]

### III.A. The Adult Copper Requirement

Turnlund et al.[211] measured copper balance of young men using a formula diet. Less than 31 to 34% of the energy in this diet was from conventional foods when 2400 calories were consumed each day. The rest of the energy was supplied by 369 g of nutrient formula: albumen, corn syrup solids, cottonseed oil, minerals, starch, and sucrose. Depending on energy requirements, men who required more energy received proportionately less food. The experimental data and interpretations of results are germane only to similar people eating a similar diet. The work is irrelevant to the measurement of the copper requirement of people who eat food.

#### *1. How Much Copper is Too Little?*

We have attempted to deplete men of copper in four separate experiments since 1980 and have been puzzled by the fact that only three of eight men responded to the diet low in copper: one with hypercholesterolemia[37] and two with decreased glucose tolerance.[53] Recent comparisons and data evaluation reveal that the three were more likely ($p < 0.002$, significant difference in proportions) than the five to lose more copper in feces and urine than they consumed no matter what their copper intake: control, depletion, or repletion.[212,213] These three men seemed to utilize copper differently from the others; perhaps they differed in their copper stores or their heredity.

To date, more than 31 men and women have responded to diets low in copper with potentially harmful changes in lipids,[37,38] glusose tolerance,[53] blood pressure[52] or electrocardiograms.[37,55]

### *2. How Much Zinc is Too Much?*

More than 76 men and women have responded to high doses of zinc with potentially harmful changes in lipid metabolism.[63-66]

## III.B. Usual Copper Intakes

Daily copper intake of men determined by atomic absorption spectrometry averaged 0.75 mg.[214] Agreement between calculated and analyzed values were poor, with the former being substantially higher.

## III.C. Modifying Factors

In 1975 I suggested that the stimulus of exercise causes an incorporation of zinc into newly synthesized bone and muscle and that this zinc may be diverted from the liver with a resulting benefit to cholesterol metabolism.[57] Conversely, atrophy of these organs from inactivity releases extra zinc. Krebs et al.[215] estimated that bed-rest releases 2.06 mg of zinc from bone and muscle per day and suggested that although total body losses of zinc are increased during rest, the amount of zinc available to other cells may actually increase. Their estimate of zinc mobilization agrees well with mine—2.29 mg/day.

Rats fed cholesterol are said to have changes in fatty acid desaturases in liver microsomes.[216] Some of the experiments were done using diets containing both cholesterol and cholic acid—a classic regimen recently shown to increase the requirement of the animals for dietary copper.[199] There seems little doubt that copper deficiency can affect desaturase enzymes, but agreement is lacking on the details and directions of all of the changes.[187,209,217-222]

Carville and Strain found that lactose can modify several responses of rats to a diet low in copper.[223] (See Section IV.E.9.) Koo and Lee[224] reviewed the hypocholesterolemic effect of zinc deficiency and studied its mechanism.

## IV.B. Anatomical Similarities

Smooth muscle cell migration is important in the early pathogenesis of the atherosclerotic lesion.[225] Short stature (less than 68 inches) was associated with death from coronary heart disease by Paffenbarger et al.[226] Barker et al.[227] found that men in Hertfordshire, England with the lowest weights at birth and at 1 year had the highest death rates from ischemic heart disease, and cited data from England, Norway, and Finland on an inverse relationship between adult height and cardiovascular mortality. Redman et al.[228] have extended the description of the ventricular aneurysms found in copper deficiency.[18,19,105-107,113] They also found that the sum of three heart dimensions was increased by deficiency.[228] Echocardiography of elderly subjects of the Framingham study by Levy et al.[229] revealed that increased ventricular mass was a predictor of coronary heart disease, and this measurement was more sensitive than electrocardiography.

### IV.C. Chemical and Physiological Similarities

Copper in plasma of spontaneously hypertensive rats had a different pattern of change during the development of hypertension than bromine, rubidium, selenium, or zinc. Copper differences from normotensive rats also were greater.[230] Aalbers[231] found less copper in the anterior wall of the left ventricle in people who had died from ischemic heart disease than in people who died of other causes; sodium was higher in heart disease. We found cardiac sodium increased in rats deficient in copper.[232] Rats deficient in copper and fed salt[136] have a decrease in both aldosterone and plasma renin of approximately 94%.[233] Dermatan sulfate is the major glycosaminoglycan that accumulates in atherosclerotic plaques.[234] Dermatan sulfate also is increased in the aortas of copper deficient rats.[235] Carr and Lei[236] found rats deficient in copper have an increase in plasma volume.

### IV.D. Physiological Methods and Other Results

Dietary fructose induced hypertension in rats again.[237]

### IV.E. Links with Other Hypotheses

McCormick[238] refers to the concept of the multifactorial aetiology of coronary heart disease as a dangerous delusion that provides an inadequate basis for mass prevention strategies giving rise to false promises and unreal expectations. He agrees that current therapeutic interventions do not lengthen life[239-241] and is dubious that changes in diet are responsible for recent secular trends in mortality.[242]

In a unique study, over 12 years, of 7053 men and 8346 women, Isles et al.[243] found no relationship between plasma cholesterol and all-cause mortality. More cancer at low concentrations balanced more heart disease at high concentrations.

Stehbens[244] is more pessimistic on diagnostic accuracy than I[5] and doubts that coronary heart disease increased early in this century or is decreasing now. Zarling et al.[245] and Battle et al.[246] confirm a frequent lack of agreement between the clinical diagnosis of acute myocardial infarct and autopsy findings. Lie summarizes some of these discrepancies.[247]

### *1. The Lipid Hypothesis*

Stehbens identifies some inconsistencies in the lipid hypothesis on the origin of atherosclerosis.[248] Holtzman[249] emphasizes that data from Framingham[250] reveal that even among men in the highest quintile of serum cholesterol, 69% are free of coronary heart disease during the next 20 years. These data have implications of public policy and illustrate the fact that ischemic (coronary) heart disease is not merely a problem of lipid metabolism.[4] Atherosclerosis occurs without risk factors.[251]

The Framingham study[249,250] also revealed that nearly 9% of men in the lowest quintile of cholesterol had heart disease within 20 years. Since the Lipid Research Clinics trial,[252] the idea that a 1% decrease in cholesterol will produce a 2% decrease in risk has been widely advertised. If one considers the upper and lower boundaries of the cholesterol quintile,[249,250] 193 and 256 mg/dl, respectively, the concept that risk decreases faster than cholesterol during a drug intervention program seems reasonable. However, if one realized that members of each quintile generally did not have cholesterol values equal to those that defined the boundaries of the groups, then it seems that the decline in risk to be expected is more similar to the decline in cholesterol if subjects of intervention are similar to people of Framingham. Others[253,254] have joined Palumbo[54] in being critical of the concept that risk falls more rapidly than cholesterol. It may be interesting to examine the cholesterol values of those who died in Framingham; were men in the upper half of each quintile more likely to die than men in the lower halves?

Stehbens[248] also mentions that diets varying in carbohydrate, fat, and protein also are different in other constituents. A statistical association between the amount of fat and the ratio of zinc to copper has been found for 71 common foods.[255] Thus, modification of diets to lower fat intake is likely to improve copper metabolism.[5] A 40% decrease in dietary fat produced a 7.4% decrease in the ratio of zinc to copper.[256] Althougth cautious interpretation is desirable because copper values were calculated rather than measured,[4,50,56,86,214] this change may be important because on a molar basis copper is a more powerful determinant of cholesterolemia than dietary fat.[56]

### *2. Alcoholic Beverages*

The experiment on the drinking of beer by rats deficient in copper[172,257] also may explain the lower concentration of cholesterol in human plasma in summer because more beer is consumed in summer than in winter.

### *4. Magnesium Deficiency and Metabolism*

Magnesium deficiency in rats decreases liver copper and increases liver zinc[258] without affecting liver calcium, magnesium, phosphorus, potassium, or sodium. Liver iron also was decreased by deficiency. The effect of dietary magnesium on liver copper and zinc was similar to that of calcium,[176] confirming my prediction.[58]

### 5. *Sucrose Excess and Disease*

There also is a link between high sucrose intakes and the endocrinology of ischemic heart disease. A decrease in dietary sucrose caused a decrease in estradiol-17β in plasma of young men.[259] Phillips[260,261] has provided evidence that increased plasma estrogen is associated with increased risk of myocardial infarction. Yudkin has reviewed more than a dozen abnormalities seen in heart disease that can be produced by sucrose.[262]

### 9. *Injury by Free Radicals*

Cunnane et al.[209] perfused rat hearts *in vitro* and found that those from animals deficient in copper contracted less forcibly than hearts from supplemented rats after a period of anoxia.

Copper deficiency decreased the superoxide dismutase activity of granulocytes and lymphocytes (but not of erythrocytes or macrophages) of rats.[263] A later report by the authors[223] revealed that copper deficiency also decreased erythrocyte activity. Lactose can decrease further the antioxidant responses to diet low in copper.

Women supplemented with zinc (50 mg/day) were found to have decreased activity of copper-zinc superoxide dismutase in erythrocytes.[264] Rats fed a diet marginal in copper had decreases in activities of several antioxidant enzymes.[265]

### 10. *Thrombosis and Ischemic Heart Disease*

A recent review[266] mentions the inactivation of endothelium-derived relaxing factor (nitric oxide) by superoxide and that thromboxane $A_2$ may intensify the effects of endothelium-derived contracting factors. Endothelium-dependent contractions become more prominent with hypertension (see Section IV.A.5).

### 11. *Irrelevant Correlations*

Uza and Vlaicu[267] found increased copper in serum of patients with atherosclerosis obliterans. Although no measurements of plasma lipids were reported, one should not infer that organ copper was normal. Another disparity between plasma copper and organ copper has been found.[268]

## ACKNOWLEDGMENTS

I thank L. K. Johnson for the dietary frequency calculations and K. Grembowski for typing the manuscript.

## REFERENCES


1. **Eckstein, G.,** *The Body has a Head,* Harper & Row, New York, 1969, 124.
2. **Arey, L. B.,** *Developmental Anatomy,* W. B. Saunders, Philadelphia, 1954, 106, 351.
3. **Anon.**, *Vital Statistics of the United States,* Vol. 2, Mortality, Part A, 1984; DHHS Pub. No.(PHS) 87-1122, Public Health Service, U.S. Government Printing Office, Hyattsville, MD, 1987, Tables 1-6, 1-22, 2-5, 2-7.
4. **Klevay, L. M.,** The role of copper, zinc, and other chemical elements in ischemic heart disease, in *Metabolism of Trace Metals in Man,* Vol. 1, Rennert, O. M and Chan, W-Y, Eds., CRC Press, Boca Raton, FL, 1984, 129.
5. **Klevay, L. M.,** Ischemic heart disease: a major obstacle to becoming old, *Clin. Ger. Med.,* 3, 361, 1987.
6. **Spiekerman, R. E., Brandenburg, J. T., Achor, R. W. P., and Edwards, J. E.,** The spectrum of coronary heart disease in a community of 30,000, *Circulation,* 25, 57, 1962.
7. **Scotti, T. M.,** Heart, in *Pathology,* 7th ed., Anderson, W. A. D. and Kissane, J. M., Eds., C. V. Mosby, St. Louis, 1977, 737.
8. **Klevay, L. M.,** The ratio of zinc to copper in milk and mortality due to coronary heart disease: an association, in *Trace Substances in Environmental Health-VIII,* Hemphill, D. D., Ed., University of Missouri Press, Columbia, 1974, 9.
9. **Osborn, G. R.,** Stages in development of coronary disease observed from 1,500 young subjects. Relationship of hypotension and infant feeding to aetiology, in *Le Rôle de la Paroi Artérielle dans l'Athérogénèse,* Lenègre, M. J., Scebat, L., and Renais, J., Eds., Centre National de la Recherche Scientifique, Paris, 1968, 93.
10. **Klevay, L. M. and Allen, K. G. D.,** Vitamin $B_6$, copper, and atherosclerosis, *Lancet,* 1, 1209, 1977.
11. **Levine, C. I and Murray, J. C.,** The aetiological role of maternal vitamin-$B_6$ deficiency in the development of atherosclerosis, *Lancet,* 1, 628, 1977.
12. **Wissler, R. W.,** Principles of the pathogenesis of atherosclerosis, in *Heart Disease,* 2nd ed., Braunwald, E., Ed., W. B. Saunders, Philadelphia, 1984, 1183.
13. **Klevay, L. M.,** Hypercholesterolemia in rats produced by an increase in the ratio of zinc to copper ingested, *Am. J. Clin. Nutr.,* 26, 1060, 1973.
14. **Klevay, L. M.,** Elements of ischemic heart disease, *Perspect. Biol. Med.,* 20, 186, 1977.
15. **Klevay, L. M.,** The role of copper and zinc in cholesterol metabolism, in *Advances in Nutritional Research,* Vol 1., Draper, H. H., Ed., Plenum Press, New York, 1977, 227.
16. **Klevay, L. M., Petering, H. G., and Stemmer, K. L.,** A controlled environment for trace metal experiments on animals, *Environ. Sci. Technol.,* 5, 1196, 1971.
17. **Gallagher, C. H., Judah, J. D., and Rees, K. R.,** The biochemistry of copper deficiency. I. Enzymological disturbances, blood chemistry and excretion of amino-acids, *Proc. R. Soc. London Ser. B.,* 145, 134, 1956.
18. **Klevay, L. M.,** Clofibrate hypocholesterolemia associated with increased hepatic copper, *Drug Nutr. Interact.,* 2, 131, 1983.
19. **Klevay, L. M.,** Aspirin hypocholesterolemia associated with increased microsomal copper in liver, *Nutr. Res.,* 6, 1281, 1986.
20. **Carpenter, K. J., Gotsis, A., and Hegsted, D. M.,** Estimation of total cholesterol in serum by a micro method, *Clin. Chem.,* 3, 233, 1957.
21. **Klevay, L. M.,** Atrial thrombosis, abnormal electrocardiograms and sudden death in mice due to copper deficiency, *Atherosclerosis,* 54, 213, 1985.
22. **Klevay, L. M.,** The biotin requirement of rats fed 20% egg white, *J. Nutr.,* 106, 1643, 1976.
23. **Klevay, L. M.,** Hypercholesterolemia due to ascorbic acid, *Proc. Soc. Exp. Biol. Med.,* 151, 579, 1976.

24. **Moore, R. J. and Klevay, L. M.,** Effect of copper deficiency on blood pressure and plasma and lung angiotensin-converting enzyme activity in rats, *Nutr. Res.*, 8, 489, 1988.
25. **Klevay, L. M.,** This week's citation classic, *Curr. Contents Clin. Med.*, 15, 20, 1987.
26. **Lei, K. Y.,** Cholesterol metabolism in copper-deficient rats, *Nutr. Rep. Int.*, 15, 597, 1977.
27. **Allen, K. G. D. and Klevay, L. M.,** Cholesterolemia and cardiovascular abnormalities in rats caused by copper deficiency, *Atherosclerosis*, 29, 81, 1978.
28. **Harvey, P. W. and Allen, K. G. D.,** Decreased plasma lecithin: cholesterol acyltransferase activity in copper-deficient rats, *J. Nutr.*, 111, 1855, 1981.
29. **O'Dell, B. L.,** Copper-zinc interaction, effect of excess dietary zinc on copper status, Proc. N. Z. Workshop on Trace Elements in New Zealand, University of Otago, Dunedin, 1981, 157.
30. **Nielsen, F. H., Zimmerman, T. J., and Shuler, T. R.,** Interactions among nickel, copper, and iron in rats. Liver and plasma content of lipids and trace elements, *Biol. Trace Element Res.*, 4, 125, 1982.
31. **Resier, S., Ferretti, R. J., Fields, M., and Smith, J. C., Jr.,** Role of dietary fructose in the enhancement of mortality and biochemical changes associated with copper deficiency in rats, *Am. J. Clin. Nutr.*, 38, 214, 1983.
32. **Wu, B. N., Medeiros, D. M., Lin, K-N., and Thorne, B. M.,** Long term effect of dietary copper and sodium upon blood pressure in the Long-Evans rat, *Nutr. Res.*, 4, 305, 1984.
33. **Jones, D. G.,** Effects of dietary copper depletion on acute and delayed inflammatory responses in mice, *Res. Vet. Sci.*, 37, 205, 1984.
34. **Lefevre, M., Keen, C. L., Lönnerdal, B., Hurley, L. S., and Schneeman, B. O.,** Copper deficiency-induced hypercholesterolemia: effects on HDL subfractions and lipoprotein receptor activity in the rat, *J. Nutr.*, 116, 1735, 1986.
35. **Valsala, P. and Kurup, P. A.,** Investigations on the mechanism of hypercholesterolemia observed in copper deficiency in rats, *J. Biosci.* 12, 137, 1987.
36. **Lynch, S. M. and Strain, J. J.,** Effects of dietary copper deficiency on hepatic antioxidant enzymes in male and female rats, *Nutr. Rep. Int.*, 37, 1127, 1988.
37. **Klevay, L. M., Inman, L., Johnson, L. K., Lawler, M., Mahalko, J. R., Milne, D. B., Lukaski, H. C., Bolonchuk, W., and Sandstead, H. H.,** Increased cholesterol in plasma in a young man during experimental copper depletion, *Metabolism*, 33, 1112, 1984.
38. **Reiser, S., Powell, A., Yang, C-Y, and Canary, J. J.,** Effect of copper intake on blood cholesterol and its lipoprotein distribution in men, *Nutr. Rep. Int.*, 36, 641, 1987.
39. **Frimpong, N. A. and Magee, A. C.,** Effects of dietary copper and zinc on serum lipid parameters in young male rats, *Nutr. Rep. Int.*, 35, 551, 1987.
40. **Sinthusek, G. and Magee, A. C.,** Relationships of dietary zinc/copper ratios to plasma cholesterol and liver trace mineral deposition in young rats fed saturated and unsaturated fats, *Nutr. Res.*, 4, 841, 1984.
41. **Anon.,** *Nutrient Requirements of Laboratory Animals,* National Academy of Sciences, Washington, D.C., 1978, 23.
42. **Sherman, A. R.,** Copper and iron deficiencies: effects on serum lipids and tissue minerals, *Nutr. Res.*, 1, 363, 1981.
43. **Prohaska, J. R., Korte, J. J., and Bailey, W. R.,** Serum cholesterol levels are not elevated in young copper-deficient rats, mice or brindled mice, *J. Nutr.*, 115, 1702, 1985.
44. **Lau, B. W. C. and Klevay, L. M.,** Plasma lecithin: cholesterol acyltransferase in copper deficient rats, *J. Nutr.*, 111, 1698, 1981.
45. **Lau, B. W. C. and Klevay, L. M.,** Postheparin plasma lipoprotein lipase in copper-deficient rats, *J. Nutr.*, 112, 928, 1982.
46. **Lei, K. Y., Ed.,** Cholesterol metabolism in copper-deficient rats, in *Role of Copper in Lipid Metabolism,* CRC Press, Boca Raton, FL, 1990, chap. 2.

47. **Anon.**, *Recommended Dietary Allowances,* Food and Nutrition Board, National Research Council, 9th ed., National Academy of Sciences, Washington, D.C., 1980, 151.
48. **Anon.**, *Recommended Dietary Allowances,* Food and Nutrition Board, National Research Council, 8th ed., National Academy of Sciences, Washington, D.C., 1974, 95.
49. **Cartwright, G. E.,** Copper metabolism in human subjects, in *Copper Metabolism,* McElroy, W. D. and Glass, B., Eds., Johns Hopkins University Press, Baltimore, 1950, 274.
50. **Klevay, L. M., Reck, S. J., Jacob, R. A., Logan, G. M., Jr., Munoz, J. M., and Sandstead, H. H.,** The human requirement for copper. I. Healthy men fed conventional, American diets, *Am. J. Clin. Nutr.,* 33, 45, 1980.
51. **Klevay, L. M.,** Dietary requirements for trace elements in humans, *Trace Element Analytical Chemistry in Medicine and Biology,* Vol. 4, Bätter, P. and Schramel, P., Eds., Walter deGruyter, Berlin, 1987, 43.
52. **Lukaski, H. C., Klevay, L. M., and Milne, D. B.,** Effects of dietary copper on human autonomic cardiovascular function, *Eur. J. Appl. Physiol.,* 58, 74, 1988.
53. **Klevay, L. M., Canfield, W. K., Gallagher, S. K., Henriksen, L. K., Lukaski, H. C., Bolonchuk, W., Johnson, L. K., Milne, D. B., and Sandstead, H. H.,** Decreased glucose tolerance in two men during experimental copper depletion, *Nutr. Rep. Int.,* 33, 371, 1986.
54. **Palumbo, P. J.,** National Cholesterol Education Program; Does the emperor have any clothes?, *Mayo Clin. Proc.,* 63, 88, 1988.
55. **Reiser, S., Smith, J. C., Jr., Mertz, W., Holbrook, J. T., Scholfield, D. J., Powell, A. S., Canfield W. K., and Canary, J. J.,** Indices of copper status in humans consuming a typical American diet containing either fructose or starch, *Am. J. Clin. Nutr.,* 42, 242, 1985.
56. **Klevay, L. M.,** Dietary copper: a powerful determinant of cholesterolemia, *Med. Hypoth.,* 24, 111, 1987.
57. **Klevay, L. M.,** Coronary heart disease: the zinc/copper hypothesis, *Am. J. Clin. Nutr.,* 28, 764, 1975.
58. **Klevay, L. M.,** The influence of copper and zinc on the occurrence of ischemic heart disease, *J. Environ. Pathol. Toxicol.,* 4, 281, 1980.
59. **Klevay, L. M.,** Ischemic heart disease: updating the zinc/copper hypothesis, in *Nutrition and Heart Disease,* Naito, H. K., Ed., S.P. Medical & Scientific Books, New York, 1982, 61.
60. **Klevay, L. M.,** Interactions of copper and zinc in cardiovascular disease, *Ann. N.Y. Acad. Sci.,* 355, 140, 1980.
61. **Klevay, L. M.,** Copper and ischemic heart disease, *Biol. Trace Element Res.,* 5, 245, 1983.
62. **Underwood, E. J.,** *Trace Elements in Human and Animal Nutrition,* 4th ed., Academic Press, New York, 1977, 71, 95, 231.
63. **Chandra, R. K.,** Excessive intake of zinc impairs immune responses, *JAMA,* 252, 1443, 1984.
64. **Hooper, P. L., Visconti, L., Garry, P. J., and Johnson, G. E.,** Zinc lowers high-density lipoprotein-cholesterol levels, *JAMA,* 244, 1960, 1980.
65. **Black, M. R., Medeiros, D. M., Brunett, E., and Welke, R.,** Zinc supplements and serum lipids in young adult white males, *Am. J. Clin. Nutr.,* 47, 970, 1988.
66. **Goodwin, J. S., Hunt, W. C., Hooper, P., and Garry, P. J.,** Relationship between zinc intake, physical activity, and blood levels of high-density lipoprotein cholesterol in a healthy elderly population, *Metabolism,* 34, 519, 1985.
67. **Freeland-Graves, J. H., Friedman, B. J., Han, W.-H., Shorey, R. L., and Young, R.,** Effect of zinc supplementation on plasma high-density lipoprotein cholesterol and zinc, *Am. J. Clin. Nutr.,* 35, 988, 1982.
68. **Crouse, S. F., Hooper, P. L., Atterbom, H. A., and Papenfuss, R. L.,** Zinc ingestion and lipoprotein values in sedentary and endurance-trained men, *JAMA,* 252, 785, 1984.

69. **Pachotikarn, C., Medeiros, D. M., and Windham F.,** Effect of oral zinc supplementation on plasma lipids, blood pressure, and other variables in young adult white males, *Nutr. Rep. Int.* 32, 373, 1985.
70. **Samman, S. and Roberts, D. C. K.,** The effect of zinc supplements on lipoproteins and copper status, *Atherosclerosis,* 70, 247, 1988.
71. **Fischer, P. W. F. and Giroux, A.,** Effect of zinc supplementation on the copper status and cholesterol levels of Cynomolgus monkeys, *Nutr. Res.,* 7, 499, 1987.
72. **Burke, D. M., DeMicco, F. J., Taper, L. J., and Ritchey, S. J.,** Copper and zinc utilization in elderly adults, *J. Gerontol.,* 36, 558, 1981.
73. **Greger, J. L., Zaikis, S. C., Abernathy, R. P., Bennett, O. A., and Huffman, J.,** Zinc, nitrogen, copper, iron and manganese balance in adolescent females fed two levels of zinc, *J. Nutr.,* 108, 1449, 1978.
74. **Buchet, J. P., Lauwerys, R., Vandevoorde, A., and Pycke, J. M.,** Oral daily intake of cadmium, lead, manganese, copper, chromium, mercury, calcium, zinc and arsenic in Belgium: a duplicate meal study, *Food Chem. Toxicol.,* 21, 19, 1983.
75. **Bunker, V. W., Hinks, L. J., Lawson, M. S., and Clayton, B. E.,** Assessment of zinc and copper status of healthy elderly people using metabolic balance studies and measurement of leucocyte concentrations, *Am. J. Clin. Nutr.,* 40, 1096, 1984.
76. **Bunker, V. W., Hinks, L. J., Stansfield, M. F., Lawson, M. S., and Clayton, B. E.,** Metabolic balance studies for zinc and copper in housebound elderly people and the relationships between zinc balance and leukocyte concentrations, *Am. J. Clin. Nutr.,* 46, 353, 1988.
77. **Gibson, R. S., Martinez, O. B., and MacDonald, A. C.,** The zinc, copper, and selenium status of a selected sample of Canadian elderly women, *J. Gerontol.,* 40, 296, 1985.
78. **Krebs, J. M., Schneider, V. S., and LeBlanc, A. D.,** Zinc, copper and nitrogen balance during bed rest and fluoride supplementation in healthy adult males, *Am. J. Clin. Nutr.,* 47, 509, 1988.
79. **Moser, P. B., Reynolds, R. D., Acharya, S., Howard, M. P., Andon, M. B., and Lewis, S. A.,** Copper, iron, zinc, and selenium dietary intake and status of Nepalese lactating women and their breast-fed infants, *Am. J. Clin. Nutr.,* 47, 729, 1988.
80. **Patterson, K. Y., Holbrook, J. T., Bodner, J. E., Kelsay, J. L., Smith, J. C., Jr., and Veillon, C.,** Zinc, copper and manganese intake and balance for adults consuming self-selected diets, *Am. J. Clin. Nutr.,* 40, 1397, 1984.
81. **Taper, L. J., Oliva, J. T., and Ritchey, S. J.,** Zinc and copper retention during pregnancy: the adequacy of prenatal diets with and without dietary supplementation, *Am. J. Clin. Nutr.,* 41, 1184, 1985.
82. **Thomas, A. J., Bunker, V. W., Hinks, L. J., Sodha, N., Mullee, M. A., and Clayton, B. E.,** Energy, protein, zinc and copper status of twenty-one elderly inpatients: analyzed intake and biochemical indices, *Br. J. Nutr.,* 59, 181, 1988.
83. **Wyttenbach, A., Bajo, S., Tobler, L., and Zimmerli, B.,** The concentration of 19 trace elements in the Swiss diet, *Trace Element Analytical Chemistry in Medicine and Biology,* Brätter, P. and Schramel, P., Eds., Walter deGruyter, Berlin, 1987, 4, 169.
84. **Rivai, I. F., Suzuki, S., Koyama, H., Hyodo, K., Djuangsih, N., and Soemarwoto, O.,** Copper content of foods of Java Island and estimation of daily copper intake, *Bull. Environ. Contam. Toxicol.,* 41, 114, 1988.
85. **Soman, S. D., Panday, V. K., Joseph, K. T., and Raut, S. J.,** Daily intake of some major and trace elements, *Health Phys.,* 17, 35, 1969.
86. **Klevay, L. M.,** Cholesterotropic and cuprotropic chemicals, *Proc. 5th Int. Symp. Trace Element Metabolism in Man and Animals,* Mills, C. F., Bremner, I., and Chesters, J. K., Eds., Commonwealth Agricultural Bureaux, Farnham Royal, U.K., 1985, 180.
87. **Klevay, L. M.,** Magnesium, calcium, copper, and zinc in meals. Correlations related to the epidemiology of ischemic heart disease, *Biol. Trace Element Res.,* 4, 95, 1983.

88. **Gibson, R. S and Scythes, C. A.,** Trace element intakes of women, *Br. J. Nutr.*, 48, 241, 1982.
89. **Holden, J. M., Wolf, W. R., and Mertz, W.,** Zinc and copper in self-selected diets, *J. Am. Diet. Assoc.*, 75, 23, 1979.
90. **Klevay, L. M., Reck, S. J., and Barcome, D. F.,** Evidence of dietary copper and zinc deficiencies, *JAMA*, 241, 1916, 1979.
91. **Klevay, L. M.,** Changing patterns of disease: some nutritional remarks, *J. Am. Coll. Nutr.*, 3, 149, 1984.
92. **Klevay, L. M.,** An appraisal of current human copper nutriture, in *Inflammatory Diseases and Copper*, Sorenson, J. R. J., Ed., The Humana Press, Clifton, NJ, 1982, 123.
93. **Klevay, L. M.,** Dietary cholesterol lowers liver copper in rabbits, *Biol. Trace Element Res.*, 16, 51, 1988.
94. **Hopkins, P. N. and Williams, R.,** A survey of 246 suggested coronary risk factors, *Atherosclerosis*, 40, 1, 1981.
95. **Strasser, T.,** Coronary risk factors revisited, *World Health Forum*, 3, 85, 1982.
96. **Insull, W., Jr.,** Coronary Risk Handbook, American Heart Association, New York, 1973, 3.
97. **Kannel, W. B.,** Some lessons in cardiovascular epidemiology from Framingham, *Am. J. Cardiol.*, 37, 269, 1976.
98. **Keil, H. L. and Nelson, V. E.,** The rôle of copper in carbohydrate metabolism, *J. Biol. Chem.*, 106, 343, 1934.
99. **Cohen, A. M., Teitelbaum, A., Miller, E., Ben-Tor, V., Hirt, R., and Fields, M.,** Effect of copper on carbohydrate metabolism in rats, *Isr. J. Med. Sci.*, 18, 840, 1982.
100. **Hassel, C. A., Marchello, J. A., and Lei, K. Y.,** Impaired glucose tolerance in copper-deficient rats, *J. Nutr.*, 113, 1081, 1983.
101. **Fields, M., Ferretti, R. J., Smith, J. C., Jr., and Reiser, S.,** Effect of copper deficiency on metabolism and mortality in rats fed sucrose or starch diets, *J. Nutr.*, 113, 1335, 1983.
102. **Johnson, W. T., Nordlie, R. C., and Klevay, L. M.,** Glucose-6-phosphatase activity in copper-deficient rats, *Biol. Trace Element Res.*, 6, 369, 1984.
103. **Klevay, L. M.,** An increase in glycosylated hemoglobin in rats deficient in copper, *Nutr. Rep. Int.*, 26, 329, 1982.
104. **Klevay, L. M. and Viestenz, K. E.,** Abnormal electrocardiograms in rats deficient in copper, *Am. J. Physiol.*, 240, H185, 1981.
105. **Viestenz, K. E. and Klevay, L. M.,** A randomized trial of copper therapy in rats with electrocardiographic abnormalities due to copper deficiency, *Am. J. Clin. Nutr.*, 35, 258, 1982.
106. **Kopp, S. J., Klevay, L. M., and Feliksik, J. M.,** Physiological and metabolic characterization of a cardiomyopathy induced by chronic copper deficiency, *Am. J. Physiol.*, 245, H855, 1983.
107. **Klevay, L. M., Milne, D. B., and Wallwork, J. C.,** Comparison of some indices of copper deficiency in growing rats, *Nutr. Rep. Int.*, 31, 963, 1985.
108. **Fields, M., Lewis, C., Scholfield, D. J., Powell, A. S., Rose, A. J., Reiser, S., and Smith, J. C.,** Female rats are protected against the fructose induced mortality of copper deficiency, *Proc. Soc. Exp. Biol. Med.*, 183, 145, 1986.
109. **Anon.,** Sex differences in the survival of copper-deficient rats, *Nutr. Rev.*, 45, 250, 1987.
110. **Prohaska, J. R. and Heller, L. J.,** Mechanical properties of the copper-deficient rat heart, *J. Nutr.*, 112, 2142, 1982.
111. **Fields, M., Ferretti, R. J., Smith, J. C., Jr., and Reiser, S.,** Effect of dietary carbohydrates and copper status on blood pressure of rats, *Life Sci.*, 34, 763, 1984.
112. **Medeiros, D. M., Lin, K-N., Liu, C-C.F., and Thorne, B. M.,** Pre-gestation dietary copper restriction and blood pressure in the Long-Evans rat, *Nutr. Rep. Int.*, 30, 559, 1984.

113. **Borg, T. K., Klevay, L. M., Gay, R. E., Siegel, R., and Bergin, M. E.,** Alteration of the connective tissue network of striated muscle in copper deficient rats, *J. Mol. Cell. Cardiol.,* 17, 1173, 1985.
114. **Klevay, L. M.,** Hypertension in rats due to copper deficiency, *Nutr. Rep. Int.,* 35, 999, 1987.
115. **Medeiros, D. M.,** Hypertension in the Wistar-Kyoto rat as a result of post-weaning copper restriction, *Nutr. Res.,* 7, 231, 1987.
116. **Owen, C. A., Jr.,** *Physiological Aspects of Copper,* Noyes Publications, Park Ridge, NJ, 1982, 25.
117. **Kannel, W. B., Castelli, W. P., and Gordon, T.,** Cholesterol in the prediction of atherosclerotic disease, *Ann. Intern. Med.,* 90, 85, 1979.
118. **Newland, H.,** Hyperuricemia in coronary, cerebral and peripheral arterial disease: an explanation, *Med. Hypotheses,* 1, 152, 1975.
119. **Klevay, L. M.,** Hyperuricemia in rats due to copper deficiency, *Nutr. Rep. Int.,* 22, 617, 1980.
120. **Bennetts, H. W., Harley, R., and Evans, S. T.,** Studies on copper deficiency of cattle: the fatal termination ("Falling disease"), *Aust. Vet. J.,* 18, 50, 1942.
121. **Bennetts, H. W. and Hall, H. T. B.,** "Falling disease" of cattle in the south-west of western Australia, *Aust. Vet. J.,* 15, 152, 1939.
122. **Carnes, W. H., Coulson, W. F., and Albino, A. M.,** Intimal lesions in muscular arteries of young copper-deficient swine, *Ann. N.Y. Acad. Sci.,* 127, 800, 1965.
123. **Maximov, A. A. and Bloom, W.,** *A Textbook of Histology,* 6th ed., W. B. Saunders, Philadelphia, 1952, 211.
124. **Hunsaker, H. A., Morita, M., and Allen, K. G. D.,** Marginal copper deficiency in rats. Aortal morphology of elastin and cholesterol values in first-generation adult males, *Atherosclerosis,* 51, 1, 1984.
125. **Hill, K. E. and Davidson, J. M.,** Induction of increased collagen and elastin biosynthesis in copper-deficient pig aorta, *Arteriosclerosis,* 6, 98, 1986.
126. **Ross, R. and Glomset, J. A.,** Atherosclerosis and the arterial smooth muscle cell, *Science,* 180, 1332, 1973.
127. **Ross, R. and Glomset, J. A.,** The pathogenesis of atherosclerosis, *N. Engl. J. Med.,* 295, 420, 1976.
128. **Morris, J. N., Kagan, A., Pattison, D. C., Gardner, M. J., and Raffle, P. A. B.,** Incidence and prediction of ischaemic heart-disease in London busmen, *Lancet,* ii, 553, 1966.
129. **Castillo-Duran, D. and Uauy, R.,** Copper deficiency impairs growth of infants recovering from malnutrition, *Am. J. Clin. Nutr.,* 47, 710, 1988.
130. **Penttilä, O., Neuvonen, P. J., Vouri, E., Himberg, J-J., Siltanen, P., Järvinen, A., and Merikallio, E.,** Auricular myocardial cation concentrations in certain heart diseases in man, *Trace Elements in Med.,* 3, 47, 1986.
131. **Zama, N. and Towns, R. L. R.,** Cardiac copper, magnesium, and zinc in recent and old myocardial infarction, *Biol. Trace Elem. Res.,* 10, 201, 1986.
132. **Fregly, M. J.,** Sodium and potassium, in *Present Knowledge in Nutrition,* 5th ed., Olson, R. E., Broquist, H. P., Chichester, C. O., Darby, W. J., Kolbye, A. C., Jr., and Stalvey, R. M., Eds., Nutrition Foundation, Washington, DC, 1984, 439.
133. **Achuff, S. C. and Ross, R. S.,** Congestive heart failure and/or enlarged cardiac silhouette: approach to the patient and management, in *The Principles and Practice of Medicine,* 21st ed., Harvey, A.McG., Johns, R. J., McKusick, V. A., Owens, A. H., Jr., and Ross, R. S., Eds., Appleton-Century-Crofts, 1984, 191.
134. **Bulkley, B. H., Becker, L. C., and Flaherty, J. T.,** Myocardial infarction, in *The Principles and Practice of Medicine,* 21st ed., Harvey, A.Mc.G., Johns, R. J., McKusick, V. A., Owens, A. H., Jr., and Ross, R. S., Eds., Appleton-Century-Crofts, 1984, 257.
135. **Kaplan, N. M.,** Endocrine hypertension, in *Williams Textbook of Endocrinology,* 7th ed., Wilson, J. D. and Foster, D. W., Eds., W. B. Saunders, Philadelphia, 1985, 966.

136. **Moore, R. J., Lukaski, H. C., Hall, C. B., and Klevay, L. M.,** Fluid retention and edema formation in Cu-deficient rats fed a high-NaCl diet, *FASEB J.*, 2, A1422, 1988.
137. **Meneely, G. R., Tucker, R. G., Darby, W. J., and Auerbach, S. H.,** Chronic sodium chloride toxicity in the albino rat. II. Occurrence of hypertension and of a syndrome of edema and renal failure, *J. Exp. Med.*, 98, 71, 1953.
138. **Clegg, M. S., Ferrell, F., and Keen, C. L.,** Hypertension-induced alterations in copper and zinc metabolism in Dahl rats, *Hypertension,* 9, 624, 1987.
139. **Owen, C. A., Jr.,** *Copper Deficiency and Toxicity,* Noyes Publications, Park Ridge, NJ, 1981, 50.
140. **Blackett, P. R., Lee, D. M., Donaldson, D. L., Fesmire, J. D., Chan, W. Y., Holcombe, J. H., and Rennert, O. M.,** Studies of lipids, lipoproteins, and apolipoproteins in Menkes' disease, *Ped. Res.*, 18, 864, 1984.
141. **Owen, C. A., Jr.,** *Wilson's Disease,* Noyes Publications, Park Ridge, NJ, 1981, 1.
142. **Hill, G. M., Brewer, G. J., McGinnis, M. M., and Block, W. D.,** Lipid parameters in Wilson's disease patients on zinc therapy, *Trace Elements in Man and Animals 6,* Hurley, L. S., Keen, C. L., Lönnerdal, B., and Rucker, R. B., Eds., 1988, 235.
143. **Jacob, R. A., Klevay, L. M., and Logan, G. M., Jr.,** Hair as a biopsy material. V. Hair metal as an index of hepatic metal in rats: copper and zinc, *Am. J. Clin. Nutr.*, 31, 477, 1978.
144. **Harvey, P. W., Hunsaker, H. A., and Allen, K. G. D.,** Dietary L-histidine-induced hypercholesterolemia and hypocupremia in the rat, *J. Nutr.*, 111, 639, 1981.
145. **Hennekens, C. H. et al.,** Preliminary report: findings from the aspirin component of the ongoing Physician's Health Study, *N. Eng. J. Med.*, 318, 262, 1988.
146. **Oliver, M. F. et al.,** A co-operative trial in the primary prevention of ischaemic heart disease using clofibrate, *Br. Heart J.*, 40, 1069, 1978.
147. **Ahrens, R. A., Demuth, P., Lee, M. K., and Majkowski, J. W.,** Moderate sucrose ingestion and blood pressure in the rat, *J. Nutr.*, 110, 725, 1980.
148. **Jacob, R. A., Baesler, L. G., Klevay, L. M., Lee, D. E., and Wherry, P. L.,** Hypercholesterolemia in mice with meat anemia, *Nutr. Rep. Int.*, 16, 73, 1977.
149. **Hart, E. B., Steenbock, H., Waddell, J., and Elvehjem, C. A.,** Iron in nutrition. VII. Copper as a supplement to iron for hemoglobin building in the rat, *J. Biol. Chem.*, 77, 797, 1928.
150. **Shields, G. S., Coulson, W. F., Kimball, D. A., Carnes, W. H., Cartwright, G. E., and Wintrobe, M. M.,** Studies on copper metabolism. XXXII. Cardiovascular lesions in copper-deficient swine, *Am. J. Pathol.*, 41, 603, 1962.
151. **Beebe, C. G., Schemmel, R., and Mickelsen, O.,** Blood pressure of rats as affected by diet and concentration of NaCl in drinking water, *Proc. Soc. Exp. Biol. Med.*, 151, 395, 1976.
152. **Preuss, M. B. and Preuss, H. G.,** The effects of sucrose and sodium on blood pressures in various substrains of Wistar rats, *Lab. Invest.*, 43, 101, 1980.
153. **Young, J. B. and Landsberg, L.,** Effect of oral sucrose on blood pressure in the spontaneously hypertensive rat, *Metabolism,* 30, 421, 1981.
154. **Buñag, R. D., Tomita, T., and Sasaki, S.,** Chronic sucrose ingestion induces mild hypertension and tachycardia in rats, *Hypertension,* 5, 218, 1983.
155. **Michaelis, O. E., IV, Martin, R. E., Gardner, L. B., and Ellwood, K. C.,** Effect of dietary carbohydrates on systolic blood pressure of normotensive and hypertensive rats, *Nutr. Rep. Int.*, 23, 261, 1981.
156. **Bialkowska, M., Hoser, A., Szostak, W. B., Dybczynski, R., Sterlinski, S., Nowicka, G., Majchrzak, J., Kaczorowski, J., and Danko, B.,** Hair zinc and copper concentration in survivors of myocardial infarction, *Ann. Nutr. Metab.*, 31, 327, 1987.
157. **Chamberlin, T. C.,** The method of multiple working hypotheses, *Science,* 148, 754, 1965, (reprinted from 7 February 1890 issue).
158. **Kuhn, T. S.,** *The Structure of Scientific Revolutions,* 2nd ed., University of Chicago Press, Chicago, 1970, 95.

159. **Medawar, P. B.,** *Advice to a Young Scientist,* Harper & Row, New York, 1979, 39.
160. **Barboriak, J. J., Rimm, A. A., Anderson, A. J., Schmidhoffer, M., and Tristani, F. E.,** Coronary artery occlusion and alcohol intake, *Br. Heart J.,* 39, 289, 1977.
161. **Yano, K., Rhoads, G. G., and Kagan, A.,** Coffee, alcohol and risk of coronary heart disease among Japanese men living in Hawaii, *N. Engl. J. Med.,* 297, 405, 1977.
162. **Hennekens, C. H., Rosner, B., and Cole, D. S.,** Daily alcohol consumption and fatal coronary heart disease, *Am. J. Epidemiol.,* 107, 196, 1978.
163. **Turner, T. B., Bennett, V. L., and Hernandez, H.,** The beneficial side of moderate alcohol use, *Johns Hopkins Med. J.,* 148, 53, 1981.
164. **Klatsky, A. L., Friedman, G. D., and Siegelaub, A. B.,** Alcohol and mortality, *Ann. Intern. Med.,* 95, 139, 1981.
165. **Gordon, T., Kagan, A., Garcia-Palmieri, M., Kannel, W. B., Zukel, W. J., Tillotson, J., Sorlie, P., and Hjortland, M.,** Diet and its relation to coronary heart disease and death in three populations, *Circulation,* 63, 500, 1981.
166. **Kozarevic, D., Demirovic, J., Gordon, T., Kaelber, C. T., McGee, D., and Zukel, W. J.,** Drinking habits and coronary heart disease, *Am. J. Epidemiol.,* 116, 748, 1982.
167. **Gordon, T. and Kannel, W. B.,** Drinking and mortality: the Framingham Study, *Am. J. Epidemiol.,* 120, 97, 1984.
168. **Gordon, T. and Kannel, W. B.,** Drinking habits and cardiovascular disease: the Framingham Study, *Am. Heart J.,* 105, 667, 1983.
169. **Colditz, G. A., Branch, L. G., Lipnick, R. J., Willett, W. C., Rosner, B., Posner, B., and Hennekens, C. H.,** Moderate alcohol and decreased cardiovascular mortality in an elderly cohort, *Am. Heart J.,* 109, 886, 1985.
170. **Shaper, A. G., Phillips, A. N., Pocock, S. J., and Walker, M.,** Alcohol and ischaemic heart disease in middle aged British men, *Br. Med. J.,* 294, 733, 1987.
171. **Stampfer, M. J., Colditz, M. B., Willett, W. C., Speizer, F. E., and Hennekens, C. H.,** A prospective study of moderate alcohol consumption and the risk of coronary disease in women, *N. Engl. J. Med.,* 319, 267, 1988.
172. **Klevay, L. M.,** Beer increases the longevity of rats fed a diet deficient in copper, *Trace Elements in Man and Animals 6,* Hurley, L. S., Keen, C. L., Lönnerdal, B. and Rucker, R. B., Eds., 1988, 453.
173. **Gordon, D. J., Trost, D. C., Hyde, J., Whaley, F. S., Hannon, P. J., Jacobs, D. R., Jr., and Ekelund, L-G.,** Seasonal cholesterol cycles: the Lipid Research Clinics Coronary Primary Prevention Trial placebo group, *Circulation,* 76, 1224, 1987.
174. **Kobayashi, J.,** On geographical relationship between the chemical nature of river water and death-rate from apoplexy, *Ber. Ohara Inst.,* 11, 12, 1957.
175. **Sharrett, A. R. and Feinleib, M.,** Water constituents and trace elements in relation to cardiovascular disease, *Prev. Med.,* 4, 20, 1975.
176. **Romasz, R. S., Lemmo, E. A., and Evans, J. L.,** Diet calcium, sex and age influences on tissue mineralization and cholesterol in rats, in *Trace Substances in Environmental Health-XI,* Hemphill, D. D., Ed., University of Missouri Press, Columbia, 1977, 289.
177. **Seelig, M. S.,** *Magnesium Deficiency in the Pathogenesis of Disease: Early Roots of Cardiovascular, Skeletal, and Renal Abnormalities,* Plenum Press, New York, 1980.
178. **Anon.**, Magnesium deficiency and ischemic heart disease, *Nutr. Rev.,* 46, 311, 1988.
179. **Yudkin, J.,** Patterns and trends in carbohydrate consumption and their relation in disease, *Proc. Nutr. Soc.,* 23, 149, 1964.
180. **Yudkin, J.,** Dietetic aspects of atherosclerosis, *Angiology,* 17, 127, 1966.
181. **Reaven, G. M.,** Parma Symposium: current controversies in nutrition, *Am. J. Clin. Nutr.,* 47, 1078, 1988.
182. **Cleave, T. L. and Campbell, G. D.,** *Diabetes, Coronary Thrombosis, and the Saccharine Disease,* John Wright & Sons, Bristol, 1966.
183. **Burkitt, D. P. and Trowell, H. C.,** *Refined Carbohydrate Foods and Disease,* Academic Press, London, 1975.

184. **Klevay, L. M. and Van Soest, P. J.,** Positive correlations between dietary fibers and copper or zinc in breakfast cereals, *Fed. Proc.,* 40, 854, 1981.
185. **McCord, J. M.,** Oxygen-derived free radicals in postischemic tissue injury, *N. Engl. J. Med.,* 312, 159, 1985.
186. **Southorn, P. A. and Powis, G.,** Free radicals in medicine, II. Involvement in human disease, *Mayo Clin. Proc.,* 63, 390, 1988.
187. **Owen, C. A., Jr.,** *Biochemical Aspects of Copper,* Noyes Publications, Park Ridge, NJ, 1982, 78, 120.
188. **Southorn, P. A. and Powis, G.,** Free radicals in medicine. I. Chemical nature and biologic reactions, *Mayo Clin. Proc.,* 63, 381, 1988.
189. **Paynter, D. I., Moir, R. J., and Underwood, E. J.,** Changes in activity of the Cu-Zn superoxide dismutase enzyme in tissues of the rat with changes in dietary copper, *J. Nutr.,* 109, 1570, 1979.
190. **Saari, J. T.,** Chronic treatment with dimethyl sulfoxide protects against cardiovascular defects of copper deficiency, *Proc. Soc. Exp. Biol. Med.,* 190, 121, 1989.
191. **Dubick, M. A., Hunter, G. C., Casey, S. M., and Keen, C. L.,** Aortic ascorbic acid, trace elements, and superoxide dismutase activity in human aneurysmal and occlusive disease, *Proc. Soc. Exp. Biol. Med.,* 184, 138, 1987.
192. **Chandler, A. B., Eurenius, K., McMillan, G. C., Nelson, C. B., Schwartz, C. J., and Wessler, S.,** The thrombotic process in atherogenesis, *Adv. Exp. Med. Biol.,* 104, 1978.
193. **Schwartz, C. J., Chandler, A. B., Gerrity, R. G., and Naito, H. K.,** Clinical and pathological aspects of arterial thrombosis and thromboembolism, *Adv. Exp. Med. Biol.,* 104, 111, 1978.
194. **Mustard, J. F., Packham, M. A., and Kinlough-Rathbone, R.,** Platelets, thrombosis and atherosclerosis, *Adv. Exp. Med. Biol.,* 104, 127, 1978.
195. **Mitchell, L. L., Allen, K. G. D., and Mathias, M. M.,** Copper deficiency depresses rat aortae superoxide dismutase activity and prostacyclin synthesis, *Prostaglandins,* 35, 977, 1988.
196. **Allen, K. G. D.,** *Role of Copper in Lipid Metabolism,* Lei, K. Y., Ed., CRC Press, Boca Raton, FL, 1990, chap. 10.
197. **Vanhoutte, P. M.,** The endothelium—modulator of vascular smooth-muscle tone, *N. Engl. J. Med.,* 319, 512, 1988.
198. **Solomons, N. W.,** On the assessment of zinc and copper nutriture in man, *Am. J. Clin. Nutr.,* 32, 856, 1979.
199. **Klevay, L. M.,** Metabolic interactions among cholesterol, cholic acid and copper, *Nutr. Rep. Int.,* 26, 405, 1982.
200. **Kincaid, R. L.,** Toxicity of ammonium molybdate added to drinking water of calves, *J. Dairy Sci.,* 63, 608, 1980.
201. **Evans, G. W., Cornatzer, N. F., and Cornatzer, W. E.,** Mechanism for hormone-induced alterations in serum ceruloplasmin, *Am J. Physiol.,* 218, 613, 1970.
202. **Chappuis, P., Jacqueson, A., Ducimetière, P., and Rousselet, F.,** Magnesium, copper and zinc: relationships with ischemic heart disease risk factors, in *Magnesium Deficiency,* Halpern, M. J. and Durlach, J., Eds., S. Karger, Basel, 1985, 142.
203. **Kromhout, D., Wibowo, A. A. E., Herber, R. F. M., Dalderup, L. M., Heerdink, H., deLezenne Coulandor, C., and Zielhuis, R. L.,** Trace metals and coronary heart disease risk indicators in 152 elderly men (the Zutphen study), *Am. J. Epidemiol.,* 122, 378, 1985.
204. **Thuillier-Juteau, Y., Jaudon, M. C., Clavel, J. P., Delattre, J., and Galli, A.,** Zinc et cuivre sériques dans l'hypercholestérolémie, *Pathol. Biol.,* 35, 387, 1987.
205. **Kok, F. J., van Duijn, C. M., Hofman, A., Van Der Voet, G. B., De Wolff, F. A., Paays, C. H. Ch., and Valkenburg, H. A.,** Serum copper and zinc and the risk of death from cancer and cardiovascular disease, *Am. J. Epidemiol.,* 128, 352, 1988.

206. **Meissner, D., Leonhardt, W., Schulze, J., and Julius, U.,** Correlations between parameters of lipoprotein metabolism and magnesium, copper, and zinc, Hormones, Lipoproteins & Atherosclerosis, Proc. Sat. Symp. 28th Int. Congr. Physiol. Sci. Bratislava, Czechoslovakia 1980, *Adv. Physiol Sci. Ser.*, 35, 91, 1981.
207. **Tiber, A. M., Sakhaii, M., Joffe, C. D., and Ratnaparkhi, M. V.,** Relative value of plasma copper, zinc, lipids and lipoproteins as markers for coronary artery disease, *Atherosclerosis,* 62, 105, 1986.
208. **Koo, S. I., Lee, C. C., and Norvell, J. E.,**Effect of copper deficiency on the lymphatic absorption of cholesterol, plasma chylomicron clearance, and post heparin lipase activities, *Proc. Soc. Exp. Biol. Med.,* 188, 410, 1988.
209. **Cunnane, S. C., McAdoo, K. R., and Karmazyn, M.,** Copper intake affects rat heart performance during ischemia-reperfusion: possible relation to altered lipid and fatty acid metabolism, *Prostaglandins Leukotrienes and Essential Fatty Acids,* 34, 61, 1988.
210. **Yount, N. Y., McNamara, D. J., Al-Othman, A. A., and Lei, K. Y.,** Effect of copper deficiency on rat hepatic 3-hydroxy-3-methyl-glutaryl-coenzyme A reductase, *FASEB J.*, 3, A357, 1989.
211. **Turnlund, J. R., Keyes, W. R., Anderson, H. L., and Acord, L. L.,** Copper absorption and retention in young men at three levels of dietary copper by use of the stable isotope $^{65}$Cu, *Am. J. Clin. Nutr.*, 49, 870, 1989.
212. **Milne, D. B., Johnson, P. E., and Klevay, L. M.,** Effect of copper intake on balance, absorption, and status indices of copper in men and women, in preparation.
213. **Bancroft, H.,** *Introduction to Biostatistics,* P. B. Hoeber, Inc., New York, 1957, 115.
214. **Rawson, J. R. and Medeiros, D. M.,** Macronutrient, copper, and zinc intakes of young adult males as determined by duplicate food samples and diet records, *Nutr. Rep. Int.*, 39, 1149, 1989.
215. **Krebs, J. M., Schneider, V. S., and LeBlanc, A. D.,** Zinc, copper, and nitrogen balances during bed rest and fluoride supplementation in healthy adult males, *Am. J. Clin. Nutr.*, 47, 509, 1988.
216. **Anon.,** Dietary cholesterol and liver microsomal fatty acid desaturases, *Nutr. Rev.*, 46, 397, 1988.
217. **Wahle, K. W. J. and Davies, N. T.,** Effect of dietary copper deficiency in the rat on fatty acid composition of adipose tissue and desaturase activity of liver microsomes, *Br. J. Nutr.*, 34, 105, 1975.
218. **Balevska, P. S., Russanov, E. M., and Stanchev, P. I.,** Copper deficiency-induced changes in fatty acid composition of mitochondrial and microsomal membranes of rat liver, *Biol. Trace Element Res.*, 8, 211, 1985.
219. **Cunnane, S. C., Horrobin, D. F., and Manku, M. S.,** Contrasting effects of low or high copper intake in rat tissue lipid essential fatty acid composition, *Ann. Nutr. Metab.*, 29, 103, 1985.
220. **Cunnane, S. C., McAdoo, K. R., and Prohaska, J. R.,** Lipid and fatty acid composition of organs from copper-deficient mice, *J. Nutr.*, 116, 1248, 1986.
221. **Cunnane, S. C.,** The profile of long-chain fatty acids in serum phospholipids: a possible indicator of copper status in humans, *Am. J. Clin. Nutr.*, 48, 1475, 1988.
222. **Johnson, S. B., Kramer, T. R., Briske-Anderson, M., and Holman, R. T.,** Fatty acid pattern of tissue phospholipids in copper and iron deficiencies, *Lipids,* 24, 141, 1989.
223. **Carville, D. G. M. and Strain, J. J.,** The effect of copper deficiency on blood antioxidant enzymes in rats fed sucrose or sucrose and lactose diets, *Nutr. Rep. Int.*, 39, 25, 1989.
224. **Koo, S. I. and Lee, C. C.,** Cholesterol and apolipoprotein distribution in plasma high-density-lipoprotein subclasses from zinc-deficient rats, *Am. J. Clin. Nutr.*, 50, 73, 1989.
225. **Wissler, R. W., Vesselinovitch, D., and Davis, H. R.,** Cellular components of the progressive atherosclerotic process, in *Atherosclerosis Biology and Clinical Science,* Olsson, A. G., Ed., Churchill Livingstone, Edinburgh, 1987, 57.

226. **Paffenbarger, R. S., Notkin, J., Krueger, D. E., Wolf, P. A., Thorne, M. C., LeBauer, E. J., and Williams, J. L.,** Chronic disease in former college students. II. Methods of study and observations on mortality from coronary heart disease, *Am. J. Publ. Health,* 56, 962, 1966.
227. **Barker, D. J. P., Osmond, C., Winter, P. D., Margetts, B., and Simmonds, S. J.,** Weight in infancy and death from ischaemic heart disease, *Lancet,* 2, 577, 1989.
228. **Redman, R. S., Fields, M., Reiser, S., and Smith, J. C., Jr.,** Dietary fructose exacerbates the cardiac abnormalities of copper deficiency in rats, *Atherosclerosis,* 74, 203, 1988.
229. **Levy, D., Garrison, R. J., Savage, D. D., Kannel, W. B., and Castelli, W. P.,** Left ventricular mass and incidence of coronary heart disease in an elderly cohort, *Ann. Intern. Med.,* 110, 101, 1989.
230. **Berthelot, A., Luthringer, C., and Exinger, A.,** Trace elements during the development of hypertension in the spontaneously hypertensive rat, *Clin. Sci.,* 72, 515, 1987.
231. **Aalbers, Th. G.,** Cardiovascular diseases and trace elements, Drukkerij Blok En Zonen, Dieren, Netherlands, 1984, 206.
232. **Klevay, L. M., and Halas, E. S.,** Effect of restraint and copper deficiency on blood pressure and mortality of adult rats, *FASEB J.,* 3, A1062, 1989.
233. **Moore, R. J., Hall, C. B., Carlson, E. C., Lukaski, H. C., and Klevay, L. M.,** Acute renal failure and fluid retention and kidney damage in copper-deficient rats fed a high-NaCl diet, *J. Lab. Clin. Med.,* 113, 516, 1989.
234. **Fuster, V. and Kottke, B. A.,** Atherosclerosis A. Pathogenesis, pathology, and presentation of atherosclerosis, in *Cardiology: Fundamentals and Practice,* Brandenburg, R. O., Fuster, V., Giuliani, E. R., and McGoon, D. C., Eds., Year Book Medical Publ., Chicago, 1987, 951.
235. **Radhakrishnamurthy, B., Ruiz, H., Dalferes, E. R., Jr., Klevay, L. M., and Berenson, G. S.,** Composition of proteoglycans in the aortas of copper-deficient rats, *Proc. Soc. Exp. Biol. Med.,* 190, 98, 1989.
236. **Carr, T. P. and Lei, K. Y.,** *In vivo* apoprotein catabolism of high density lipoproteins in copper-deficient, hypercholesterolemic rats, *Proc. Soc. Exp. Biol. Med.,* 191, 370, 1989.
237. **Reaven, G. M., Ho, H., and Hoffman, B. B.,** Attenuation of fructose-induced hypertension in rats by exercise training, *Hypertension,* 12, 129, 1988.
238. **McCormick, J.,** The multifactorial aetiology of coronary heart disease: a dangerous delusion, *Perspect. Biol. Med.,* 32, 103, 1988.
239. **Klevay, L. M.,** Cholesterol reduction: safety, other concerns, *Ann. Intern. Med.,* 107, 421, 1987.
240. **Klevay, L. M.,** Cost-effectiveness of antihyperlipemic therapy, *JAMA,* 259, 1811, 1988.
241. **Oliver, M. F.,** Reducing cholesterol does not reduce mortality, *J. Am. Coll. Cardiol.,* 12, 814, 1988.
242. **McCormick, J. and Skrabanek, P.,** Coronary heart disease is not preventable by population interventions, *Lancet,* 2, 839, 1988.
243. **Isles, C. G., Hole, D. J., Gillis, C. R., Hawthorne, V. M., and Lever, A. F.,** Plasma cholesterol, conorary heart disease, and cancer in the Renfrew and Paisley survey, *Br. Med. J.,* 298, 920, 1989.
244. **Stehbens, W. E.,** An appraisal of the epidemic rise of coronary heart disease and its decline, *Lancet,* 1, 606, 1987.
245. **Zarling, E. J., Sexton, H., and Milnor, P., Jr.,** Failure to diagnose acute myocardial infarction, *JAMA,* 250, 1177, 1983.
246. **Battle, R. M., Pathak, D., Humble, C. G., Key, C. R., Vanatta, P. R., Hill, R. B., and Anderson, R. E.,** Factors influencing discrepancies between premortem and postmortem diagnoses, *JAMA,* 258, 339, 1987.

247. **Lie, J. T.,** Atherosclerosis B. Pathology of coronary artery disease, in *Cardiology: Fundamentals and Practice,* Brandenburg, R. O., Fuster, V., Giuliani, E. R., and McGoon, D. C., Eds., Year Book Medical Publ., Chicago 1987, 972.
248. **Stehbens, W. E.,** Diet and atherogenesis, *Nutr. Rev.,* 47, 1, 1989.
249. **Holtzman, N. A.,** Genetic variation in nutritional requirements and susceptibility to disease: policy implications, *Am. J. Clin. Nutr.,* 48, 1510, 1988.
250. **Kannel, W. B. and Gordon, T.,** The search for an optimum serum cholesterol, *Lancet,* 2, 374, 1982.
251. **Anon.,** Coronary heart disease without risk factors, *Nutr. Rev.,* 47, 18, 1989.
252. **Rifkind, B. M.,** The Lipid Research Clinics coronary primary prevention trial results. I. Reduction in incidence of coronary heart diseases, *JAMA,* 251, 351, 1984.
253. **Scott, F. I., Jr.,** A risky experiment and some serious implications, *Am. Clin. Lab.,* 8, 4-6, Nov. 1989.
254. **Smith, R. L.,** Dietary lipids and heart disease: the contriving of a relationship, *Am. Clin. Lab.,* 8, 26, Nov. 1989.
255. **Klevay, L. M.,** An association between the amount of fat and the ratio of zinc to copper in 71 foods: inferences about the epidemiology of coronary heart disease, *Nutr. Rep. Int.,* 9, 393, 1974.
256. **Dougherty, R. M., Fong, A. K. H., and Iacono, J. M.,** Nutrient content of the diet when fat is reduced, *Am. J. Clin. Nutr.,* 48, 970, 1988.
257. **Klevay, L. M. and Moore, R. J.,** Beer mitigates some effects of copper deficiency in rats, *Am. J. Clin. Nutr.,* in press.
258. **Kimura, M. and Itokawa, Y.,** Inefficient utilization of iron and minerals in magnesium deficient rats, in *Magnesium in Health and Disease,* Itokawa, Y. and Durlach, J., Eds., John Libby, London, 1989, 95.
259. **Yudkin, J. and Eisa, O.,** Dietary sucrose and oestradiol concentration in young men, *Ann. Nutr. Metab.,* 32, 53, 1988.
260. **Phillips, G. B.,** Evidence for hyperoestrogenaemia as a risk factor for myocardial infarction in men, *Lancet,* 2, 14, 1976.
261. **Phillips, G. B.,** Relationship between serum sex hormones and glucose, insulin, and lipid abnormalities in men with myocardial infarction, *Proc. Natl. Acad. Sci. U.S.A.,* 74, 1729, 1977.
262. **Yudkin, J.,** Sucrose, coronary disease, diabetes, and obesity: do hormones provide a link?, *Am. Heart J.,* 115, 493, 1988.
263. **Carville, D. G. M. and Strain, J. J.,** The effect of a low copper diet on blood cholesterol and enzymic antioxidant defense mechanisms in male and female rats, *Int. J. Vit. Nutr. Res.,* 58, 456, 1988.
264. **Yadrick, M. K., Kenney, M. A., and Winterfeldt, E. A.,** Iron, copper, and zinc status: response to supplementation with zinc or zinc and iron in adult females, *Am, J. Clin, Nutr.,* 49, 145, 1989.
265. **Lynch, S. M. and Strain, J. J.,** Effects of copper deficiency on hepatic and cardiac antioxidant enzyme activities in lactose- and sucrose-fed rats, *Br. J. Nutr.,* 61, 345, 1989.
266. **Furchgott, R. F. and Vanhoutte, P. M.,** Endothelium-derived relaxing and contracting factors, *FASEB J.,* 3, 2007, 1989.
267. **Uza, G. and Vlaicu, R.,** Serum zinc and copper in patients with atherosclerosis and thromboangiitis obliterans, *Biol. Tr. Element Res.,* 20, 197, 1989.
268. **Kennedy, M. L., Failla, M. L., and Smith, J. C., Jr.,** Influence of genetic obesity on tissue concentrations of zinc, copper, manganese and iron in mice, *J. Nutr.,* 116, 1432, 1986.

# INDEX

## A

# D

E

## F



## G

## H

## J



## K



## L

## M

## R



## S

## W



## X



## Y



## Z